AF356442

Assessment, Reconstruction and Decision Procedures for the Preservation of Existing Structures after Earthquakes

Assessment, Reconstruction and Decision Procedures for the Preservation of Existing Structures after Earthquakes

Editors

Mislav Stepinac
Chiara Bedon
Marco Francesco Funari
Tomislav Kišiček
Ufuk Hancilar

Basel • Beijing • Wuhan • Barcelona • Belgrade • Novi Sad • Cluj • Manchester

Editors

Mislav Stepinac
University of Zagreb
Zagreb
Croatia

Chiara Bedon
University of Trieste
Trieste
Italy

Marco Francesco Funari
University of Surrey
Gulidford
United Kingdom

Tomislav Kišiček
University of Zagreb
Zagreb
Croatia

Ufuk Hancilar
Bogazici University
Bebek
Turkey

Editorial Office
MDPI
St. Alban-Anlage 66
4052 Basel, Switzerland

This is a reprint of articles from the Topical Collection published online in the open access journal *Buildings* (ISSN 2075-5309) (available at: https://www.mdpi.com/journal/buildings/topical_collections/Assess_Reconst_Decis_Proced_Preser_Struct_Earth).

For citation purposes, cite each article independently as indicated on the article page online and as indicated below:

Lastname, A.A.; Lastname, B.B. Article Title. *Journal Name* **Year**, *Volume Number*, Page Range.

ISBN 978-3-7258-0483-2 (Hbk)
ISBN 978-3-7258-0484-9 (PDF)
doi.org/10.3390/books978-3-7258-0484-9

Cover image courtesy of Mislav Stepinac

Contents

Preface . . . . **vii**

Qiushan Li, Kabilijiang Wumaier and Koide Osamu
Post-Earthquake Housing Reconstruction Management and Implementation in Rural Areas:
Review and Lessons from Dujiangyan, Wenchuan Earthquake
Reprinted from: *Buildings* **2023**, *13*, 2251, doi:10.3390/buildings13092251 . . . . **1**

Karlo Ožić, Ivan Markić, Antonela Moretić and Luka Lulić
The Assessment and Retrofitting of Cultural Heritage—A Case Study of a Residential Building
in Glina
Reprinted from: *Buildings* **2023**, *13*, 1798, doi:10.3390/buildings13071798 . . . . **18**

Predrag Blagojević, Svetlana Brzev and Radovan Cvetković
Seismic Retrofitting of Mid-Rise Unreinforced Masonry Residential Buildings after the 2010
Kraljevo, Serbia Earthquake: A Case Study
Reprinted from: *Buildings* **2023**, *13*, 597, doi:10.3390/buildings13030597 . . . . **38**

Giorgia Predari, Lorenzo Stefanini, Marko Marinković, Mislav Stepinac and Svetlana Brzev
Adriseismic Methodology for Expeditious Seismic Assessment of Unreinforced Masonry
Buildings
Reprinted from: *Buildings* **2023**, *13*, 344, doi:10.3390/buildings13020344 . . . . **74**

Mario Uroš, Marija Demšić, Maja Baniček and Ante Pilipović
Seismic Retrofitting of Dual Structural Systems—A Case Study of an Educational Building in
Croatia
Reprinted from: *Buildings* **2023**, *13*, 292, doi:10.3390/buildings13020292 . . . . **104**

Aida Salaman, Mislav Stepinac, Ivan Matorić and Mija Klasić
Post-Earthquake Condition Assessment and Seismic Upgrading Strategies for a
Heritage-Protected School in Petrinja, Croatia
Reprinted from: *Buildings* **2022**, *12*, 2263, doi:10.3390/buildings12122263 . . . . **138**

Assaf Shmerling and Matthias Gerdts
A Design Methodology for the Seismic Retrofitting of Existing Frame Structures
Post-Earthquake Incident Using Nonlinear Control Systems
Reprinted from: *Buildings* **2022**, *12*, 1886, doi:10.3390/buildings12111886 . . . . **166**

Zvonko Sigmund, Mladen Radujković and Josip Atalić
The Role of Disaster Risk Governance for Effective Post-Disaster Risk Management—Case of
Croatia
Reprinted from: *Buildings* **2022**, *12*, 420, doi:10.3390/buildings12040420 . . . . **188**

Anđelko Vlašić, Mladen Srbić, Dominik Skokandić and Ana Mandić Ivanković
Post-Earthquake Rapid Damage Assessment of Road Bridges in Glina County
Reprinted from: *Buildings* **2022**, *12*, 42, doi:10.3390/buildings12010042 . . . . **205**

Mattia Zizi, Jafar Rouhi, Corrado Chisari, Daniela Cacace and Gianfranco De Matteis
Seismic Vulnerability Assessment for Masonry Churches: An Overview on Existing
Methodologies
Reprinted from: *Buildings* **2021**, *11*, 588, doi:10.3390/buildings11120588 . . . . **229**

Mija Milić, Mislav Stepinac, Luka Lulić, Nataša Ivanišević, Ivan Matorić, Boja Čačić Šipoš and Yohei Endo
Assessment and Rehabilitation of Culturally Protected Prince Rudolf Infantry Barracks in Zagreb after Major Earthquake
Reprinted from: *Buildings* **2021**, *11*, 508, doi:10.3390/buildings11110508 . . . . **254**

Preface

Seismic events not only lead to the tragic loss of lives, but also cause extensive structural damage, economic downturns, and have highlighted the urgent need for societal preparedness against natural calamities. This reprint embarks on an in-depth exploration of these incidents, emphasizing the significance of understanding vulnerability and performing earthquake performance assessments.

Historically, structural assessments have relied heavily on visual inspections, non-destructive testing methods, and professional judgment. However, the past decade has witnessed remarkable technological advancements that have transformed our approach to assessing and enhancing the safety and resilience of existing structures. This collection discusses these advancements within the context of pre- and post-earthquake scenarios, including the reconstruction and renovation efforts that follow such disasters.

Through a series of studies, this reprint bridges the gap between theoretical research and practical application, offering invaluable insights into the processes involved in the preservation, maintenance, and restoration of buildings, particularly those of significant cultural and heritage value. It underscores the role of emerging technologies, innovative materials, and advanced numerical modeling techniques in shaping the future of structural engineering and disaster mitigation.

This reprint aims to compile a rich database of research and best practices focusing on the assessment, monitoring, reconstruction, maintenance, and preservation of masonry, concrete, and timber structures. Moreover, it seeks to inspire educational initiatives that aim to preserve our built heritage, ensuring that the lessons learned from past earthquakes guide our efforts in building safer, more resilient communities for the future.

Mislav Stepinac , Chiara Bedon , Marco Francesco Funari , Tomislav Kišiček , and Ufuk Hancilar
Editors

Article

Post-Earthquake Housing Reconstruction Management and Implementation in Rural Areas: Review and Lessons from Dujiangyan, Wenchuan Earthquake

Qiushan Li, Kabilijiang Wumaier * and Koide Osamu

The Hong Kong Polytechnic University-Institute for Disaster Management and Reconstruction (IDMR), Sichuan University, Chengdu 610207, China; liqiushan@scu.edu.cn (Q.L.); koide@scu.edu.cn (K.O.)
* Correspondence: kabil@scu.edu.cn

Abstract: Housing reconstruction plays a crucial role in renovating disaster-hit areas. Rural areas are considerably different from urban areas in terms of geographic environment, building size, residential culture, and social organization. Therefore, post-disaster recovery and reconstruction models for urban areas cannot be applied directly to disaster-hit rural areas. This study, based on the experience of rural housing reconstruction after the Wenchuan earthquake, identified key strategic issues in housing reconstruction that must be addressed to achieve the goal of "building back better" in the future. By taking the experience of Dujiangyan as our reference, the study found that the following strategies are important for successful housing reconstruction in rural areas: (1) actively involve disaster victims through a participatory institutional design; (2) coordinate the interests of governments, markets, and disaster victims and the functions of living, production, and ecology through a classified housing reconstruction system; and (3) activate the quota for rural collective construction land and create a new source of funding for housing reconstruction through the market circulation of urban-rural land. Additionally, in the context of urban-rural integration, changes in land use can lead to rural spatial reconstruction and sustainable regional development, providing a reference for formulating optimal post-disaster reconstruction strategies.

Keywords: housing reconstruction; earthquake disaster; recovery; rural community; Jiangyin

1. Introduction

For people in developing countries, housing is usually one of the most precious assets. In the event of disasters (especially sudden disasters), housing is usually impacted or damaged most severely and suffers the greatest losses among the overall impacts on the national economy. Therefore, many post-disaster recovery programs allocate the largest proportion of resources and give first priority to housing and infrastructure reconstruction compared with other sectors [1]. Reconstructing housing post-disaster is considerably different from reconstructing other forms of infrastructure and public facilities (particularly, interest re-allocation arising from the diversity of property rights), as it involves directly engaging with the victims. Thus, housing reconstruction has been the primary concern of victims and governments in the aftermath of disasters. Post-disaster housing reconstruction involves repairing structural damage to housing and providing necessary services and facilities to ensure the health, safety, and security of local residents [2,3]. During the reconstruction period, resettlement of victims, reinforcement of damaged housing, and construction of new housing are critical issues for post-disaster emergency management and reconstruction. Early studies of post-disaster recovery focused on describing the behavioral responses of victims and recovery and reconstruction organizations during and after post-disaster recovery [4]. By investigating the responses of individuals and communities across the recovery and reconstruction processes, Barton [5] argued that conflicts and contradictions between victims and local governments are more significant

Citation: Li, Q.; Wumaier, K.; Osamu, K. Post-Earthquake Housing Reconstruction Management and Implementation in Rural Areas: Review and Lessons from Dujiangyan, Wenchuan Earthquake. *Buildings* **2023**, *13*, 2251. https://doi.org/10.3390/buildings13092251

Academic Editors: Mislav Stepinac, Marco Francesco Funari, Chiara Bedon, Tomislav Kišiček and Ufuk Hancilar

Received: 7 August 2023
Revised: 29 August 2023
Accepted: 3 September 2023
Published: 5 September 2023

at the intermediate stage than during the emergency aid period or at later stages. Rubin suggests that well-targeted policy measures can effectively solve conflicts during the reconstruction period [6]. Alesch argues that post-disaster reconstruction should promote further development and progress in disaster-hit areas rather than merely restoring the pre-disaster state [7]. Developing countries should implement post-disaster recovery and reconstruction in conjunction with regional vulnerability analysis to overcome the vulnerabilities of disaster-hit areas [8,9]. Housing reconstruction is crucial for overall reconstruction effectiveness and involves direct stakeholders and long-term relationships with diverse parties [10]. Many housing reconstruction cases worldwide have revealed three major problems: management and coordination, funding and compensation, and major reconstruction participants [11–13]. Among them, reconstruction funding is the first prerequisite to ensuring rapid recovery in disaster-hit areas. Further, reasonable victim compensation and fund allocation are an important basis for evaluating the effectiveness of housing reconstruction and a significant factor in maintaining the order of the housing market. A statistical analysis of 209 articles [14] shows that ineffective management and coordination failure are the most frequently mentioned challenges to housing reconstruction in the existing literature because they increase the frequency of repetitions, errors, and faults during urban-rural housing reconstruction.

Since the 2004 Indian Ocean tsunami and 2005 US hurricane Katrina, academic seminars and studies on post-disaster reconstruction planning have increased sharply [15]. This, combined with the increasing international attention on global climate change, has led to an increasing number of people realizing the importance of disaster adaptability in human settlements [16,17]. One of China's major structural problems is the urban-rural dichotomy, which has resulted in a significant urban-rural gap. Rural areas are considerably different from urban areas in terms of geographic environment, building size, residential culture, and social organization. Therefore, post-disaster recovery and reconstruction models for urban areas cannot be applied directly to disaster-hit rural areas. Li et al. have investigated in detail the urban housing reconstruction policy and reconstruction pattern after the Wenchuan earthquake [18]. This study provides a thematic analysis and systematic summary of the entire process of post-disaster rural housing reconstruction in four aspects (including planning system and special policy, implementation mechanism and management process, land supply, and reconstruction effectiveness). As a supplement to the planning and management mode of housing reconstruction after major earthquakes, our findings will serve as a reference for rural post-disaster emergency management.

2. Study Area

Dujiangyan City is located in the northwestern Sichuan Plain. The Wenchuan earthquake impacted the housing of 132,487 rural households (approximately 94% of total rural households in Dujiangyan), with 75% being either collapsed or damaged. Safety appraisal shows that the severely damaged and collapsed houses collectively accounted for 58% (44.34% and 14.12%), and the moderately and slightly damaged houses collectively accounted for 35% (16.57% and 18.64%). Additionally, 73% of the impacted houses (95,130 rural households) needed to be repaired, and 27% of the impacted houses (housing 34,447 rural households) needed to be reconstructed. The rural housing was mainly distributed in mountainous areas near the Longmen Mountain Fracture Zone, with most of these areas being devastated. The damage rate of rural housing in Longchi Town, Xiang'e Township, Zipingpu Town, Yutang Town, and Hongkou Town was 100%, 88%, 94%, 87%, and 69%, respectively. Compared with the above townships/towns, the damage rate of rural housing was relatively low in the townships/towns adjacent to or along the mountainous areas of the Longmen Mountain Fracture Zone. However, compared with the townships/towns in plain areas, the damage rate of rural housing was relatively high (e.g., 25% and 41% in Qingchengshan Town and Daguan Town, respectively). The damage rate of rural housing was considerably low in most plain areas (e.g., 8%, 18%, 12%, 5%, and 11% in Juyuan, Chongyi, Shiyang, Liu-jie, and Anlong Town, respectively) but relatively high in certain plain areas (e.g., 23%, 37%, and 46% in Tianma, Xingfu, and Cuiyuehu, respectively), as shown in Figure 1.

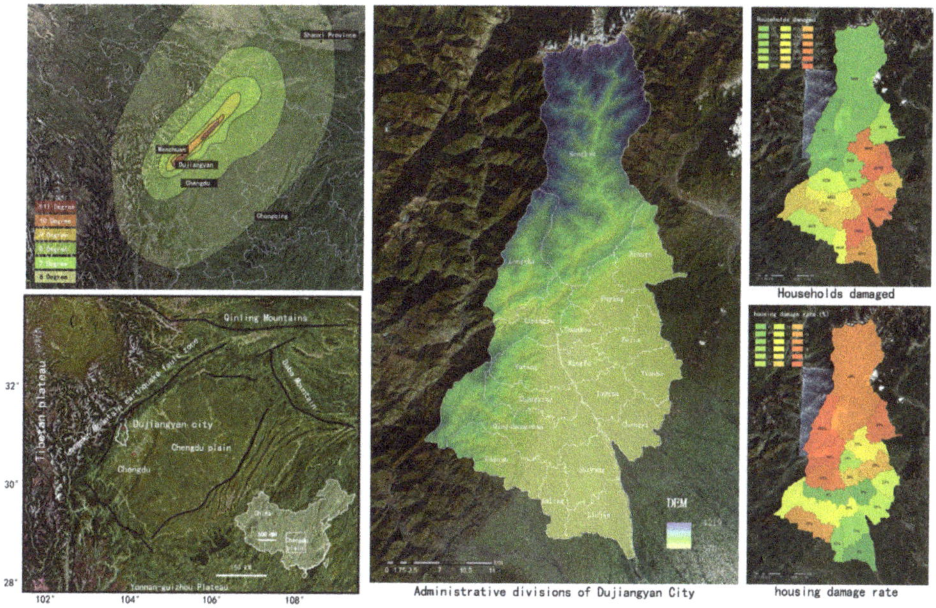

Figure 1. Regional location and spatial distribution of housing damage in the study area.

3. Differences between Urban and Rural Planning Systems

3.1. Changes in China's Legal Systems of Urban and Rural Planning

Figure 2 describes the changes in China's legal system for urban and rural planning. Before the implementation of the new planning law, urban and rural areas were planned and constructed in accordance with different laws or regulations. Specifically, urban planning and construction complied with the City Planning Law of the People's Republic of China (hereinafter referred to as the "Old Planning Law"), whereas rural planning and construction complied with the Regulations on the Planning and Construction of Villages and Towns (hereinafter referred to as the "Old Rural Regulations"). As a national law adopted by the National People's Congress, the old planning law stipulates that urban land is state-owned and may be used on a paid basis. The old rural regulations fall under administrative regulations, stipulating that rural land is collectively owned and that the right to use rural land may be transferred to governments before peasants are compensated for the requisitioned land. The new planning law was promulgated at the end of 2007 and came into force on 1 January 2008, while the old planning law became null and void. The new planning law is groundbreaking in proposing the requirement for urban-rural integrated development and management, suggesting that China's modernization and urbanization entered a new stage. Compared with the old planning law, the new planning law emphasizes urban-rural integrated planning and management, specifically the following: (1) planning industry and agriculture, cities and villages, urban residents and rural villagers as a whole; (2) performing institutional reforms and policy adjustments to promote urban-rural integration in terms of planning and construction, industrial development, market information, policy measures, ecological and environmental protection, and social undertakings and change the long-standing urban-rural dual economic structure; (3) achieving equality in policies, complementarity in industrial development, and equality in national treatment (e.g., helping peasants enjoy the same civilization and benefits as urban residents); and (4) achieving all-round, coordinated, and sustainable socioeconomic development of both urban and rural areas.

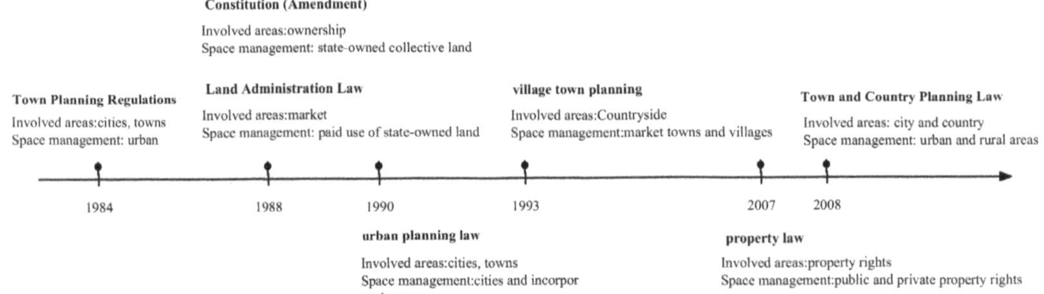

Figure 2. Changes in China's urban and rural planning legal systems.

3.2. Institutional Background of Post-Disaster Reconstruction Planning

Chengdu City has established a national-level demonstration zone for urban-rural integrated development, with a focus on reforming the rural property rights system, freeing up the circulation of rural land markets, and adjusting the spatial structure of modern ecological agriculture. These reforms provide the legal and institutional background for Dujiangyan's rural reconstruction planning. Property rights are the core of rural land systems, and the registration of land ownership and property rights is a crucial aspect of rural reconstruction. In 2007, Chengdu took the lead in implementing the rural property rights confirmation and registration system in China, which explicitly confirmed related property rights (e.g., collective land ownership, use rights of collective land, housing ownership, land contracted management rights, and use rights of forest land) and provided rural land to rural collective organizations and households [19]. According to the Decision on Deepening the Reform in Strict Land Management promulgated by the State Council in December 2004, Chengdu implemented an increase and decrease linkage system of urban-rural construction land to facilitate rural reconstruction after the Wenchuan earthquake. On the premise of expert review and approval, as well as the consent of at least 66.7% of villager representatives, the following reform measures were implemented: (1) further consolidating and allocating land through the "demolition and reconstruction" project; (2) transforming the surplus quota for collective construction land in a market-oriented way; and (3) helping rural households in demolition areas earn extra revenue from the differential rent of circulated land [20].

3.3. Institutional Connection between Post-Disaster Reconstruction Planning and Pre-Disaster Rural Planning

The Decisions of the Central Committee of the Communist Party of China on Several Major Issues in Building a Harmonious Socialist Society (promulgated in 2006) have planned to "actively promote the construction of rural communities, develop a perfect management and service system for new-type communities, and turn rural communities into well-managed, well-served, civilized, and peaceful social life communities." The concept of "rural communities" was first used in an official document released by China's central government. Dujiangyan's New Rural Community Planning (2007–2020) has planned to reduce Dujiangyan's total peasant population from 390,000 to 270,000. Specifically, (1) 120,000 peasants will be resettled in new rural communities, (2) 150,000 peasants will be settled in Linpan (Dujiangyan's traditional rural settlements) residential conservation areas, and (3) 120,000 peasants will be turned into rural citizens by developing amalgamated dwellings in urban planning areas. Before the Wenchuan earthquake, Dujiangyan's overall plan for urban-rural integration was formulated, considering a "large concentration" pattern. After the Wenchuan earthquake, the Chengdu municipal government proposed the planning principles of "three concentrations" in August 2008, specifically the following: (1) industrial parks where manufacturing enterprises are concentrated; (2) concentration of rural towns or new rural communities; and (3) large-scale concentration of agricultural land. The "three concentrations" principle aims to achieve compact, concentrated, and intensive development and promote the coordination of industrialization, urbanization, and agricultural modernization. The Dujiangyan Municipal Government further proposed four principles for rural reconstruction (development, diversity, sharing, and compatibility) and specified 15 new-type rural communities in the original rural reconstruction planning as permanent resettlement housing, which is intended to accommodate 16% of the total resettled rural population. Accordingly, Dujiangyan's overall planning of urban-rural integration was adjusted and revised to highlight a pattern of "relative concentration and small concentration" [21]. To date, "large concentration" and "relative concentration and small concentration" are the dominant patterns of rural development planning. "Large concentration" enables residents to manage the relevant infrastructure and public service facilities conveniently but increases the cost of travel between people and land and is averse to the implementation of agricultural activities. The scale of this kind of reconstructed rural community is generally large, which is not only not conducive to the development of agricultural activities but also has a certain degree of impact on the local traditional settlements and agricultural cultural landscape [22]. "Relative concentration and small concentration" inherit the characteristics of rural settlement and facilitate the development of large-scale and modern agriculture, but incur extra costs due to the capital construction of public services.

4. Rural Housing Reconstruction Planning and Its Implementation

4.1. Objective Management and the Management Process of Post-Disaster Reconstruction

In the context of the Dujiangyan region, the reconstruction of rural residences follows a bifocal managerial approach encompassing reconstruction goal management and re-construction process management. The crux of reconstruction goal management revolves around the strategic imperative of effecting a seamless culmination of housing reconstruction initiatives within a concise biennial temporal span (as depicted in Figure 3). Noteworthy temporal milestones encompass the accomplishment of transitional resettlement within the tri-monthly ambit subsequent to seismic events; the fruition of development intentions pertaining to housing reconstruction within the hexa-monthly aftermath of the seismic occurrence; and the consummation of permanent domicile construction endeavors within the biennial post-seismic timeline, notably culminating on 12 May 2010.

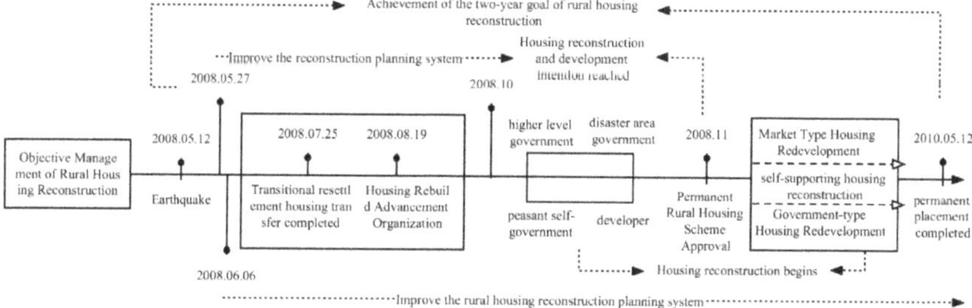

Figure 3. Flowchart on Dujiangyan's rural housing reconstruction objective management.

Facilitating the reconstruction process (as delineated in Figure 4) entails a dual-faceted approach comprising seismic disaster response strategies and housing reconstruction strategies. The former is principally geared towards damage evaluations coalesced with comprehensive safety assessments. The latter, marked by its comprehensive ambit, encapsulates a tri-level strategy. Commencing with the trajectory of planning guidance, the constitution of a Rural Recovery and Reconstruction Work Promotion Organization, as expounded in Section 4.2, synergizes the objectives of transitional settlement paradigms and fosters sustainable development, all while aligning with doctrinal positions stipulated within regulatory frameworks concerning rural housing reconstruction. This august organization duly engenders a comprehensive blueprint catering to the perpetual reconstruction of rural domiciles. Subsequently, within the rubric of implementation systems, this pivotal organization promulgates a specialized set of policies spanning four cardinal dimensions, encompassing land tenure systems, financing frameworks, property rights reform frameworks, and administrative support services, all geared towards fortifying the reconstruction enterprise. Lastly, within the terrain of development models, anchored upon a bedrock of respecting the volition of disaster survivors and safeguarding their inalienable rights, an assortment of five distinct reconstruction paradigms is presented for selection, encompassing on-site self-reconstruction, government-driven standardized construction, government-driven standardized construction coupled with a self-reconstruction facet, community-participatory standardized construction, and community-participatory standardized construction complemented by self-reconstruction.

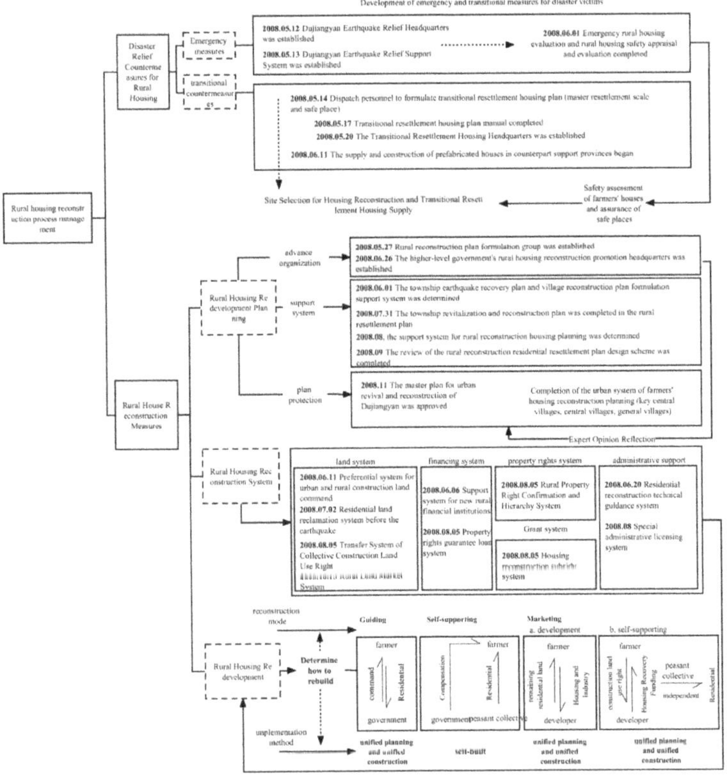

Figure 4. Flowchart on Dujiangyan's rural housing reconstruction process management.

4.2. Executive Organizations for Rural Housing Reconstruction

4.2.1. Government-Led Executive Organizations and Task Allocation

- Executive organization at the planning formulation stage

After the Wenchuan earthquake in China on 12 May 2008, local governments in the affected rural areas initiated post-disaster rural housing reconstruction planning on 25 May 2008, and were scheduled to complete it within two months. The planning included township, town and district, village, and rural housing reconstruction planning.

Dujiangyan formed a post-disaster reconstruction planning coordination team on 27 May as the supreme decision-making body for rural housing reconstruction planning, responsible for programming, outlining, management, and coordination. Planning formulation teams were established under the coordination team to receive guidance from higher-level governments. A support system for planning formulation was established on 1 June 2008, based on pairing-assistance and dispatched talents or volunteers in planning and design. The coordination team was responsible for the following: (1) determining the scale (including construction land and resettled population) and safe layout of town reconstruction, investigating the disaster status, and formulating the overall planning (for townships, towns, districts, and villages), and (2) investigating the intentions of township/town-level governments, village committees, and rural households.

- Executive organization at the planning and implementation stage

The executive organization for rural housing reconstruction was established on 5 August 2008, with the Dujiangyan Municipal Government as the main implementer. The executive organization was responsible for policymaking and announcements, as well as decision-making on work programs and objective management. The executive organization comprised seven municipal government agencies (the Coordination Bureau, Civil Affairs Bureau, Planning Bureau, Land Resources Bureau, Construction Bureau, Housing Management Bureau, and Economy Bureau) and township/town-level governments with a vertical work responsibility system (Table 1). The policy announcement covered the following content: (1) means and methods of housing reconstruction; (2) housing reconstruction policies and measures; (3) technical guidance and investment for reconstructed housing; (4) inspiration for revenue allocation to rural households or collective economic organizations; (5) social attention; (6) progress and key events in housing reconstruction; and (7) new situations and problems in the process of housing reconstruction. Policy announcements were mainly conducted through the news media, the Internet, and news conferences. The management aimed to achieve housing reconstruction and reinforcement on a village scale, according to the project ledger for rural housing reconstruction.

Table 1. The executive organization and task allocation of rural housing reconstruction.

Task Allocation	City-Level Government	Township/Town-Level Government
Administrative services	Issuing one letter and three permits [1] (i.e., a letter of site selection opinion, a planning permit of construction land, planning permit of construction engineering, and a planning permit of rural construction) for administrative licensing and re-registering land property rights (including land ownership, the use right of collective construction land, the contracted management right of agricultural land, the use right of forest land, and rural housing ownership).	Receiving and reviewing the applications for construction land and housing land at resettlement points, handling the related administrative procedures, and applying for the property rights of resettlement housing and land.
Market services	Circulation of collective construction land and increase and decrease linkage of urban-rural construction land.	Confirming, bulletining, and registering the adjustment of land property rights, adjusting the interest distribution of intensive construction land, and supervising the progress in housing land development.
Housing reconstruction	Handling the application for housing safety appraisal, disaster-hit housing subsidies, and housing reconstruction and resettlement agreements on the village or villager group scale.	Handling the selection of housing reconstruction models, investigating and reviewing the profile of rural households, and adjusting the ownership of resettlement land.

Table 1. *Cont.*

Task Allocation	City-Level Government	Township/Town-Level Government
Reconstruction planning guidance	Specifying the boundaries of urban planning areas and rural planning areas, performing field boundary measurement for collective construction land and reconstructed housing, conducting layout design and safety appraisal for post-disaster reconstructed permanent housing and centralized resettlement points, and specifying the supporting facilities for the increase and decrease linkage of urban-rural construction land.	Publicizing the superior governments' related policies to rural households or collective economic organizations and providing guidance to the circulation of collective construction land.
Reconstruction planning formulation	Providing unified guidance to reconstruction planning and design, architectural landscape, architectural quality, infrastructure, and public facilities.	-
Technical guidance and supervision	Providing technical guidance and supervision, including construction quality supervision, safety management, and engineering completion inspection and acceptance for reconstructed housing, completion inspection and acceptance for land circulation and development projects, and inspection and acceptance for cultivated land development on the township, village, or villager group scale.	Application management, construction supervision, and completion inspection and acceptance records for small-sized engineering projects.

[1] The Urban and Rural Planning Law expressly stipulates the setting of planning permission as well as the bulletining and disclosure of the implementation procedure, condition, process, and results of planning permission. Articles 36, 37, 38, and 40 stipulate that the related systems (e.g., letter of site selection opinion, planning permit of construction land, and planning permit of construction engineering) are implemented for urban planning in China; Article 41 stipulates that the rural construction planning permit system is implemented for rural planning in China.

4.2.2. Villager Self-Governed Executive Organization and Task Allocation

On the premise of respect for victims' willingness and villager self-governance, five housing reconstruction models (in situ reconstruction, unified planning and self-construction, unified planning and unified construction, monetized resettlement, and market-oriented development) were available for victims or rural collective economic organizations in accordance with the opinions on the implementation of rural housing reconstruction (released by Dujiangyan Municipal Government in August 2008). On a village or villager group scale, grass roots (e.g., township/town-level) administrative organizations publicized the post-disaster relief and resettlement policy and investigated the victims' basic information and need for housing reconstruction.

Except for in situ reconstruction, the planning programs for the four other housing reconstruction models must be based on the premise of close communication between design institutions and victims, and the implementation programs must be based on the opinions of village cadres and victim representatives. A victim discussion system was established for the planning and implementation programs, for example, according to the unanimous opinion of the village meetings, villager representative meetings, village council meetings, and courtyard meetings. The discussion results were published on the bulletin boards of villages or villager groups to solve the problem of interest adjustment (2009, field survey). The earthquake victims made decisions about related issues (e.g., land consolidation, industrial development, living environment governance, and concentration mode of collective construction land) within villages or villager groups. Unified planning and self-construction were mainly implemented by village committees, and in situ reconstruction was mainly implemented by rural households or joint-tenancy parties.

Property management was performed for villager groups' self-resettlement points based on the principles of autonomy, voluntarism, and self-government and through democratic consultation, and villager group committees addressed the difficulties and contradictions faced by property management at centralized resettlement points.

4.2.3. Technical Guidance from Dispatched Experts and Task Allocation

In mid-August 2008, a technical guidance system for housing reconstruction planning and design was implemented under the leadership of the Construction Committee of the Chengdu Municipal Planning Bureau. Under the technical guidance system, experts from Chengdu's more than 100 registered planning and architectural design institutions constituted the executive organization for rural housing reconstruction planning and were dispatched to the earthquake-hit areas to provide guidance for reconstruction planning. The following tasks were allocated to them: (1) providing technical guidance to rural planning and construction in various aspects (e.g., safe layout, construction standard for public and municipal facilities in rural resettlement points, architectural configurations, built environments, architectural landscapes, and architectural structures); (2) meeting the requirements for the four principles of planning and design (including diversity of regional characteristics, sharing of infrastructure, integration of production, living and ecological integration, and development); and (3) providing rural planning and construction atlas (e.g., drawings for environmental design, architectural configuration, residential building, and aseismic structure design) to guide peasants' self-built housing. In particular, to satisfy the housing needs of peasants, the experts must fully understand the disaster severity and victims' willingness for housing reconstruction, consider diverse factors (e.g., cost-effectiveness and practicality of building space design, household size, blood relations, and family relatives), design appropriate apartment layouts, and ensure equality in the design of construction land and construction area.

4.3. Special Systems and Policies for Rural Housing Reconstruction

4.3.1. Cash Subsidy and In-Kind Relief System for Rural Housing Reconstruction

Cash subsidies for Dujiangyan's rural housing reconstruction were sourced from post-disaster reconstruction funds allocated by the central and local governments of different levels (including the people's governments of Sichuan Province, Chengdu City, and Dujiangyan City) and granted to rural households whose housing was considered to meet the standard for damaged housing. Depending on the severity of the housing damage, household income, and household composition, the subsidy standard was set at CNY 20,000 per general rural household and CNY 23,000 per poor rural household. Cash subsidies were advertised and confirmed in the villages or village groups concerned before they were granted. Rural households were not eligible to receive cash subsidies if they had no earthquake-damaged housing and were centrally resettled; rural households were eligible to receive transferred subsidies separately if they were resettled from geologically hazardous areas. By using the pairing-assistance funds or social donations, in-kind aid was used to centrally resettle and take care of the disadvantaged groups (e.g., orphans, solitary elderly, and solitary disabled people, as well as homeless, vagrant, and five-guarantee peasants) in welfare houses, elderly nursing homes, and happy civil dwellings. According to the field surveys in 2009 and 2010, in-kind aid mostly existed in the centralized resettlement peasant communities and in situ, self-reliantly reconstructed areas where related service facilities were not available.

4.3.2. Special Loan System for Rural Housing Reconstruction

In August 2008, Dujiangyan implemented a special loan system for rural housing reconstruction among rural households and rural collective economic organizations that selected self-reliant in situ reconstruction, unified planning, and self-construction. As a newly established and solely state-owned enterprise with corporate property management rights, Dujiangyan Property Rights Circulation Financing Guarantee Co. Ltd. provided loan guarantees for rural housing reconstruction. Meanwhile, an administrative registration and licensing system was implemented to protect the legal rights and interests of the loan guarantor and reduce the risk of loans (government interview records, 2009/3). The financial institutions of Dujiangyan granted special loans for rural housing reconstruction, with a maximum loan amount of CNY 60,000, a maximum repayment period of eight years, and a preferential loan rate of 60% of the benchmark loan rate for three years subsequent to the Wenchuan earthquake. Additionally, rural households registered with the special loan system before 2009 were exempt from the administrative handling fee. For rural households who paid off their loan principal before the end of 2009, Dujiangyan's special funds for post-disaster reconstruction would pay their loan interest in full; for rural households who paid off their loan principal before the end of 2010, Dujiangyan's special funds for post-disaster reconstruction would pay 5% of their loan interest [23].

4.3.3. Innovative Rural Land Systems

Property rights circulation system for rural land. In October 2008, Chengdu established the Chengdu Rural Property Rights Exchange, with subordinate branches established in Dujiangyan City and towns under its jurisdiction. The exchange was renamed the "Urban-rural Property Rights Exchange" in 2009 and provided one-stop market services for the property rights circulation of urban-rural land. The property rights circulation system for rural land aimed to subsidize the conversion of agricultural land into state-owned construction land and restrict the transfer of rural collective construction land. This system was China's first land marketization system, institutionalizing the free circulation of agricultural land and the marketization of rural collective construction land.

Increase and decrease the linkage system of urban-rural construction land. The increase–decrease linkage system of urban-rural construction land mainly has the following features: (1) the increase or increase in rural collective construction land linked with the change of land use; (2) higher priority for the earthquake-hit areas in Chengdu; (3) the obligatory to demolish existing buildings and reclaim land in the original collective construction land; (4) the earthquake-hit rural households were eligible to take advantage of the housing, infrastructure, and public facilities reconstructed according to the rural housing reconstruction planning; and (5) the surplus quota for collective construction land was used for the development of earthquake-hit areas or large cities' peripheral areas.

Property rights re-registration and reconfirmation of rural land. A city-level liaison meeting system is developed to coordinate, manage, and supervise the property rights re-registration of city-wide rural reconstruction land and implement integrated property rights management of urban and rural land. Hence, it is necessary to develop a two-level (town-level and township-level) management system that provides administrative services for property right re-registration (e.g., definition of property rights, property right registration, policy announcement, and certificate handling), clarifies the land revenue relationship, and issues certificates for five rights (land ownership, use right of collective construction land, contracted management right of agricultural land, use right of forest land, and rural housing ownership). Official statistics showed that in 2009, Dujiangyan totally issued 2259 certificates on collective land ownership, 91,744 certificates on use rights of collective construction land, 74,995 certificates on use rights of forest land, and 54,192 certificates on rural housing ownership.

Contracted management rights re-registration and reconfirmation of agricultural land. Villages, villager groups, and rural households jointly assist the land management authorities to demarcate the farmland of each rural household. Subsequently, the results of farmland demarcation are signed and confirmed by rural households if they are bulletined without any objection. Finally, rural households are granted the certificate of contracted management rights if villages and villager groups file an application for registration and confirmation with the township/town-level land management authorities, and such an application is reviewed and approved by Dujiangyan's rural land management authorities. The above procedure is applicable to the area adjustment and contract extension regarding farmland contracted management rights. It is of considerable significance for ascertaining the total amount of agricultural land, simplifying administrative management, and publishing market information on agricultural land circulation.

Use the right registration and confirmation of collection of construction land for rural housing reconstruction. Eligible investors and village or villager group collective organizations may apply for the construction land use right of jointly built housing, which is China's first property rights recognition system. Eligible real estate developers and village or villager group collective organizations may apply for the collective construction land use right of centrally rebuilt housing (including the collective construction land use right of centralized peasants' residential areas and new urban development areas).

Ownership registration and confirmation of peasants' reconstructed housing. The ownership of buildings in reconstruction areas should be registered and confirmed. Depending on the method of housing reconstruction, there are three types of ownership: ownership of scattered rural households' self-constructed housing; ownership of resettlement housing in centralized residential areas; and ownership of jointly constructed peasant housing. Jointly constructed peasant housing includes scattered rural households' jointly constructed housing and jointly reconstructed housing, with their ownership owned by rural households and the joint-construction party.

4.3.4. Classified Housing Reconstruction System

Table 2 describes the effectiveness of Dujiangyan's classified rural housing reconstruction policy. The first type is self-reliant housing reconstruction, including in situ reconstruction (e.g., in situ self-reconstruction and in situ joint reconstruction) and unified planning and self-construction. The second type is government-led unified planning and unified construction. As agreed, rural collective economic organizations transfer the surplus quota for collective construction land generated during centralized resettlement to governments, in return for which governments provide collective economic organizations with reconstructed housing and living environments. Governments are the main implementers of unified planning and unified construction. The third type is market-oriented unified planning and unified construction. With the consent of at least 66.7% of total rural households and in accordance with the bidding contract, rural collective economic organizations may transfer the surplus quota for collective construction land to real estate developers, in return for which disaster victims are provided with reconstructed housing and living environments.

4.3.5. Centralized Resettlement System

According to the principle of "three concentrations," Dujiangyan's overall planning of post-disaster reconstruction defines three types of centralized resettlement points: (1) new-type community of the key central village level or large rural resettlement point with at least 300 rural households; (2) new-type community of the central village level or medium rural resettlement point with 100 to 300 rural households; and (3) a resettlement point of the basic village level with less than 100 rural households. Table 3 describes their spatiotemporal distribution characteristics. Among them, the largest single centralized resettlement point has 1017 rural households, 1.6 times the average village size before the Wenchuan earthquake (635 households per village). Centralized resettlement enables governments to acquire the urban development rights to 4,860,000 m^2 of construction land and the ownership of such land. When earthquake victims are resettled in urban-type, centrally managed housing, they are registered as urban households and entitled to enjoy community management.

Table 2. Implementation effect of housing reconstruction in rural areas of Dujiangyan rural area.

Category	Reconstruction Method	Implementing Subject	Reconstruction Funds						Living Environment			Housing Reconstruction Achievements				
			Own Funds	Subsidies	Pothook Project	Land Transfer	Loan	Contribution	Infrastructure Renovation	Public Facility Renovation	Management of Original Residential Land	Placement Point (Location)	Land Area (mu)	Floor Space (m²)	Number of Resettlement Households	Number of Households Destroyed by Disaster (Including)
Self-reliant	In situ reconstruction	Farmer	✔	✔			✔	✔			Reserve					1994
	Unified planning and self-construction	Farmers' Committee	✔	✔	✔	✔	✔	✔	✔	✔	Return to cultivation	85		850,400	7444	6262
Government-led	Unified planning and unified construction	Government		✔	✔			✔	✔	✔	Return to cultivation	48		3,043,000		16,009
Market-oriented	Unified planning and unified construction	Developer				✔			✔	✔	Land consolidation	31		1,340,000		6249
	Unified planning and self-construction	Farmers' Committee enterprise	✔	✔		✔	✔	✔	✔	✔	Land consolidation	39	960.12	481,300	4233	3455
	Joint reconstruction	Investor				✔			✔	✔	Land consolidation	2	26	78,000	168	93
Monetizing	Resettlement	Farmer		✔							Land consolidation					395
Total												205	11,477.2	5,792,700	41,870.9	34,447

Table 3. Implementation effect of a centralized resettlement system for housing reconstruction in Dujiangyan rural area.

Resettlement Type	Reconstruction Method	Placement Point (Location)	Construction Area (mu)	Floor Space (mu)	Internal Ratio (%)		Overall Ratio (%)		Resettlement Households		Affected Person Ratio (%)	Scale of Centralized Resettlement		Per Capita Area (m²)	
					Land Area	Floor Space	Construction Area	Floor Space	Transferred Households	Destroyed Households		Average Number of Resettlement Households	Placement Ratio (%)	Land Area	Floor Space
≥300	UPUC [1] (government-led)	24	4012	251	64	68	36	45	24,410	10,821	44	12,017	46.4	32	30
	UPUC (market-oriented)	12	1970	110	32	30	18	20	11,153	4641	42	797	21.2	35	29
	UPSC [2] (market-oriented)	1	70	5	1	1	1	1	350	426	122	350	0.7	39	42
	UPSC (government-led)	1	185	4	3	1	2	1	347	537	155	347	0.7	105	354
	sub-total	40	5237	370	100	100	56	66	36,260	16,425	45	907	68.9	34	30
100–300	UPUC (government-led)	20	1306	47	38	36	12	8	3796	3471	91	190	7.2	67	36
	UPUC (market-oriented)	6	473	19	14	14	4	3	1390	1143	82	232	2.6	67	40
	UPSC (market-oriented)	9	531	23	16	18	5	4	2183	1876	86	243	4.2	48	31
	UPSC (government-led)	22	1100	41	32	32	10	7	3491	2878	77	159	6.6	62	35
	sub-total	57	3409	130	100	100	30	23	10,860	9168	84	191	20.6	62	35
≤100	UPUC (government-led)	3	145	3	9	5	1	1	204	257	126	68	0.4	139	42
	UPUC (market-oriented)	11	149	6	10	9	1	1	516	465	90	47	1	57	32
	UPSC (market-oriented)	26	194	14	12	23	2	3	1151	967	84	44	2.2	33	36
	UPSC (government-led)	62	1074	40	69	64	10	7	3606	3237	90	58	6.9	58	32
	sub-total	102	1561	62	100	100	14	11	5477	4926	90	54	10.4	56	34
	Total	199	11,207	562			100	100	52,597	30,519	58	264	100	42	31

[1] UPUC = unified planning and unified construction; [2] UPSC = unified planning and self-construction.

5. Characteristics and Benefits of Dujiangyan's Rural Housing Reconstruction

5.1. Self-Reliant Housing Reconstruction

Rural households that selected self-reliant in situ reconstruction, unified planning, and self-construction collectively accounted for 17% (9438 households). In situ reconstruction was mainly implemented by rural households, was unplanned and completely self-reliant, and was not provided with public infrastructure but retained the use rights of the original housing land. Unified planning and self-construction were mainly implemented by village committees, which implied that public infrastructure would be built as planned and original housing land would be reclaimed as farmland. For unified planning and self-construction, the development scale was relatively small; the resettled rural households accounted for 14%, the covered land area accounted for 21%, the floor area accounted for 15%, and the average floor area was 114 m^2 per household. Here, the reconstruction of Anlong Town is taken as an example. Specifically, 13% (872 households) of its total rural households registered before the Wenchuan earthquake were resettled in seven reconstructed residential areas, and 40% of the 872 rural households selected government-led unified planning and self-construction, as typified by Hejia Village and Dongyi Village. Dujiangyan Urban and Rural Planning Institute provided planning and design services for the resettlement points, adopting the Linpan-style layout of Western Sichuan, in which courtyards, gardens, and lanes were interlaced, forming unique spatial characteristics (retaining the original scattered residence and forming a relatively centralized residence pattern).

5.2. Government-Led Housing Reconstruction

Numerous rural households (28,720 households, 54%) selected government-led unified planning and unified construction. For government-led unified planning and unified construction, the covered land area accounted for 48%, the floor area accounted for 53%, and the average floor area was 106 m^2 per household. Government-led unified planning and unified construction were mainly implemented by local governments and characterized by unified planning and construction based on government-led market attributes. Governments could recover the surplus quota for collective construction land generated by the centralized development of construction land, in return for which disaster victims were provided with reconstructed housing, infrastructure, and public facilities. Additionally, disaster victims' original housing land was reclaimed as farmland. Here, the housing reconstruction of Xiang'e Township is taken as an example. Government-led unified planning and unified construction were adopted with the consent of 94.5% of peasant voters. With the raised housing reconstruction fund of approximately 600 million yuan, 16 new rural communities were built to resettle 3425 rural households, saving nearly 4000 mu collective construction land. Furthermore, land preparation was performed to promote the development of modern ecological agriculture (e.g., kiwi fruit, green tea, shoot bamboo, and three types of medicinal herbs [Eucommia ulmoides, Magnolia officinalis Rehd et Wils, and Cortex Phellodendri Chinensis]) and form a large-scale distinctive agricultural base (approximately 10,000 mu), transforming the traditional modes of life and production into modern modes of life and production.

5.3. Market-Oriented Housing Reconstruction

Rural households that selected market-oriented unified planning and unified construction, unified planning and self-construction, or joint construction collectively accounted for 31%. These housing reconstruction models were mainly implemented by real estate developers or village committees. In market-oriented unified planning and self-construction or joint construction, rural collective organizations or households transferred their surplus quota for collective construction land to real estate developers and received reconstructed housing, infrastructure, and public facilities. For market-oriented unified planning and unified construction and unified planning and self-construction or joint construction, the resettled rural households accounted for 31% (17,460 households), the covered land area accounted for 31%, and the floor area accounted for 33% [24]. Moreover, housing reconstruction was accompanied by land preparation. Here, Shiqiao Village in Qingchengshan Town is taken as an example. With the consent of at least 66.7% of the villagers of Shiqiao Village, market-oriented unified planning and unified construction were adopted, and the surplus quota for collective construction land was publicly transferred to decide on housing reconstructors. Real estate developers provided village committees with reconstructed housing (hereinafter referred to as "new communities") that was self-governed by rural households and rural collective economic organizations. Thus, all 11 villager groups and 910 rural households (2000 people) registered before the Wenchuan earthquake were resettled in one new community. After the original housing land was reclaimed as farmland and Linpan land was reclaimed as forest land, the surplus quota for collective construction land and forest land might be circulated in the market to raise funds for housing reconstruction. For example, the 110,000 m^2 collective construction land in Shiqiao Village was circulated and used to develop "Happy Farmhouse" rural tourism.

6. Discussion and Conclusions

The primary issue with rural housing reconstruction is that the property rights subject of rural land is unclear, and the disposal of rural land ownership is not governed by the land property rights system [25]. In other words, peasants do not truly own land, and their property rights in rural areas cannot be sufficiently guaranteed. In the post-disaster housing reconstruction process, implementing key land management systems such as the land planning system, land requisition system, land property rights system, and collective construction land system, as well as efficiently using, supervising, and planning rural land, is challenging. The goal of Dujiangyan's post-disaster rural reconstruction planning is to achieve population concentration and land circulation through expert technical support, villager participation, administrative guidance, and information disclosure. Currently, post-disaster housing reconstruction models continuously affect subsequent rural planning and development. The recent survey of rural planning in Chengdu (including Pidu District, Dujiangyan City, and Chongzhou City) shows that villagers universally preferred government-led unified planning and construction for resettlement housing construction, and they had increasing basic life needs. Owing to various reasons (e.g., significant differences in fiscal situation between local governments and implementation of the Linpan repair and preservation project—one of Chengdu's ten major projects for happy and good life), the housing reconstruction model varied across regions, and local governments considered how to address the relationship between rural housing construction and preservation of traditional residential modes. Nevertheless, local governments have universally implemented a certain scale of centralized resettlement so that the surplus quota for construction land can be freed up to facilitate further land consolidation and industrial development. According to follow-up surveys, the various housing reconstruction model options lead to cultural inappropriateness due to a lack of understanding of local needs by implementing agencies. It is reflected in the size and style of the house, the design of the interior and surrounding space of the house, the choice of building materials, and the infrastructure services, which are all obviously inconsistent, especially the typical regiment barracks layout in the resettlement area. This is consistent with earlier observations that cultural gaps and inappropriate questions have persisted for a long time. From early survey studies [26–29] to more recent observations [30–32], there are still widespread failures and obstacles in post-disaster housing reconstruction. To address this issue, it is necessary to allocate rights and responsibilities through institutional reform, exercise administrative power according to law, and use the surplus quota for housing land intensively to generate revenue for victims. For example, planning guidance can be provided to facilitate housing reconstruction planning and land use control, increase, and decrease linkage of urban-rural construction land and rural land circulation (market services), land use change permits for the surplus quota for housing land (land development guidance), and land property right re-registration (protection of housing rights and interests). This will help promote the market circulation of urban-rural construction land and provide a new source of funding for rural reconstruction.

To sum up, post-Wenchuan earthquake reconstruction is different from conventional general planning and construction. It requires a complete and mature plan to be formed in a short period of time, which places high demands on the implementation of reconstruction. Before the Wenchuan earthquake, China's urban and rural planning systems were undergoing dramatic changes. Rather than posing a challenge, post-disaster rural housing reconstruction presents an opportunity for further reform. During the reconstruction process, the central and local governments also promulgated and implemented a series of related policies according to actual needs and properly handled the relationship between rural people and housing and between people and land. Today, in the upsurge of rural revitalization, a series of post-Wenchuan earthquake reconstruction measures have laid a solid foundation for rural development, and their internal logic and experience are worth studying and further summarizing.

Author Contributions: Conceptualization, Q.L. and K.W.; methodology, Q.L.; software, Q.L.; validation, Q.L., K.W. and K.O.; formal analysis, Q.L. and K.W.; investigation, Q.L., K.W. and K.O.; resources, Q.L. and K.W.; data curation, Q.L. and K.W.; writing—original draft preparation, Q.L.; writing—review and editing, K.W.; visualization, Q.L.; supervision, K.O.; project administration, Q.L.; funding acquisition, Q.L. All authors have read and agreed to the published version of the manuscript.

Funding: This research was funded by the basic scientific research projects of central universities, grant number 20826041E4186.

Data Availability Statement: The data involved in this paper are obtained from public information and the author's field research.

Acknowledgments: We gratefully acknowledge the support provided by the Dujiangyan government and villagers. Thanks for the data and relative sources offered by the government, and thanks to all reviewers for their helpful and valuable comments and suggestions. Funding from the Sichuan University Foundation is gratefully acknowledged.

Conflicts of Interest: The authors declare no conflict of interest.

References

1. Lyons, M. Building back better: The large-scale impact of small-scale approaches to reconstruction. *World Dev.* **2009**, *37*, 385–398. [CrossRef]
2. Lang, H. Community housing in post disaster area on Nias islands, Indonesia: Responding to community needs. In Proceedings of the 4th International i-Rec Conference, Christchurch, New Zealand, 30 April–2 May 2008; University of Canterbury: Christchurch, New Zealand, 2008.
3. Wang, J.; Nagarajaiah, S.; Shi, W.; Zhou, Y. Seismic performance improvement of base-isolated structures using a semi-active tuned mass damper. *Eng. Struct.* **2022**, *271*, 114963. [CrossRef]
4. Comerio, M.C. *Disaster Hits Home: New Policy for Urban Housing Recovery*; University of California Press: Berkeley, CA, USA, 1998.
5. Barton, A.H.; Fogleman, C.W.; Drabek, T.E. Communities in disaster: A sociological analysis of collective stress situations. *Am. Sociol. Rev.* **1969**, *35*, 150.
6. Rubin, C.B. Recovery from disaster. In *Emergency Management: Principles and Practice for Local Government*; Drabek, T.E., Hoetmer, G.J., Eds.; International City Management Association: Washington, DC, USA, 1991; pp. 224–262.
7. Alesch, D.J. Complex urban systems and extreme events: Toward a theory of disaster recovery. In Proceedings of the 1st International Conference on Urban Disaster Reduction, Kobe, Japan, 18–22 January 2005.
8. Clinton, W.J. *Lessons Learned from Tsunami Recovery: Key Propositions for Building Back Better*; Office of the United Nations Secretary-General's Special Envoy for Tsunami Recovery: New York, NY, USA, 2006.
9. Kennedy, J.; Ashmore, J.; Babister, E.; Kelman, I. The meaning of "build back better": Evidence from post-tsunami Aceh and Sri Lanka. *J. Contingencies Crisis Manag.* **2008**, *16*, 24–36. [CrossRef]
10. Lloyd-Jones, T. Building back better: How action research and professional networking can make a difference to disaster reconstruction and risk reduction. In Proceedings of the RIBA Research Symposium, London, UK, 2007.
11. Chang, Y.; Wilkinson, S.; Potangaroa, R.; Seville, E. Resourcing challenges for post-disaster housing reconstruction: A comparative analysis. *Build. Res. Inf.* **2010**, *38*, 247–264. [CrossRef]
12. Pearce, L. Disaster management and community planning, and public participation: How to achieve sustainable hazard mitigation. *Nat. Hazards* **2003**, *28*, 211–228. [CrossRef]

13. Steinberg, F. Housing reconstruction and rehabilitation in Aceh and Nias, Indonesia—Rebuilding lives. *Habitat. Int.* **2007**, *31*, 150–166. [CrossRef]
14. Bilau, A.A.; Witt, E.; Lill, I. Analysis of measures for managing issues in post-disaster housing reconstruction. *Buildings* **2017**, *7*, 29. [CrossRef]
15. Safapour, E.; Kermanshachi, S.; Pamidimukkala, A. Post-disaster recovery in urban and rural communities: Challenges and strategies. *Int. J. Disaster Risk Reduc.* **2021**, *64*, 102535. [CrossRef]
16. Yi, H.; Yang, J. Research trends of post disaster reconstruction: The past and the future. *Habitat. Int.* **2014**, *42*, 21–29. [CrossRef]
17. Guo, Y. Urban resilience in post-disaster reconstruction: Towards a resilient development in Sichuan, China. *Int. J. Disaster Risk Sci.* **2012**, *3*, 45–55. [CrossRef]
18. Charles, S.H.; Chang-Richards, A.Y.; Yiu, T.W. A systematic review of factors affecting post-disaster reconstruction projects resilience. *Int. J. Disaster Resil. Built Environ.* **2022**, *13*, 113–132. [CrossRef]
19. Li, Q.; Umaier, K.; Koide, O. Research on post-Wenchuan earthquake recovery and reconstruction implementation: A case study of housing reconstruction of Dujiangyan city. *Prog. Disaster Sci.* **2019**, *4*, 100041. [CrossRef]
20. Cheng, L.G.; Zhang, Y.; Liu, Z.B. Does rural land rights confirmation promote rural land circulation in China? *Manag. World* **2016**, *1*, 88–98.
21. Kabilijiang, W.; Shi, Y. *Project Management for Post-Disaster Urban Infrastructure Reconstruction after the Wenchuan Earthquake*; Sichuan University Press: Chengdu, China, 2015; pp. 105–107.
22. Zhu, L. Study of the Increase and Decrease Linkage Policy of Urban-Rural Construction Land. Ph.D. Dissertation, Southwest University, Chongqing, China, 2010.
23. Li, G.Y.; Wang, M.; Wu, J.Z. Preliminary study of the Chengdu pattern of urban-rural coordinated development. *Chin. J. Syst. Sci.* **2010**, *18*, 67–71.
24. Luo, D.L. Legal Choice in Housing Construction after the Wenchuan Earthquake. Ph.D. Dissertation, Southwestern University of Finance and Economics, Chengdu, China, 2009.
25. Liu, Y.Q.; Su, C.G.; Long, H.L.; Hou, X.G. Innovation in China's rural land management system in the context of urban-rural integration. *Econ. Geogr.* **2013**, *33*, 138–144.
26. Davis, I. *Shelter after Disaster*; Oxford Polytechnic: Oxford, UK, 1978.
27. Skinner, R. Peru: Low-income housing. In *Mimar 38: Architecture in Development*; Concept Media Ltd.: London, UK, 1992; pp. 52–55.
28. Delaney, P.; Shrader, E. *Gender and Post-Disaster Reconstruction: The Case of Hurricane Mitch in Honduras and Nicaragua*; World Bank: Washington, DC, USA, 2000.
29. Boen, T.; Jigyasu, R. Cultural considerations for post disaster reconstruction post-tsunami challenges. *UNDP Conf.* **2005**, 1–10.
30. Gharaati, M.; Davidson, C. Who knows best? An overview of reconstruction after the earthquake in Bam, Iran. In Proceedings of the 4th International i-Rec Conference, Christchurch, New Zealand, 30 April–2 May 2008.
31. Pamidimukkala, A.; Kermanshachi, S.; Safapour, E. Challenges in post-disaster housing reconstruction: Analysis of Urban vs. Rural communities. In Proceedings of the Creative Construction e Conference, Opatija, Croatia, 28 June–1 July 2020.
32. Wang, L.; Nagarajaiah, S.; Zhou, Y.; Shi, W. Experimental study on adaptive-passive tuned mass damper with variable stiffness for vertical human-induced vibration control. *Eng. Struct.* **2023**, *280*, 115714. [CrossRef]

Article

The Assessment and Retrofitting of Cultural Heritage—A Case Study of a Residential Building in Glina

Karlo Ožić [1], Ivan Markić [2], Antonela Moretić [1,*] and Luka Lulić [1]

[1] Department for Structures, Faculty of Civil Engineering, University of Zagreb, 10000 Zagreb, Croatia; kozic@grad.hr (K.O.); llulic@grad.hr (L.L.)
[2] PRONA—GRAD, 10090 Zagreb, Croatia; ivan.markic@prona-grad.hr
* Correspondence: amoretic@grad.hr

Abstract: The focus of the study is on the renovation of a specific case study, which is a 19th century building under cultural heritage protection. It highlights the particular challenges faced by civil engineers in the structural renovation of buildings that are under heritage protection. Preserving the identity of these buildings limits the available methods for strengthening their seismic capacity. At the beginning, information about the seismic activity and the different post-earthquake evaluation procedures are presented to identify the damage and take appropriate further steps. Then, basic information about the building is given and supported by graphic attachments. In the following, the methods and materials are explained, focusing on in situ testing with the semi-destructive flat-jack method and the analysis of the structure with the nonlinear method implemented in the software. Subsequently, the obtained results are presented and discussed, accompanied by graphics. An approach for strengthening the structure is presented, which includes a combination of traditional methods and innovative solutions suitable for the preservation of cultural heritage. The discussion and conclusions emphasize the importance of assessing and retrofitting existing masonry structures due to their vulnerability, especially in earthquake-prone areas. Finally, this article also provides insights into the local context, cultural significance, and historical background of the building, along with the specific retrofitting solutions employed to address its unique requirements.

Keywords: earthquake; structural strengthening; cultural heritage; nonlinear static analysis; in situ tests

check for
updates

Citation: Ožić, K.; Markić, I.; Moretić, A.; Lulić, L. The Assessment and Retrofitting of Cultural Heritage—A Case Study of a Residential Building in Glina. *Buildings* **2023**, *13*, 1798. https://doi.org/10.3390/buildings13071798

Academic Editors: Alberto Maria Avossa and Marco Di Ludovico

Received: 7 June 2023
Revised: 3 July 2023
Accepted: 12 July 2023
Published: 14 July 2023

1. Introduction

On 28 December 2020, at 6 h and 28 min, a strong earthquake with an epicenter near Petrinja occurred, with an intensity of 5 degrees on the Richter scale; this preceded a devastating earthquake on 29 December 2020, at 12 h and 19 min, with an epicenter 5 km southwest of Petrinja, at a depth of 11.5 km and a magnitude of 6.2 on the Richter scale (the intensity at the epicenter was VIII-IX degrees on the EMS scale) (Figure 1). After the earthquake, until today, the area of the City of Petrinja and its surroundings, including Glina and Sisak, have been hit by a series of minor and medium earthquakes, i.e., there was increased tectonic activity that resulted in a series of smaller earthquakes. The greatest material damage was recorded in the center of the town of Petrinja [1]. The earthquakes in Petrinja were also significantly felt in the area of Zagreb, which "swayed" more due to the greater epicentral distance that enabled the development of surface waves; these effects were felt more strongly than those of a recent Zagreb earthquake in 2020 [2].

It is well known that traditional masonry structures in Europe are vulnerable to earthquakes due to their inherent characteristics, such as inappropriate or nonexistent connections between wall and floor and roof structures, the absence of vertical and horizontal confining elements, uneven stiffness distribution, and poor load-bearing capacity [3–5]. The lack of flexibility and the inability to absorb seismic energy make masonry structures highly susceptible to damage during seismic events [6]. Additionally, most of the buildings in the city center have an expired service life, which means the degradation of their mechanical properties should be taken into account [7].

Additionally, the construction methods and materials used in masonry structures often do not meet modern seismic design standards. The seismic vulnerability [8] of masonry structures is also influenced by factors such as quality of construction, maintenance, and renovation history. As a result, many historic masonry structures in Europe are at risk of collapse during earthquakes. Therefore, it is essential to assess and retrofit existing masonry structures to improve their seismic resistance and ensure their safety during seismic events.

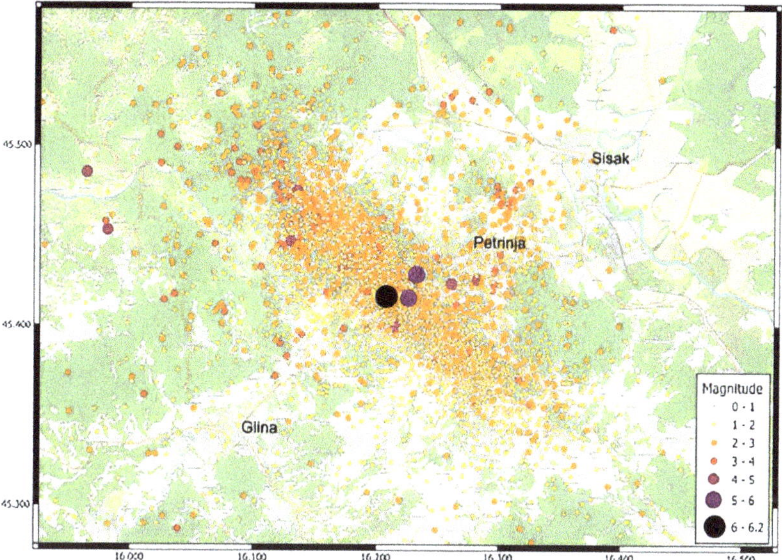

Figure 1. Map of the epicenter of the earthquake near Petrinja in the period from 29 December 2020 at 12:19 p.m. to 30 December 2020 at 10:00 p.m. [9].

To do so, thorough experimental tests and detailed geometrical and structural surveys are needed to overcome complex mechanical and geometrical issues. Additionally, we need to model a reliable simulation of the mechanical response of existing masonry buildings. Firstly, to accomplish this, well-established assessment procedures are required. An essential part of these assessment procedures is the reduction of epistemic uncertainty through gathering additional information [10]. This type of uncertainty can be reduced to a certain point, unlike aleatory uncertainties, which are defined as internal randomness of phenomena [11]. To minimize uncertainties, various inspection techniques are feasible, such as visual assessment, destructive, semi-destructive, and non-destructive methods (NDT), as well as collecting data with structural health monitoring (SHM) [12]. In the field of collecting data, some new technologies such as drone imaging and laser scanning could aid in completing a comprehensive assessment process [13]. Laser scanning and unmanned aerial vehicles (UAV) can be used in crisis management for damage detection, crack identification, and assessment of cultural heritage. Producing digital twins [14,15] is also vital for preserving the current state of the building, and this method can also be used for reconstruction purposes.

Understanding and identifying the key structural vulnerabilities of a building requires a precise evaluation of its seismic performance. Simulating the dynamic behavior of masonry structures accurately is crucial for this purpose [16]. In terms of analysis approaches, the response of masonry structures can be investigated in two main ways [17]: incremental-iterative analyses and limit analysis-based solutions. Incremental-iterative analyses are classified as either nonlinear static (pushover) analysis or nonlinear dynamic (time history) analysis. On the other hand, limit analysis-based solutions are either a lower-bound limit analysis (a static theorem) or an upper-bound limit analysis (a kinematic theorem). Regarding modelling strategies for masonry structures, four different categories are defined: block-based models [18], continuum models [18], macro-element models [19], and geometry-based models [20]. Each modelling strategy has its limitations and specific area of application. Therefore, the most suitable modelling strategy depends on the features and the complexity of the structure under investigation, the output required, the data available, and the expertise level.

This paper presents the procedure of a detailed inspection of a building under cultural heritage protection that was damaged in the 2020 Petrinja earthquake. An incremental-iterative analysis, i.e., a nonlinear static (pushover) analysis was the approach used for the renovation of the case study, while a macro-element model, that is, the equivalent frame model, was used as the modelling strategy [21]. In the sections below, the analysis approach and modelling strategies for the renovation of the case study are presented in detail, as well as the assessment procedure. This paper also shares the valuable lessons learned from the case study, including successes, failures, and practical recommendations for future retrofitting projects involving cultural heritage buildings. These insights can inform and guide professionals in the field, and contribute to the development of guidelines and standards for the conservation and retrofitting of similar structures.

2. The Case Study

The case study building (Figures 2 and 3), which is the subject of an ongoing condition assessment study, was inspected after a devastating series of earthquakes in Petrinja and its surroundings. The subject of this study is an isolated structure located at Trg bana Josipa Jelačića 21 in Glina. The building is a combined private and public property, and is a protected individual cultural asset of the Registry of Cultural Assets of the Republic of Croatia. The available existing documentation of the building shows that the building dates back to the 19th century. After a thorough reconstruction, the building took on its present form in 2000.

Figure 2. South-west facade of case study building.

Figure 3. North-east facade of case study building.

The building has a "U" shaped plan (Figures 4 and 5). The main orientation, that is, the longer side of the building, is in the north-west–south-east direction. The external dimensions of the building are about 13.4 × 32 m, with two wings, one 6.2 × 6.1 m and the other 5.7 × 8.0 m. The total height of the building is 17 m. The floor plan of each storey is 514 m². The building consists of a ground floor, first and second floors, and an attic (Figure 6). The building initially served as an educational institution (gymnasium), while today, it is used for residential and commercial purposes. The condition of the building before the earthquake was satisfactory, and the building was regularly maintained. The last reconstruction of the building was carried out in 2000. The building is built of solid brick of the old format (29 × 14 × 7 cm), as was used at the end of the 19th century. The thickness of the load-bearing walls varies within floors, and ranges between 45–75 cm. On the ground floor, the thickness is 75 cm. On the first floor, it varies from 65 cm to 45 cm, and the thickness decreases with height; on the 2nd floor, the thickness is 45 cm, and the partitioning walls are 20–30 cm thick. The floor structures differ on each storey. The floor structure of the ground floor consists of masonry vaults, except for the floor structure located in the wing, which is a semi-precast masonry/concrete floor system (with a Fert ceiling). On the first floor, a semi-precast masonry/concrete floor system (with a Fert ceiling) and timber floors with a concrete layer are found. The floor structure of the second floor consists of timber floors and Fert ceilings.

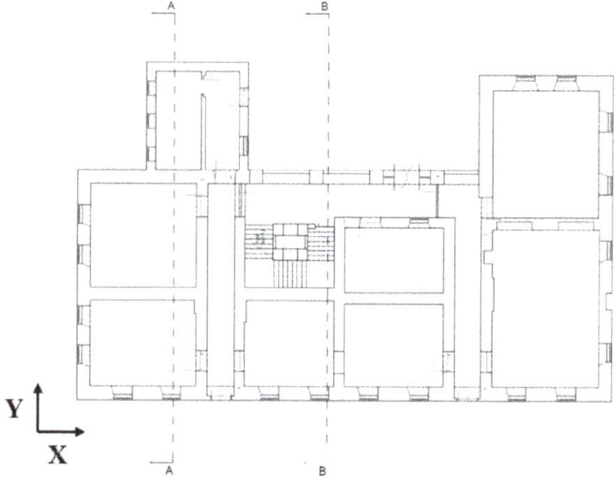

Figure 4. Ground storey floor plan.

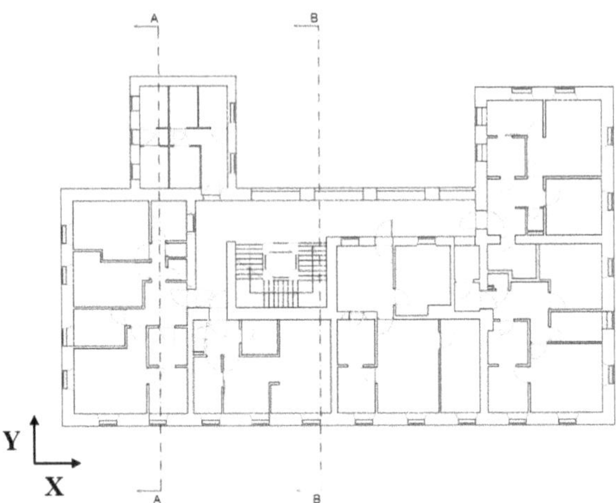

Figure 5. First and second storey floor plan.

Figure 6. Transversal building sections A and B.

3. Methods and Materials

3.1. Assessment Procedure

Shortly after the earthquake, an assessment procedure was devised for rapid, preliminary assessment of buildings damaged in the earthquake [17]. The idea was to assess the usability of all damaged buildings without further endangering engineers in the field, as some buildings were heavily damaged. This type of assessment also provided preliminary feedback about the structures to their owners and occupants. Following a quick visual inspection of the load-bearing elements and deciding on the level of damage, each building was classified into one of six possible categories: U1, Usable without limitations (green label); U2, Usable with recommendations of 25 (green label); PN1, Temporarily unusable—detailed inspection needed (yellow label); PN2, Temporary unusable—emergency interventions needed (yellow label); N1, Unusable due to external impacts (red label); and N2, Unusable due to damage (red label) [22].

A rapid, preliminary assessment of the building was carried out on 31 December 2020. The building was classified as temporarily unusable (NP, yellow label) with a recommendation of urgent repair. The following recommendations were given: urgent repair of the roof, the removal of damaged chimneys, the remediation of moisture in the foundations, and the fastening of cornices at the corners. A further detailed assessment, as presented below, was also recommended.

3.2. Detailed Assessment Results

After a rapid inspection, a detailed inspection of the building was carried out in May 2022. All instances of damage, structural and non-structural (Figure 7), were photographed and described in reports on the floors of the building and the individual rooms in which they are located, and proposed measures for their repair were given. A detailed recording of the existing condition was also made for the purpose of creating digital floor plans of the building.

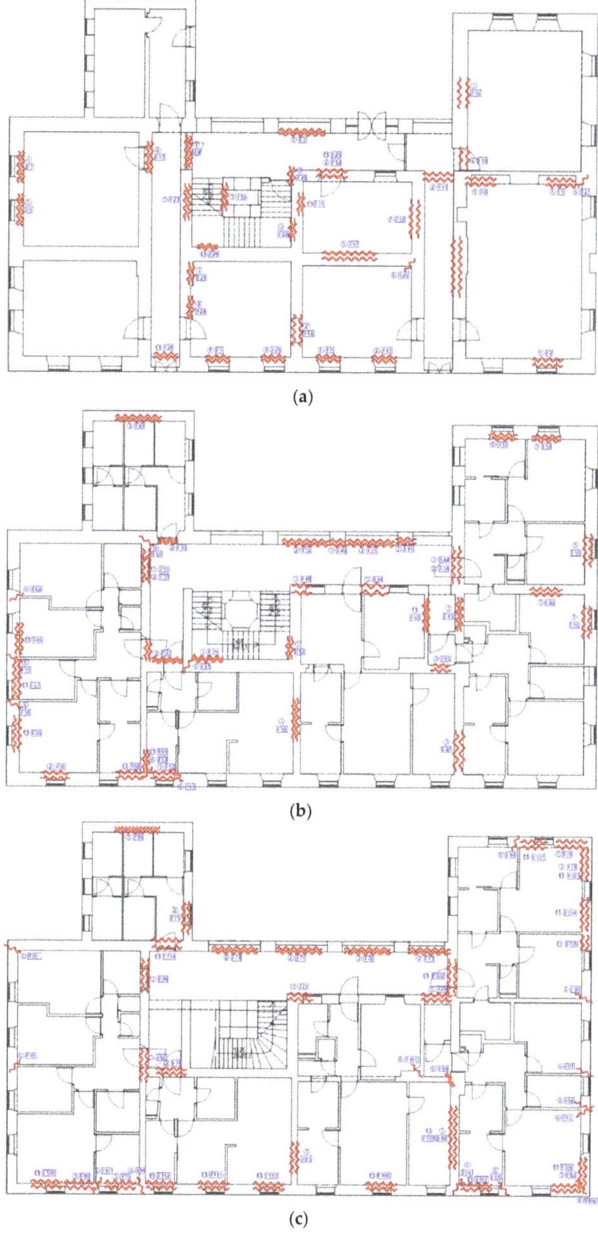

(a)

(b)

(c)

Figure 7. Damage patterns (red marks): (**a**) ground floor, (**b**) first floor, and (**c**) second floor.

Detailed inspections of the residential building (former gymnasium building) revealed the following recorded instances of damage: on the ground floor, there is visible damage in the form of cracks in the wall coverings (Figure 8), arches, vaults and ceilings, as well as separation and localized falling off of the plaster. Minor local damage to structural elements (walls, columns, and arches) is also visible. Damage is visible on all floors in the form of cracks and falling plaster on the walls. Minor local damage to the walls is also visible, and there are locally visible cracks at the junctions of load-bearing walls and ceilings (Figure 9). At the edges of the building, oblique cracks are visible on the load-bearing walls, and can also be seen on the facade of the building. In addition to the previously documented damage to the construction elements of the building, damage to other elements was observed. The partition walls are mostly damaged. Other instances of damage include the chipping of plaster on structural and non-structural elements, damage to the finishing coverings on the walls and floors, and the collapse of part of the cornice and other decorative elements (Figure 10). The roof was damaged in several places, and the roof covering near the chimneys collapsed (Figure 11).

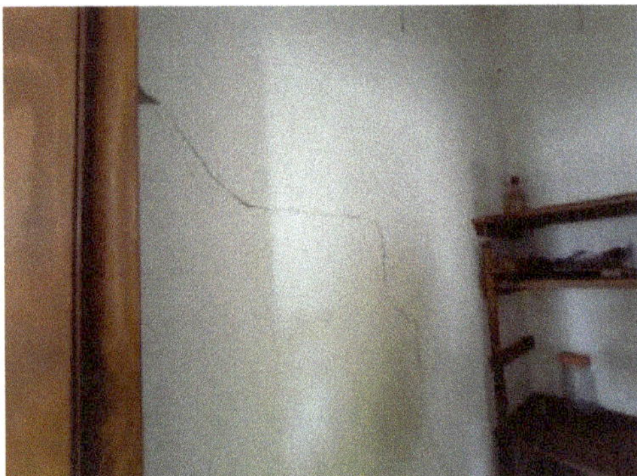

Figure 8. Cracks on the load-bearing walls on the first floor.

Figure 9. Cracks at the junctions of load-bearing walls.

Based on the detailed assessment, the entirety of the building necessitates rehabilitation and retrofitting measures. Subsequently, it is imperative to conduct a comprehensive examination encompassing both static and dynamic analyses of the existing structural conditions. Before that, it is necessary to carry out investigative work to determine the characteristics of the walls and other necessary data for the analysis of the structure.

Figure 10. Damage to the facade.

Figure 11. Collapsed chimney and damaged roof covering

3.3. In-Situ Tests

The building in question is two centuries old, and documentation of the original project is very scarce. In such existing structures, the quality of the material degrades over time, especially if it is not protected from the weather. In earthquake-prone areas, such conditions are additionally unfavorable, considering the inherent weaknesses of URM structures, such as high mass, low ductility, and energy dissipation [23]. Therefore, it is necessary to perform in situ testing of the masonry [24] as part of the assessment process. In this case study, a well-known semi-destructive method for masonry testing was used. The method mentioned is the flat-jack method. Since there are no guidelines within the European standards, the test can be performed following the guidelines given in the ASTM [25] and RILEM [26] standards. The method is divided into three phases, and provides results on the vertical stress state, the modulus of elasticity, and the shear strength of the masonry (Figure 12). Experiences from an extensive test campaign conducted after the recent devastating earthquakes in 2020 in Croatia are presented in [27]. The results from that campaign are presented in [28].

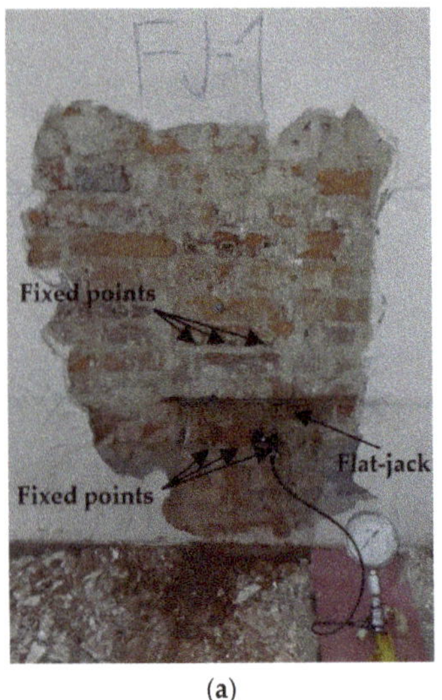

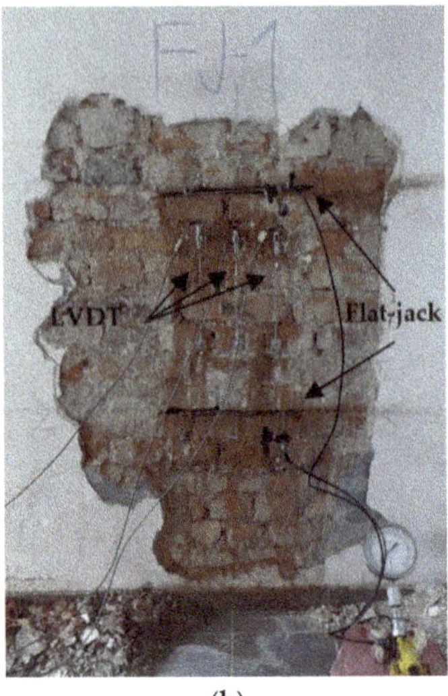

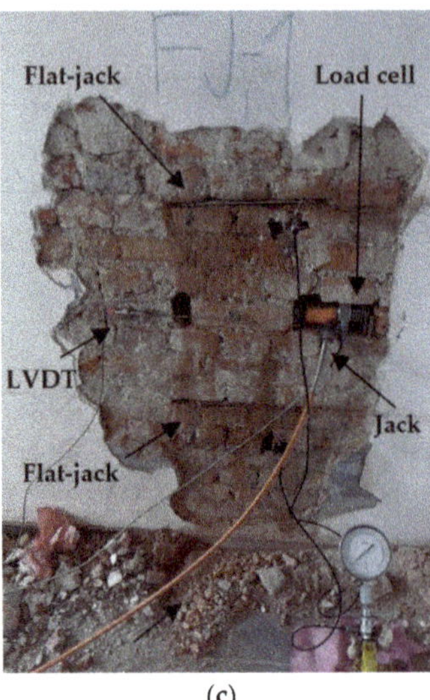

Figure 12. Flat-jack method: (**a**) single flat-jack test, (**b**) double flat-jack test, and (**c**) shear test.

The principle of the single flat-jack test is to partially relieve the compressive stress in the wall by removing the mortar from the horizontal bed joint. This is achieved by making the opening with an eccentric circular saw. The stress is then compensated by means of a flat-jack inserted in the opening. The stress is gradually increased until the original stress and strain state is established, which is verified by measuring the displacements perpendicular to the opening. Measurements are made with a portable extensometer, and fixed measurement points glued to the bricks can be seen in Figure 12a. The results of the single flat-jack tests are shown in Table 1.

Table 1. Single flat-jack test results.

Floor	Mark	t (cm)	h (cm)	Km	Ka	p (bar)	σ_0 (N/mm^2)
Ground floor	FJ-1	60	63	0.759	0.909	6.4	0.44
First floor	FJ-2	60	65	0.761	0.920	3.1	0.21

To determine the stress–strain behavior of masonry in compression and the modulus of elasticity of masonry, it is necessary to use another flat-jack placed above the first one (a double flat-jack test). Flat-jacks are inserted into parallel horizontal openings and pressure is applied gradually. In this phase, the displacement gauges (LVDTs) are placed vertically between the flat-jacks, as in Figure 12b. Simultaneously with the application of vertical pressure with the flat-jacks, the displacement is measured (Figure 13), which allows the stress–strain behavior and modulus of elasticity to be determined. For reference, see [27]. The resulting values for the double flat-jack tests are listed in Table 2.

Table 2. Double flat-jack test results.

Floor	Mark	E (N/mm^2)
Ground floor	FJ-1	831
First floor	FJ-2	648

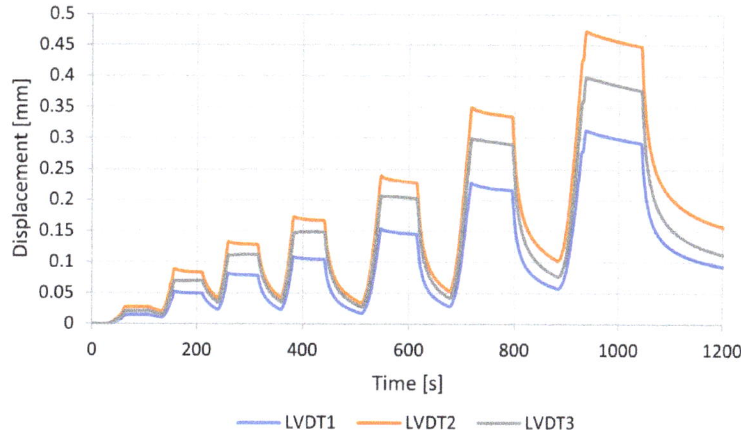

Figure 13. Measured displacement during double flat-jack test.

After the double flat-jack test, the shear (shove) test follows in the same location. Both flat-jacks with an additional horizontal hydraulic jack are used for this test (Figure 12c). The displacement gauges used in the previous phases are removed and a horizontal displacement gauge is mounted. In this phase, a horizontal brick (oriented so that a mayor dimension of the brick is parallel to the wall) is pushed until failure, that is, until it slips. This process is repeated for several vertical stress levels (Figure 14). Flat-jacks allow the control of the vertical stress at the test location. As a result, this provides the relationship between the shear stress and the vertical compressive stress. From this relationship, a best-fitting line can be drawn. The angle of this line represents the coefficient of friction, while the intersection with the vertical axis represents the initial shear strength of the masonry. The results of this test phase are shown in Table 3.

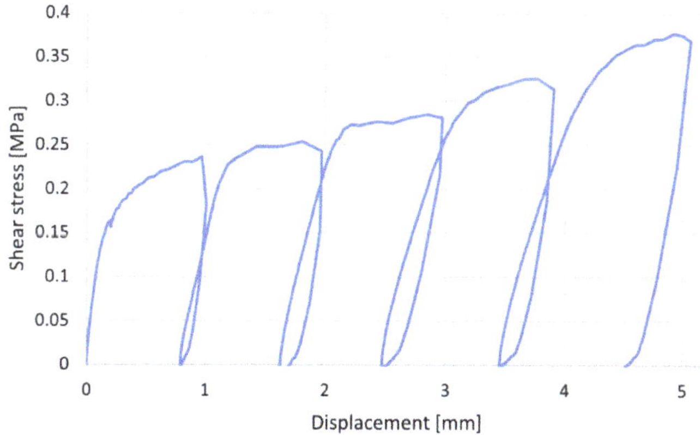

Figure 14. Change in shear stress during the test (blue line).

Table 3. Shear test results.

Floor	Mark	μ	f_{vo} (N/mm^2)
Ground floor	FJ-1	0.36	0.15
First floor	FJ-2	0.51	0.22

The results show that the quality of the masonry is relatively good, as the values obtained are within the expected range for buildings from this period of construction. Further non-destructive testing methods [29,30] can be used to validate the results. For example, in determining the modulus of elasticity, the sonic method has proven to be potentially useful. Another nondestructive method based only on visual inspection provides an approximate range of values for the shear strength, compressive strength, and modulus of elasticity of the masonry. This is the MQI method, which is explained in more detail in [31].

4. Structural Strengthening

After the earthquake, the Law on the Reconstruction of Earthquake-Damaged Buildings in the City of Zagreb, Krapina-Zagorje County, Zagreb County, Sisak-Moslavina County and Karlovac County [32] was passed. It provided a legal framework for post-earthquake rehabilitation. The law in question established four renovation levels based on factors such as the extent of damage, the building's purpose, and the investor's financial capacity. Level 1 focuses on restoring the structure's original resistance from prior to the earthquake, while levels 2, 3, and 4 aim to achieve satisfactory earthquake resistance for a specified return period. The law specifies the distribution of renovation costs, with 60% covered by the Republic of Croatia using the state budget, 20% by the counties, and the remaining 20% by the owners (excluding finishings like façade installation, parquet, plastering, and painting, which are entirely financed by the owners).

The building In question is, as already mentioned, under cultural heritage protection, and it was built at a time when seismic regulations had not yet been issued. This means its structural system does not have sufficient seismic resistance [33]. To achieve this, intrusive and detailed interventions are necessary, such as construction of a new replacement structural system (shotcrete). The construction of a new structural system would violate the cultural identity of the building; therefore, a renovation to level 3 was proposed in agreement with the investor. For level 3, the relevant return period is 225 years.

To assess the seismic resistance of the structure, a pushover analysis has been carried out. This is a nonlinear static method, meaning it considers the nonlinear properties of the material and the redistribution of the forces in the structure. The method is suitable for the evaluation of existing structures, as it helps to identify potential weaknesses or deficiencies in structural systems. Additionally, it offers a reasonable balance between accuracy and computational efficiency. The basic assumption of the method is that the structure vibrates in the first mode, which is the main drawback of the method, because it is mainly applicable to regular buildings that have a dominant response in the first mode of vibration. It does not provide reliable results for tall and irregular buildings. A monotonically increasing static load represents the distribution of forces expected during a seismic event. The analysis is performed incrementally, starting from the linear elastic response and gradually introducing nonlinear behavior. At each increment, the structural response is calculated by considering the equilibrium between the applied loads and internal forces. As the structure exhibits nonlinear behavior, the load distribution within the structure is likely to change. The structure was modelled in the software 3Muri [34], which applies macro-element approach. The macro-element approach discretizes masonry elements to primary vertical elements (piers) and secondary horizontal elements (spandrels). The longitudinal and transversal walls are connected to each other and to the floor structures by rigid nodes. As has already been mentioned, the floor structures vary on each storey. On the ground and first floor, the floor structures are set as rigid diaphragms, while on the second floor, rigid and flexible diaphragms are modelled. Although the connections are set as rigid, the load transfer differs depending on the diaphragm type.

The structural capacity is described by pushover curves, whereas the structure is a multi-degrees-of-freedom (MDoF) system. The curve displays the correlation between the base shear and control displacement. The seismic demand is given in terms of the elastic response spectrum for a single-degree-of-freedom (SDoF) system. To compare the seismic capacity and demand, both must be in the same format. The pushover curve is firstly bilinearized. The bilinearization is performed by adapting the initial stiffness so that it equals a ratio of 70% of the maximum base shear and the corresponding displacement. The base shear is calculated based on the principle of equal fracture energies. The curve is then transformed from a multi-degrees-of-freedom (MDOF) to an equivalent single-degree-of-freedom (SDOF) system by means of Γ. The curve is finally converted into ADRS format, by dividing the base shear value F^* by the equivalent mass of the SDOF m^*. The calculation of target displacement differs for the structures with short periods ($T^* < T_C$) and for the structures with medium or long periods ($T^* > T_C$). The described method is known as the N2 method [35], and it is described in detail in Eurocode 1998-1 [36]. Except for pushover curves, 3Muri software also presents the results using risk index α, which is the ratio of capacity and demand in terms of acceleration. If the risk index for a certain analysis equals $\alpha \geq 1.0$, the seismic capacity is equal to or higher than the demand.

Elastic response spectra have been calculated for acceleration values, referencing three different return periods—475, 225, and 95 years—and amplified for the soil type C (Table 4). The return periods of 475 and 225 years refer to the "Significant damage" limit state, while the return period of 95 years refers to the "Damage limitation" limit state. Soil type C (deep deposits of compacted or medium-compacted sand, gravel, or hard clay with a thickness of several tens to hundreds of meters) is assumed based on the available data.

Table 4. PGA values.

Limit State	Return Period (Year)	Peak Ground Acceleration
Significant damage	475	0.15 g
Significant damage	225	0.11 g
Damage limitation	95	0.07 g

Geometric characteristics were derived from original plans and via visual inspection. To determine the material properties (Table 5.), in situ tests were carried out. The shear modulus is taken as a percentage of the elastic modulus. The specific weight is assumed according to the type of masonry. The compressive strength of the masonry was obtained using formula (3.1) from EN1996, and available data on the strength of the bricks and mortar of the structure, which were tested earlier. The roof structure was considered a non-structural element in the model, since it does not contribute significantly to the global resistance of the structure; however, its mass was taken into account in the calculation of the seismic force [37,38]. The base restraints have been set as fixed.

Table 5. Masonry material characteristics.

Material Characteristics	Value
Modulus of elasticity (N/mm^2)	740
Shear modulus (N/mm^2)	300
Specific weight (kN/m^3)	18
Mean compressive strength (N/mm^2)	2.8
Shear strength (N/mm^2)	0.18
Characteristic compressive strength (N/mm^2)	2.2

The pushover curves are shown in Figure 15. It is noted that the pushover curves vary within the same direction. The asymmetry of the building (both in plan and height) causes a higher shear capacity for the $-X$ (negative) orientation versus the X (positive) orientation. Additionally, it is noted a higher shear capacity is accomplished for the uniform load distribution, in both the X and Y direction.

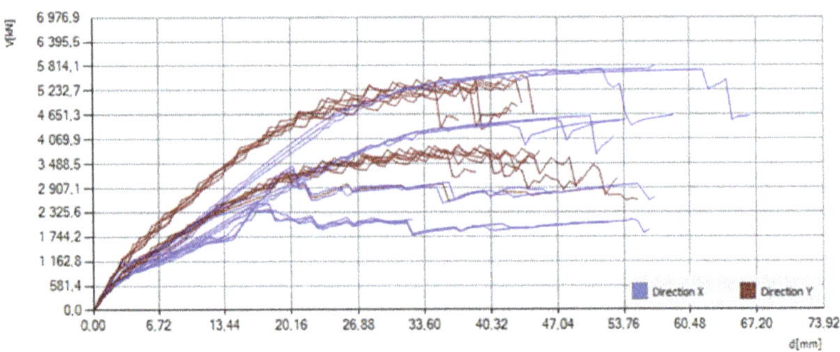

Figure 15. Pushover curves of the existing structure in the current condition.

A visualization of the damage from the critical analyses is displayed in Figure 16. The images highlight the critical parts of the structure, which correspond to the actual damage scenarios (damaged spandrels and load-bearing walls in the longitudinal and transversal direction of the central part). Figure 17. displays the capacity curves in the X and Y direction, which are derived from the pushover curves with the lowest risk indices. The radial lines defined by the slope of the first part of the capacity curve represent the elastic structural period T*. The intersecting point of the period and the seismic demand is the elastic displacement of the SDoF. It is visible that the seismic capacity is equal in both directions. Still, the seismic demand is lower for the Y direction than the X direction. The latter justifies the higher risk indices in the Y direction, which are shown in Table 6. Accordingly, the structure does not satisfy the seismic requirements.

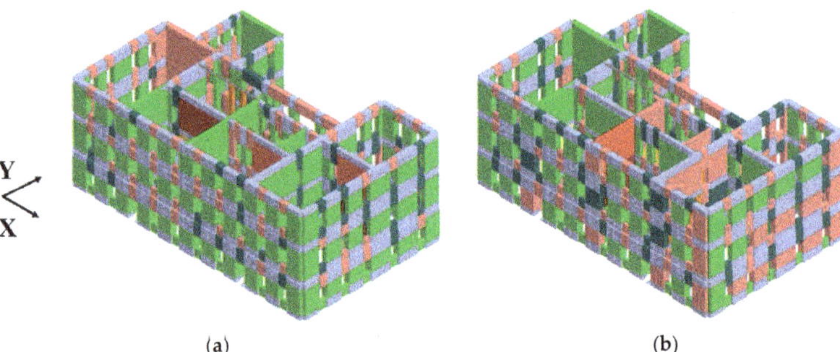

Figure 16. Damage visualization for the most critical analysis in the (**a**) X direction and (**b**) Y direction.

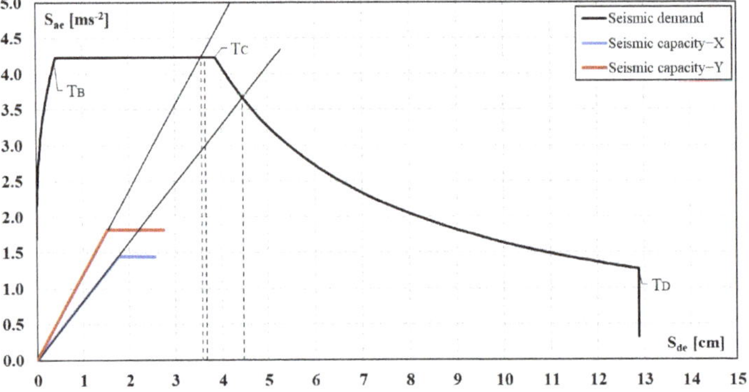

Figure 17. Seismic demand and capacity of the existing structure.

Table 6. Risk indices of the existing structure in the current condition.

Limit State	Return Period (Years)	α (X Direction)	α (Y Direction)
Significant damage	475	0.424	0.565
Significant damage	225	0.606	0.807
Damage limitation	95	0.819	0.891

The common method of strengthening in Croatia is concrete jacketing (shotcrete). It has an immense impact on the global behavior of the structure, as it affects the structural stiffness significantly. Since this technique is quite invasive, in agreement with the conservators, it was concluded that the optimal solution [39,40] is strengthening by applying an FRCM (fabric-reinforced cementitious matrix) system [41], as seen in Figure 18. For the composite, carbon fibers were used, since they are frequently used in practice in Croatia due to their material characteristics and cost efficiency. The material is characterized by low weight, high tensile strength, and corrosion and fire resistance. However, compared to shotcrete, it is not as affordable, and it requires highly skilled craftsmen. Nevertheless, since the criterion of preserving the cultural identity of the structure is a priority, and also defined by the law, strengthening using the FRCM system is a preferable solution due to its non-invasiveness. Additionally, from the environmental point of view, FRCM is a more favorable solution compared to shotcrete [42]. It causes far fewer emissions of carbon dioxide, which contributes to the global trend of reducing carbon costs [43]. The composite's characteristics have been provided by the manufacturer and are displayed in Table 7. Additionally, the existing wooden floors were coupled by introducing a 6 cm thick concrete layer. By adding the concrete layer, the in-plane stiffness of the floor structure increases, which makes the floor structure act as a rigid diaphragm [29]. This causes the distribution of internal forces proportionately to the stiffness of each element, and in case of potential future earthquake events, it ensures the translation of the floor structure without in-plane deformation.

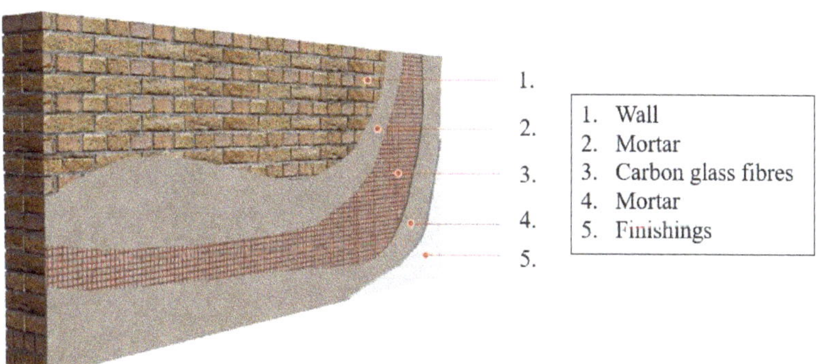

Figure 18. Application of the FRCM system.

Table 7. FRCM system characteristics [44].

FRCM System	Value
Fiber thickness t_f (mm)	0.045
Modulus of elasticity E (N/mm^2)	194,000
Conventional strain limit (%)	0.91

The required number of layers was calculated according to the values of internal forces. The resistance of the element was calculated, including the shear and bending failure. Shear failure includes both diagonal and sliding failure, while the bending failure is characterized by crushing in the compression area. The minimum value of the three is considered the resistance of the element. If the internal forces are higher than the resistance, the FRCM composite has been applied to the wall. The contribution of the FRCM system to the resistance of the wall layer was calculated. The number of layers is increased until a resistance greater than the internal forces is achieved. The FRCM composite was applied to the walls in the software, which caused the redistribution of the internal forces. Hence, the calculation of the strengthening is an iterative procedure. The proposed solution is displayed in Figures 19 and 20. The walls on which the FRCM system is applied are marked according to the legend, and denoted l/n. l represents the number of layers applied per side of the wall, while n indicates if the FRCM system is applied to a single side of the wall (n = 1) [45,46] or on both sides (n = 2). Although the system is placed asymmetrically regarding the floor plan, the solution does not cause additional torsion, as the fibers have a low individual weight, and 0.5 cm of the cementitious matrix is applied per layer of the fibers. The system exclusively affects the in-plane capacity.

Figure 19. Strengthened structure of the ground floor.

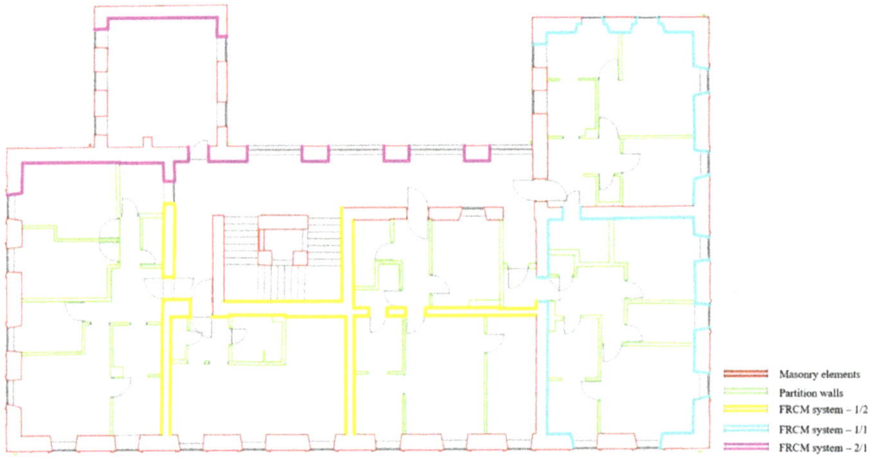

Figure 20. Strengthened structure of the first and second floor.

The results are shown in terms of risk indices, pushover curves and capacity curves (Table 8, Figures 21 and 22). In total, 24 analyses were performed. The base shear value varies depending on the pushover analysis (load distribution, eccentricity load direction, and orientation). The discrepancies are high due to the plan and height irregularities. The aim was to satisfy 24 analyses, so the increase in the capacity also varies. The average increase for the X direction is 23–53%, while for the Y direction, it is 33–40%. The suggested solution caused the increase in seismic capacity to satisfy the level 3, $\alpha_{225} > 1.0$. The significant increase in shear capacity in the X direction was noted; this was expected, since the FRCM system was mainly applied to the walls set in the longitudinal X direction. Both displacement capacity and ductility in the X direction increased. The displacement capacity increased in the Y direction, but the ductility remained the same.

Table 8. Risk indices of the strengthened structure.

Limit State	Return Period (Years)	α (X Direction)	α (Y Direction)
Significant damage	475	1.099	0.781
Significant damage	225	1.451	1.163
Damage limitation	95	1.563	1.309

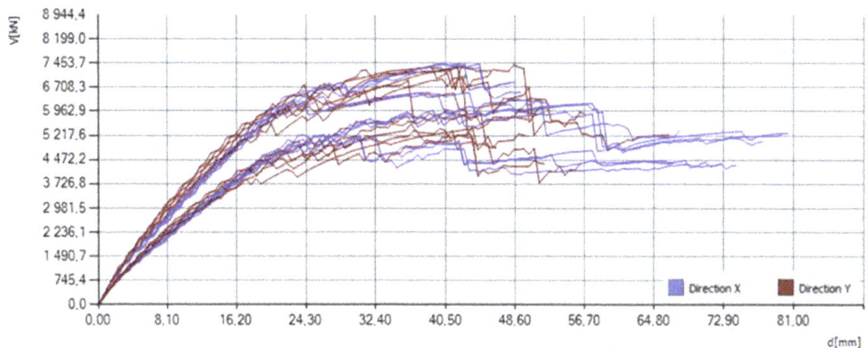

Figure 21. Pushover curves of the strengthened structure.

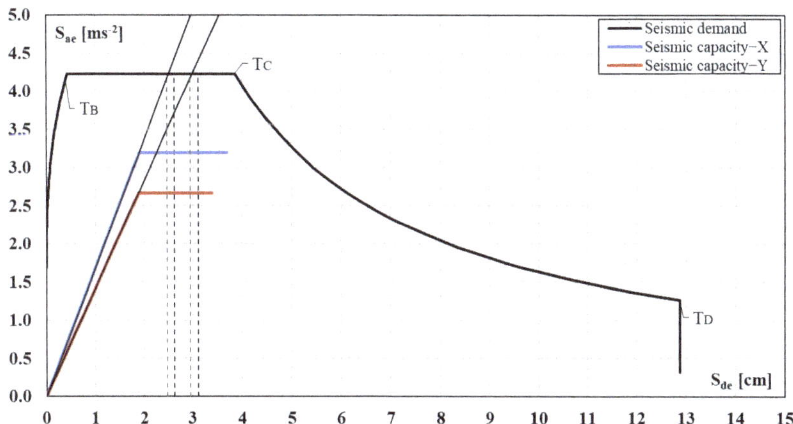

Figure 22. Seismic demand and capacity of the strengthened structure.

The dynamic parameters are displayed in Table 9, before and after strengthening, for the corresponding SDoF systems.

Table 9. Structural periods.

Direction	Existing Structure	Strengthened Structure
X	0.69	0.48
Y	0.58	0.52

Risk indices vary for each direction, which is displayed in Figure 23. It shows the comparison of risk indices for the limit state "Significant damage", and the return period of 225 years for the existing and strengthened structures. It is visible that the capacity of the strengthened structure increased overall, regardless of the direction, orientation, and distribution of the forces.

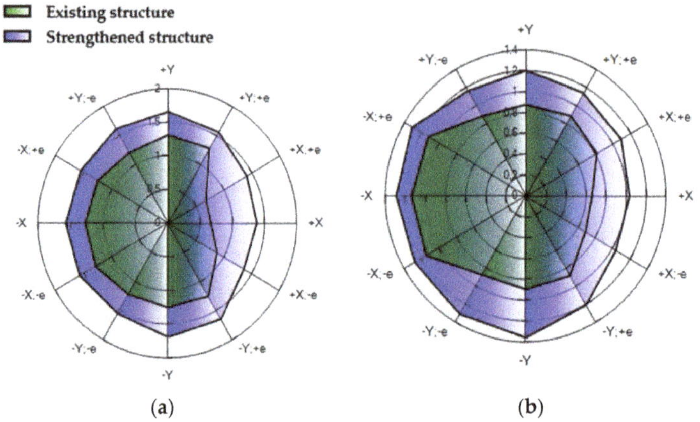

Figure 23. Comparison of the risk indices of the models of the existing structure and the structure strengthened using FRCM: (**a**) uniform distribution; (**b**) distribution proportionate to the static forces.

5. Conclusions

In conclusion, this scientific article explored the assessment and retrofitting of cultural heritage through a case study of a residential building in Glina. The study aimed to preserve and enhance the architectural and historical significance of the building while ensuring its safety, functionality, and sustainability.

Through a comprehensive assessment, various aspects of the building were examined, including its structural integrity, material characteristics, and cultural value. This holistic approach allowed for a thorough understanding of the building's characteristics and the challenges ahead. The retrofitting process involved implementing appropriate conservation and restoration strategies that respected the building's original features and cultural context. Using 3Muri software and non-linear static seismic analysis, the assessment of the building's behavior during an earthquake revealed that strengthening measures are necessary to enhance its seismic resistance. These interventions were carefully designed to strike a balance between preserving the building's heritage value and meeting contemporary standards.

The case study in Glina serves as a valuable example of the successful integration of heritage preservation and sustainable retrofitting practices. In the process of renovating and reinforcing the seismic resistance of the protected heritage building, it is crucial to adopt minimally invasive methods that preserve the original architectural and historical values while ensuring safety and functionality. It also highlights the importance of interdisciplinary collaboration among architects, engineers, conservators, and other stakeholders in achieving these objectives. By adopting a multidimensional approach, it is possible to safeguard cultural heritage while ensuring the long-term viability and functionality of historic buildings. This study emphasizes the significance of preserving cultural heritage as an integral part of sustainable development. The retrofitting process not only enhances the longevity of historic structures, but also contributes to the revitalization of communities and the promotion of cultural identity. It serves as a model for future projects aiming to balance heritage preservation with the demands of modern society. In future work, it will be important to emphasize the importance of different types of unique retrofitting solutions for each cultural heritage building, as every building brings its own problems and challenges.

In conclusion, the assessment and retrofitting of cultural heritage, as demonstrated through the case study of a residential building in Glina, offers valuable insights and guidelines for similar projects. By embracing innovative conservation strategies, interdisciplinary collaboration, and a deep appreciation of heritage value, we can successfully preserve and revitalize our cultural treasures for future generations to cherish and enjoy.

Author Contributions: Conceptualization, K.O. and I.M.; Data curation, A.M.; Methodology, K.O. and L.L.; Software, K.O., A.M. and L.L.; Supervision, I.M.; Validation, L.L.; Writing—original draft, K.O.; Writing—review and editing, K.O., I.M. and A.M. All authors have read and agreed to the published version of the manuscript.

Funding: This research was funded by the Croatian Science Foundation, grant number UIP-2019–04-3749 (ARES project—assessment and rehabilitation of existing structures—development of contemporary methods for masonry and timber structures), project leader: Mislav Stepinac.

Data Availability Statement: Data can be shared on request.

Conflicts of Interest: The authors declare no conflict of interest.

References

1. Government of the Republic of Croatia. *Croatia December 2020 Earthquake Rapid Damage and Needs Assessment*; The World Bank: Washington, DC, USA, 2021.
2. Stepinac, M.; Lourenço, P.B.; Atalić, J.; Kišiček, T.; Uroš, M.; Baniček, M.; Novak, M.Š. Damage classification of residential buildings in historical downtown after the ML5.5 earthquake in Zagreb, Croatia in 2020. *Int. J. Disaster Risk Reduct.* **2021**, *56*, 102140. [CrossRef]
3. Moretić, A.; Chieffo, N.; Stepinac, M.; Lourenço, P.B. Vulnerability Assessment of Historical Building Aggregates in Zagreb: Implementation of a Macroseismic Approach. *Bull Earthq. Eng.* **2022**, *21*, 2045–2065. [CrossRef]
4. Grillanda, N.; Valente, M.; Milani, G.; Formigoni, F.; Chiozzi, A.; Tralli, A. Fast Seismic Vulnerability Evaluation of Historical Masonry Aggregates through Local Analyses: An Adaptive NURBS-based Limit Analysis Approach. In Proceedings of the 12th International Conference on Structural Analysis of Historical Constructions (SAHC), Barcelona, Spain, 16–18 September 2020; pp. 1–13.
5. Valente, M. Seismic vulnerability assessment and earthquake response of slender historical masonry bell towers in South-East Lombardia. *Eng. Fail. Anal.* **2021**, *129*, 105656. [CrossRef]
6. Ferretti, E.; Pascale, G. Some of the Latest Active Strengthening Techniques for Masonry Buildings: A Critical Analysis. *Materials* **2019**, *12*, 1151. [CrossRef]
7. Holicky, M.; Sykora, M. Structural assessment of heritage buildings. *WIT Trans. Built Environ.* **2012**, *123*, 69–80.
8. Atalić, J.; Šavor Novak, M.; Uroš, M. Seismic risk for Croatia: Overview of research activities and present assessments with guidelines for the future. *Građevinar* **2019**, *71*, 923–947.
9. Department of Geophysics. Available online: http://www.pmf.unizg.hr/geof/en (accessed on 26 April 2023).
10. Ožić, K.; Skejić, D.; Lukačević, I.; Stepinac, M. Value of Information Analysis for the Post-Earthquake Assessment of Existing Masonry Structures—Case Studies. *Buildings* **2023**, *13*, 144. [CrossRef]
11. Der Kiureghian, A.; Ditlevsen, O. Aleatory or epistemic? Does it matter? *Struct. Saf.* **2009**, *31*, 105–112. [CrossRef]
12. Sýkora, M.; Diamantidis, D.; Holický, M.; Marková, J.; Rózsás, Á. Assessment of compressive strength of historic masonry using non-destructive and destructive techniques. *Constr. Build. Mater.* **2018**, *193*, 196–210. [CrossRef]
13. ARES PROJECT Assessment and Rehabilitation of Existing Structures—Development of Contemporary Methods for Masonry and Timber Structures. Available online: https://www.grad.hr/ares/ (accessed on 15 December 2021).
14. Stepinac, M.; Skokandić, D.; Ožić, K.; Zidar, M.; Vajdić, M. Condition Assessment and Seismic Upgrading Strategy of RC Structures—A Case Study of a Public Institution in Croatia. *Buildings* **2022**, *12*, 1489. [CrossRef]
15. Funari, M.F.; Hajjat, A.E.; Masciotta, M.G.; Oliveira, D.V.; Lourenço, P.B. A Parametric Scan-to-FEM Framework for the Digital Twin Generation of Historic Masonry Structures. *Sustainability* **2021**, *13*, 11088. [CrossRef]
16. Valente, M. Seismic behavior and damage assessment of two historical fortified masonry palaces with corner towers. *Eng. Fail. Anal.* **2021**, *134*, 106003. [CrossRef]

17. D'altri, A.M.; Sarhosis, V.; Milani, G.; Rots, J.; Cattari, S.; Lagomarsino, S.; Sacco, E.; Tralli, A.; Castellazzi, G.; de Miranda, S. Modeling Strategies for the Computational Analysis of Unreinforced Masonry Structures: Review and Classification. *Arch. Comput. Methods Eng.* **2019**, *27*, 1153–1185. [CrossRef]

18. D'Altri, A.M.; de Miranda, S.; Castellazzi, G.; Sarhosis, V. A 3D detailed micro-model for the in-plane and out-of-plane numerical analysis of masonry panels. *Comput. Struct.* **2018**, *206*, 18–30. [CrossRef]

19. Lagomarsino, S.; Penna, A.; Galasco, A.; Cattari, S. TREMURI program: An equivalent frame model for the nonlinear seismic analysis of masonry buildings. *Eng. Struct.* **2013**, *56*, 1787–1799. [CrossRef]

20. Fraternali, F. A thrust network approach to the equilibrium problem of unreinforced masonry vaults via polyhedral stress functions. *Mech. Res. Commun.* **2010**, *37*, 198–204. [CrossRef]

21. Bilgin, H.; Shkodrani, N.; Hysenlliu, M.; Ozmen, H.B.; Isik, E.; Harirchian, E. Damage and performance evaluation of masonry buildings constructed in 1970s during the 2019 Albania earthquakes. *Eng. Fail. Anal.* **2021**, *131*, 105824. [CrossRef]

22. Uroš, M.; Šavor Novak, M.; Atalić, J.; Sigmund, Z.; Baniček, M.; Demšić, M.; Hak, S. Post-earthquake damage assessment of buildings—Procedure for conducting building inspections. *J. Croat. Assoc. Civ. Eng.* **2021**, *72*, 1089–1115.

23. Kišiček, T.; Stepinac, M.; Renić, T.; Hafner, I.; Lulić, L. Strengthening of masonry walls with FRP or TRM. *J. Croat. Assoc. Civ. Eng.* **2020**, *72*, 937–953.

24. Stepinac, M.; Gašparović, M. A Review of Emerging Technologies for an Assessment of Safety and Seismic Vulnerability and Damage Detection of Existing Masonry Structures. *Appl. Sci.* **2020**, *10*, 5060. [CrossRef]

25. *ASTM C1197-14a*; Standard Test Method for in situ Measurement of Masonry Deformability Properties Using the Flatjack Method. ASTM International: West Conshohocken, PA, USA, 2014.

26. RILEM Technical Committee. RILEM TC 177-MDT: Masonry Durability and on-Site Testing Recommendation MDT. D. 4: In-Situ Stress Tests Based on the Flat jack. *Mater Struct.* **2004**, *37*, 491. Available online: http://www.springerlink.com/index/10.1007/BF02481588 (accessed on 15 March 2023).

27. Lulić, L.; Stepinac, M.; Bartolac, M.; Lourenço, P.B. Review of the Flat-Jack Method and Lessons from Extensive Post-Earthquake Research Campaign in Croatia. *Constr. Build. Mater.* **2023**, *384*. Available online: https://linkinghub.elsevier.com/retrieve/pii/S0950061823011200 (accessed on 5 June 2023). [CrossRef]

28. Stepinac, M.; Lulić, L.; Damjanović, D.; Duvnjak, I.; Bartolac, M.; Lourenço, P.B. Experimental Evaluation of Unreinforced Brick Masonry Mechanical Properties by the Flat-Jack Method–An Extensive Campaign in Croatia. *Int. J. Arch. Heritage* **2023**, 1–18. [CrossRef]

29. Costamagna, E.; Quintero, M.S.; Bianchini, N.; Mendes, N.; Lourenço, P.B.; Su, S.; Paik, Y.M.; Min, A. Advanced non-destructive techniques for the diagnosis of historic buildings: The Loka-Hteik-Pan temple in Bagan. *J. Cult. Herit.* **2019**, *43*, 108–117. [CrossRef]

30. Ortega, J.; Stepinac, M.; Lulić, L.; Nunez Garcia, M.; Saloustros, S.; Aranha, C. *Correlation between Sonic Pulse Velocity and Flat-Jack Tests for the Estimation of the Elastic Properties of Unreinforced Brick Masonry: Case Studies from Croatia (Under Review in Case Studies in Construction Materials, Netherlands)*; 2023.

31. Borri, A.; Corradi, M.; Castori, G.; De Maria, A. A method for the analysis and classification of historic masonry. *Bull. Earthq. Eng.* **2015**, *13*, 2647–2665. [CrossRef]

32. Law on the Reconstruction of Earthquake-Damaged Buildings in the City of Zagreb, Krapina-Zagorje County, Zagreb County, Sisak-Moslavina County and Karlovac County (NN 21/23). Available online: https://www.zakon.hr/z/2656/Zakon-o-obnovi-zgrada-oštećenih-potresom-na-području-Grada-Zagreba%2C-Krapinsko-zagorske-županije%2C-Zagrebačke-županije%2C-Sisačko-moslavačke-županije-i-Karlovačke-županije (accessed on 27 April 2023).

33. Milani, G.; Shehu, R.; Valente, M. Seismic Upgrading of a Masonry Church with FRP Composites. *Mater. Sci. Forum* **2016**, *866*, 119–123. [CrossRef]

34. S.T.A. DATA, 3Muri Program 12.5.0.2. Available online: http://www.stadata.com/ (accessed on 14 December 2021).

35. Fajfar, P.; Gašperšič, P. The N2 method for the seismic damage analysis of RC buildings. *Earthq. Eng. Struct. Dyn.* **1996**, *25*, 31–46. [CrossRef]

36. *EN 1998-1:2004*; Eurocode 8: Design of Structures for Earthquake Resistance—Part 1: General Rules, Seismic Actions and Rules for Buildings. European Committee for Standardization: Brussels, Belgium, 2004.

37. Gomes, M.I.; Lopes, M.; de Brito, J. Seismic resistance of earth construction in Portugal. *Eng. Struct.* **2010**, *33*, 932–941. [CrossRef]

38. Lulić, L.; Ožić, K.; Kišiček, T.; Hafner, I.; Stepinac, M. Post-Earthquake Damage Assessment—Case Study of the Educational Building after the Zagreb Earthquake. *Sustainability* **2021**, *13*, 6353. [CrossRef]

39. Law on the Protection and Preservation of Cultural Heritage. Available online: https://www.zakon.hr/z/340/Zakon-o-zaštiti-i-očuvanju-kulturnih-dobara (accessed on 16 February 2022).

40. Yavartanoo, F.; Kang, T.H.-K. Retrofitting of unreinforced masonry structures and considerations for heritage-sensitive construc-tions. *J. Build. Eng.* **2022**, *49*, 103993. [CrossRef]

41. Abbass, A.; Lourenço, P.B.; Oliveira, D.V. The use of natural fibers in repairing and strengthening of cultural heritage buildings. *Mater. Today: Proc.* **2020**, *31*, S321–S328. [CrossRef]

42. Moretić, A.; Stepinac, M.; Lourenço, P.B. Seismic upgrading of cultural heritage—A case study using an educational building in Croatia from the historicism style. *Case Stud. Constr. Mater.* **2022**, *17*, e01183. [CrossRef]

43. Low-Carbon Development Strategy of the Republic of Croatia until 2030 with a view to 2050. The Republic of Croatia, Ministry of Economy and Sustainable Development. Available online: https://mingor.gov.hr/o-ministarstvu-1065/djelokrug/uprava-za-klimatske-aktivnosti-1879/strategije-planovi-i-programi-1915/strategija-niskougljicnog-razvoja-hrvatske/1930 (accessed on 27 April 2023).
44. BAUMIT. Available online: https://baumit.hr/ (accessed on 27 April 2023).
45. Del Zoppo, M.; Di Ludovico, M.; Prota, A. Analysis of FRCM and CRM parameters for the in-plane shear strengthening of different URM types. *Compos. Part B Eng.* **2019**, *171*, 20–33. [CrossRef]
46. Babaeidarabad, S.; De Caso, F.; Nanni, A. URM Walls Strengthened with Fabric-Reinforced Cementitious Matrix Composite Subjected to Diagonal Compression. *J. Compos. Constr.* **2014**, *18*, 242. [CrossRef]

 *buildings*

Article

Seismic Retrofitting of Mid-Rise Unreinforced Masonry Residential Buildings after the 2010 Kraljevo, Serbia Earthquake: A Case Study

Predrag Blagojević [1], Svetlana Brzev [2,*] and Radovan Cvetković [1]

1   Faculty of Civil Engineering and Architecture, University of Niš, 18000 Niš, Serbia
2   Department of Civil Engineering, University of British Columbia, Vancouver, BC V6T 1Z4, Canada
*   Correspondence: sbrzev@mail.ubc.ca

Abstract: There is a significant building stock of post-WWII low- and mid-rise unreinforced masonry (URM) buildings in Serbia and the region (former Yugoslavia). Numerous buildings of this typology collapsed due to the devastating 1963 Skopje, Yugoslavia earthquake, causing fatalities, injuries, and property losses, as well as experienced damage in a few recent earthquakes in the region, including the 2010 Kraljevo, Serbia earthquake (M_W 5.5) and the 2020 Petrinja, Croatia earthquake (M 6.4). These buildings are three- to five-stories high, have clay brick masonry walls, and rigid floor slabs, usually with an RC ring beam at each floor level. This paper presents a case study of a URM building which was damaged due to the 2010 Kraljevo earthquake and subsequently retrofitted. A comparison of seismic analysis results, including the capacity/demand ratio and displacement/drift values, for the original and retrofitted building according to the seismic design and retrofit codes which were followed in Serbia as well as some of the neighboring countries for several decades and Eurocode 8 has been presented. The results of this study show that the selected retrofit solution that satisfied the Yugoslav seismic code requirements is not adequate according to the Eurocode 8, primarily due to significantly higher seismic demand.

Keywords: unreinforced masonry buildings; earthquake damage; seismic retrofitting; residential buildings

check for **updates**

Citation: Blagojević, P.; Brzev, S.; Cvetković, R. Seismic Retrofitting of Mid-Rise Unreinforced Masonry Residential Buildings after the 2010 Kraljevo, Serbia Earthquake: A Case Study. *Buildings* **2023**, *13*, 597. https://doi.org/10.3390/buildings13030597

Academic Editors: Antonio Formisano and Jorge Manuel Branco

Received: 2 January 2023
Revised: 30 January 2023
Accepted: 17 February 2023
Published: 24 February 2023

1. Introduction

Masonry construction technology has been traditionally used for residential construction in European countries, including Serbia and neighboring countries in the Balkan region [1]. Since the second half of the 19th century, residential and public buildings in Serbia and the region have been constructed using clay brick masonry technology. Reinforced concrete (RC) emerged as the preferred technology for the construction of mid- and high-rise buildings since the 1950s; however, masonry continued to be used for low- to mid-rise residential construction in the region. According to the 2011 Census of Serbia (referred to as "Census" in the following text) [2], low-rise single-family buildings constitute 95% of the national residential building stock, corresponding to 65.9% of the total number of housing units. Multi-family housing accounts for only 2.6% of the housing stock in terms of the number of buildings. Still, the proportion is significantly higher (33%) in terms of the number of housing units. According to the Census, 72% of all residential buildings in Serbia were constructed between 1946 (after WWII) and 1990, when Serbia was a part of the Socialist Federal Republic of Yugoslavia (SFRY), referred to as "former Yugoslavia" in this paper (note that Croatia, Slovenia, North Macedonia, Montenegro, and Bosnia and Herzegovina were also a part of the former Yugoslavia). The majority of pre-1960 multi-family residential buildings were unreinforced masonry (URM) buildings, with load-bearing masonry walls as a structural system for resisting both gravity and lateral loads. Most of URM multi-family residential buildings of post-WWII vintage have semi-prefabricated RC floor systems. Buildings of this type constitute a significant fraction of the building stock in urban areas of Serbia and neighboring countries and are the focus of this study. A few examples of typical urban URM multi-family residential buildings from Serbian cities Belgrade and Niš are shown in Figure 1a and Figure 1b respectively.

(a)   (b)

Figure 1. Examples of URM multi-family residential buildings in Serbian cities: (**a**) Belgrade and (**b**) Niš.

URM buildings are among the most seismically vulnerable building typologies and are expected to experience severe damage and/or collapse in moderate-to-strong earthquake events and cause a significant number of casualties. Masonry is a composite material characterized by a high compressive strength and at the same time, a very low tensile strength. As a result, load-bearing URM wall structures are able to sustain high gravity loads but are vulnerable to the effects of low-to-moderate earthquakes due to tensile stresses and cracking of the walls induced by seismic actions. Another deficiency of URM structures is associated with buildings with wooden floors and roofs, which act as flexible diaphragms and may cause out-of-plane damage or collapse of URM walls with deficient wall-to-floor and wall-to-roof connections. Numerous older URM buildings experienced damage due to recent earthquakes in the region, e.g., the March 2020 Zagreb, Croatia earthquake [3,4] and the December 2020 Petrinja, Croatia earthquake [5].

URM buildings with rigid diaphragms (e.g., RC floor structures), which are the scope of this study, have performed better in past earthquakes compared to otherwise similar URM buildings with flexible diaphragms. Rigid diaphragms enhance the integrity of masonry structures, which is a very important seismic resilient feature. The results of experimental research studies on URM building models subjected to shaking-table testing confirmed the beneficial effect of rigid diaphragms on the seismic performance of URM buildings [6]; however, a few URM buildings with prefabricated RC hollow core slabs collapsed in the November 2019, Albania earthquake due to inadequate wall-to-floor connections and an absence of RC slab topping [7]. It is also important to ensure a symmetrical wall layout and a sufficient number of walls in each horizontal direction in order to minimize increased seismic demands in the walls due to torsional effects in these buildings.

Several experimental and analytical studies have contributed to the state-of-the-art knowledge related to the seismic behavior of URM buildings with rigid diaphragms, which are the scope of this paper. Comprehensive experimental research studies were performed by Prof. Miha Tomaževič and his team in ZAG, Slovenia, on the seismic response of URM walls subjected to reversed cyclic loading, as well as shaking-table testing of URM building models [8]. A few analytical studies also contributed toward the understanding of global behavior and deficiencies of typical buildings, ranging from simple, practice-oriented seismic evaluation approaches to advanced non-linear seismic analysis procedures [9,10].

Buildings of this typology were exposed to several damaging earthquakes in the region, including the devastating 1963 Skopje, North Macedonia earthquake (M 6.0); the 1979 Montenegro earthquake (M_W 6.9); the 2010 Kraljevo, Serbia earthquake (M_W 5.5), and the 2020 Petrinja, Croatia earthquake (M 6.4). A widely accepted EMS-98 damage scale [11] can be used to classify the extent and type of earthquake damage in a standardized manner. EMS-98 includes the following five Damage States (DSs): DS1 (slight damage), DS2 (moderate damage), DS3 (substantial to heavy damage), DS4 (very heavy damage), and DS5 (destruction). A detailed description of the EMS-98 DSs for these buildings reported after past earthquakes in the region has been presented elsewhere [12].

The most extensive damage related to this typology was reported in the 1963 Skopje earthquake, in which many buildings of this type located in Skopje experienced severe (unrepairable) damage (DS4) or partial/full collapse (DS5) (Figure 2a). The total death toll in the earthquake was estimated at 1.500, and a significant number of fatalities was attributed to the collapse of URM multi-family residential buildings [13]. The affected buildings were designed according to standardized designs, which did not consider seismic effects. Hence, it was acceptable to include load-bearing walls in one horizontal direction only, while the lateral load-resisting system in the other horizontal direction was non-existent.

Figure 2. Seismic performance (DSs) of URM residential buildings in past earthquakes in the Balkan region: (**a**) DS5—a collapsed building in Skopje due to the 1963 Skopje earthquake (credit: Z. Milutinović); (**b**) DS4—a URM building in Petrinja experienced extensive cracking in wall piers due to the 2020 Petrinja earthquake (credit: SUZI-SAEE); (**c**) DS3—a URM building in Petrinja experienced a major shear-induced cracking in piers at the ground floor level due to the 2020 Petrinja earthquake (credit: SUZI-SAEE), and (**d**) DS2—a moderately damaged URM building which experienced moderate shear-induced cracking due to the 2010 Kraljevo earthquake (credit: P. Blagojević).

More recently, buildings of this typology were affected by the 2020 Petrinja earthquake, which was characterized by a higher magnitude than the 1963 Skopje earthquake [5]. Some buildings of this type, located in the epicentral area of the earthquake, experienced damage that could be classified as DS4 (Figure 2b) or DS3 (Figure 2c). There were no reports of severe damage or collapse associated with buildings of this type. Note that URM buildings in the Petrinja area had load-bearing walls aligned in both horizontal directions, unlike the buildings which were severely affected by the Skopje earthquake (due to load-bearing walls provided in one horizontal direction only). The main cause of damage for buildings both in the 2010 Kraljevo and the 2020 Petrinja earthquakes was high in-plane seismic demand which exceeded the shear capacity of URM walls.

Buildings of this typology were also affected by the 2010 Kraljevo earthquake, which was less severe (in terms of magnitude) than the 1963 Skopje earthquake. The damage was mostly due to in-plane seismic effects, and was mostly in the form of shear cracks observed in lower portions of the buildings (usually at the ground floor level) [14]. Structural damage observed in the affected buildings could be classified as DS3 or DS2 (Figure 2d).

Although Serbia is located in an area characterized by moderate seismic hazard, urban areas of the country were not affected by major damaging earthquakes in the past century, with the exception of the 2010 Kraljevo earthquake (M_W 5.5). A few other damaging earthquakes took place in Serbia in the last 40 years, namely the 1980 Kopaonik earthquake (M 5.8) and the 1998 Mionica earthquake (M 5.7), but they affected mostly rural areas of the country. Consequently, design and construction experience in Serbia related to repair and seismic retrofitting of buildings in post-earthquake situations has been rather limited. The 2010 Kraljevo earthquake prompted a need for the repair and retrofitting of a significant number of damaged URM residential buildings. The main objective of the post-earthquake recovery was to restore damaged building infrastructure to its original pre-earthquake condition within a relatively short time frame and with limited financial resources. An additional constraint was to minimize the impact of construction activities on building occupants. As a result, the design and execution of seismic rehabilitation projects related to residential buildings used simple retrofitting techniques which were suitable for easy on-site implementation on a large scale. RC jacketing was selected because it was a well-established technique used for the structural strengthening of URM buildings in Serbia before the 2010 earthquake.

In this paper, the authors have shared lessons on seismic retrofitting of URM buildings damaged in the 2010 Kraljevo, Serbia earthquake. The focus of this study was on the evaluation of the effectiveness of RC jacketing, a common seismic retrofitting technique for URM buildings. This study also provides an insight into differences in the results of seismic evaluation and retrofitting of masonry building according to the seismic codes from former Yugoslavia and Eurocode 8. The PTN-S code for seismic design of new structures and the PTN-R code for seismic retrofitting of existing structures were followed in Serbia and former Yugoslavia for almost 40 years. It should be noted that Eurocode 8 became the governing code for seismic design and retrofitting of buildings in Serbia in 2019. Similarly, neighboring countries in the region are either in the process of adopting Eurocode 8, or have adopted it in the last 5–10 years. A case study on a URM building in Kraljevo, which was damaged due to the 2010 earthquake and subsequently retrofitted, is presented in this paper. A comparison of seismic analysis results, including the capacity/demand ratio and displacement/drift values, has been performed for the original and retrofitted building, according to both the Yugoslav seismic design and retrofit codes and Eurocode 8. Linear elastic seismic analysis was the default procedure for ordinary buildings, such as residential buildings, according to the seismic codes from former Yugoslavia, and was used in this study. The results of this study show that the selected retrofit solution satisfied the Yugoslav seismic code requirements, but it is not adequate according to the Eurocode 8 requirements. The findings of this paper may be of particular interest to engineers in the Balkan countries, which recently adopted Eurocode 8 as the governing code for the seismic design of new buildings and the evaluation/retrofitting of existing buildings.

2. Design Code Requirements for Seismic Retrofitting of Masonry Buildings from the Serbian Code and Eurocode 8, Part 3

The first comprehensive seismic design code in the SFRY was published in 1964 [15], after the devastating 1963 Skopje earthquake. Its subsequent edition (PTN-S), issued in 1981 [16], was the governing design code in Serbia until 2019. It was reported that the PTN-S code was as advanced as other international seismic design codes at the time [17].

Eurocodes were adopted as official codes for the design, construction, and maintenance of building structures in Serbia in 2019 [18]. As a result, Eurocode 8—Part 1 [19] (also referred to as EC8-1 in this paper) is currently used for the seismic design of new buildings

(SRPS EN 1998-1/NA:2018) [20], while Eurocode 8—Part 3 [21] (also referred to as EC8-3 in this paper) has been followed in projects related to seismic assessment and retrofitting of existing buildings (SRPS EN 1998-3/NA:2018, 2018) [22].

Figure 3 presents seismic hazard maps for Serbia. Deterministic seismic hazard maps used in the SFRY were originally published in 1982 as a companion to the PTN-S code and were updated in 1987 (Figure 3a). The territory of Serbia was divided into zones VI to IX based on the MCS-64 seismic macrointensity scale (note that the borders of Serbia are shown in black color on the map). Figure 3b shows the current official seismic hazard map for Serbia developed for the design according to Eurocode 8 [23]. According to the map, Peak Ground Acceleration (PGA) values for design-level earthquake with 10% probability of exceedance in 50 years (corresponding to a 475-year return period) are largest for Southern and Central Serbia (0.25 g and 0.20 g, respectively), while other localities in the country have been assigned lower PGA values. Seismic design requirements for masonry buildings from Serbian codes and a comparison with the corresponding Eurocode 8 provisions were outlined by the authors in an earlier paper [1].

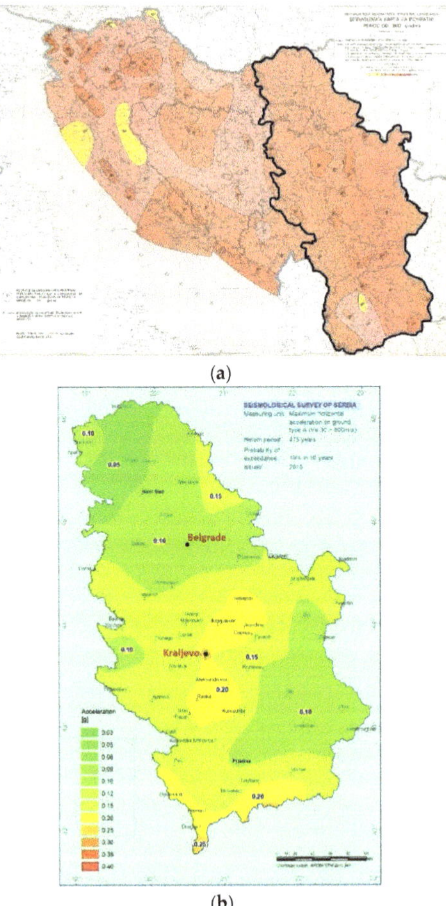

(a)

(b)

Figure 3. Seismic hazard maps: (**a**) seismic intensity map for SFRY (including Serbia) according to the MCS-64 scale published in 1987 for the 500-year return period earthquake (in compliance with the PTN-S code) and (**b**) seismic hazard map for Serbia showing the design PGA values for an earthquake with 10% probability of exceedance in 50 years, according to Eurocode 8, Part 1 (EC8-1).

Based on the experience gained after the 1979 Montenegro earthquake, the first national design code for repair, rehabilitation, and retrofitting of existing buildings was issued in 1985 (PTN-R) [24]. The code addressed structural rehabilitation and seismic retrofitting of masonry and RC buildings and the foundations.

Structural rehabilitation of existing masonry buildings was performed to improve their load-bearing capacity, and it was mandatory in one of the following cases: (i) when the walls have experienced structural damage, (ii) when the number of walls in each horizontal direction (expressed in terms of the total cross-sectional area) is inadequate, (iii) when allowable stresses in the walls have been exceeded, (iv) when there is a change of building function, and the total mass is increased by more than 10%, or (v) in case of a building renovation/extension.

According to the PTN-R code, the main performance objective was the same for retrofitted existing buildings and new buildings, that is, structural damage due to a major damaging earthquake was acceptable, but collapse had to be avoided. Seismic retrofitting provisions for a specific building were dependent on its configuration, type, and quality of the materials, the extent of damage, as well as the expected seismic performance. The code prescribed the following approaches for structural rehabilitation or seismic retrofitting of existing masonry buildings: (a) rehabilitation/retrofitting of the existing load-bearing structure, (b) reconstruction of the existing walls, (c) construction of new structural walls to enhance the seismic capacity of the existing lateral load-resisting system, and (d) retrofitting of existing floor structures and wall-to-floor connections (discussed in Section 3.7).

The PTN-R code prescribed that damaged clay brick/block masonry walls had to be either replaced (if they experienced heavy damage) or repaired by injecting cracks using cement-based grout. Masonry walls could also be strengthened by means of the following alternative techniques: (a) by applying one- or two-sided RC jackets (3–5 cm thick); (b) by constructing new RC confining elements, or (c) by post-tensioning of existing walls. In the context of RC jacketing, the code prescribed the minimum amount of distributed horizontal and vertical reinforcement, concentrated reinforcement at the wall ends, and the anchorage of vertical reinforcement into the floor/roof structure. A discussion on the wall retrofitting techniques is presented in Section 3.

A key seismic analysis requirement was to simulate the effect of both original and new (or retrofitted) structural elements by considering their deformation characteristics. The code provided prescriptive provisions related to the modelling of masonry walls retrofitted by means of RC jacketing. The thickness of a retrofitted wall had to be equal to the original wall thickness plus four times the thickness of each RC jacket. The intent of this provision was to approximately take into account the effect of RC jackets on increasing wall stiffness.

Seismic retrofitting solutions could be designed either according to the Allowable Stress Design approach or the Ultimate Limit States design approach. When the structural safety of a wall retrofitted using the RC jacketing technique was verified using the Allowable Stress Design approach, the thickness of the retrofitted wall was taken equal to the sum of thicknesses for the original masonry wall and RC jackets.

Similarly to the PTN-R code, which was developed in former Yugoslavia, Eurocode 8, specifically EC8-3 (Annex C), outlined provisions related to seismic retrofitting of masonry buildings. Acceptable seismic analysis approaches include linear static and dynamic analysis procedures. The numerical model of a masonry structure needs to consider the cracked stiffness of structural elements/walls, which may be taken as 50% of their uncracked values. It should be noted that the PTN-R code did not prescribe any stiffness reduction; that is, gross (uncracked) section properties were considered in the analysis.

EC8 prescribed nonlinear seismic analysis (nonlinear static or dynamic procedure) for irregular structures, while linear elastic seismic analysis was permitted for regular structures. It should be noted that the PTN-R code did not contain provisions related to the application of nonlinear seismic analysis procedures for the evaluation of existing buildings.

Repair and retrofitting techniques for masonry buildings, prescribed by the EC8-3 (Annex C), are similar to those prescribed by the PTN-R code and include the repair of cracks, repair and retrofitting of wall intersections, retrofitting and stiffening of horizontal diaphragms, provision of tie beams, retrofitting by means of steel ties, retrofitting of walls by means of RC jackets or steel profiles, and fiber-reinforced polymer (FRP) jackets.

In addition to the PTN-R code, which was the governing code for seismic retrofitting of buildings in Serbia until 2019, all masonry structural components (e.g., walls) had to be designed or evaluated according to the PTN-Z code which was issued in 1991 [25]. Similarly, Eurocode 6 (EN 1996-1-1:2004) [26], which is currently used in Serbia, contains design provisions for masonry buildings.

3. Seismic Retrofitting Techniques for URM Buildings

3.1. Seismic Retrofitting Objectives

Seismic retrofitting solutions should be effective in enhancing the performance of existing structures to achieve predetermined performance objectives. Performance objectives for a specific structure are either set by the design code or project-specific design criteria. In some countries, technical codes/standards for existing buildings may permit relaxed seismic performance objectives for the evaluation and retrofitting of existing buildings relative to the design of new structures, e.g., ASCE/SEI 41-17 code in the USA [27]. However, seismic design codes in former Yugoslavia did not make a distinction between new and existing buildings in terms of performance objectives.

One of the key design aspects of a seismic retrofitting project is to identify retrofitting goals. After the seismic evaluation of a building is performed and the deficiencies have been identified, a designer should be able to determine the retrofitting goals. Is the main goal of the retrofitting to enhance the lateral load-resisting capacity and/or stiffness and/or ductility of the existing structure—or perhaps a combination of those structural characteristics? An appropriate seismic retrofitting solution may be selected after the goals have been established.

Figure 4 illustrates different seismic retrofitting goals [28,29]. In many cases, the primary goal of retrofitting is to enhance the ductility of the existing structure, which may be feasible for the retrofitting of older RC structures (Figure 4a). Alternatively, stiffness and capacity enhancement (Figure 4b) may be feasible for retrofitting of an existing non-ductile structure. For example, the provision of new RC shear walls or steel bracings are common retrofitting techniques for enhancing both the capacity and stiffness of existing buildings. Stiffness, capacity, and ductility enhancement (illustrated in Figure 4c) may be feasible for existing buildings with high seismic demand, which prompts a need for increased lateral load-resisting capacity. On the other hand, stiffness may be increased due to inherent features of a selected retrofitting technique, e.g., provision of new shear walls or braces. In the context of URM structures, it is important to note that it is unlikely for a retrofitting solution to achieve a significant increase in ductility due to the brittle nature of masonry. It is expected that a typical global retrofitting solution for a URM structure should primarily be effective in increasing its lateral load-resisting capacity. Several researchers have studied different seismic retrofitting techniques for RC and masonry buildings and compared their effectiveness [10,30–36].

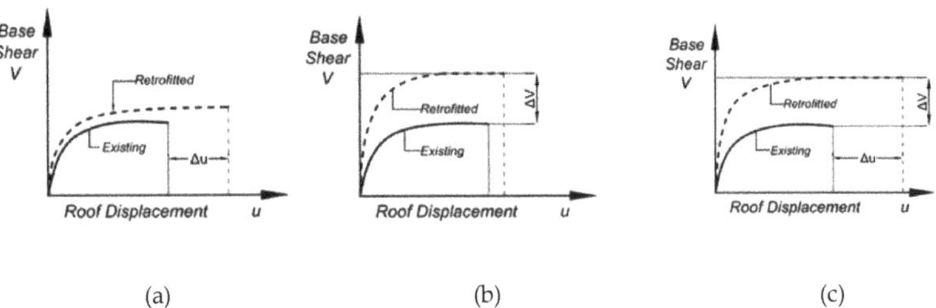

Figure 4. Seismic retrofitting goals for existing buildings: (**a**) ductility enhancement; (**b**) stiffness and strength enhancement, and (**c**) stiffness, strength and ductility enhancement [30].

3.2. Seismic Retrofitting Techniques—An Overview

Seismic retrofitting projects in the Balkan region were initiated after the 1979 Montenegro earthquake (M 6.9), which caused the damage and collapse of buildings in coastal areas of Montenegro and Croatia. Engineers and academics from all parts of the former Yugoslavia participated in the planning, design, and construction supervision of post-earthquake recovery. The earthquake also prompted a few relevant regional projects, which engaged experts from neighboring countries, such as the UNIDO-sponsored project "Building Construction Under Seismic Conditions in the Balkan Region". A series of comprehensive technical resources were produced as a result of the project, including the guidelines for seismic retrofitting of existing RC and masonry buildings [37]. Notable experimental research studies and field applications of seismic retrofitting on existing masonry buildings were performed by Prof. Miha Tomaževič and his colleagues at ZAG, Slovenia [32,38]. Comprehensive technical guidelines have recently been developed for the repair and retrofitting of masonry buildings affected by the March 2020 Zagreb, Croatia earthquake [39]. A valuable resource is available in Serbia for engineers engaged in the structural and seismic rehabilitation of buildings [40].

The most common retrofitting techniques for URM structures include: (i) retrofitting of existing masonry walls, (ii) construction of new RC walls attached to the existing masonry walls, (iii) retrofitting of wall connections, as well as the wall-to-floor connections, and (iv) retrofitting of the existing floor and/or roof structures. In some cases, retrofitting of existing foundations may be required, when shear and/or flexural capacity of the retrofitted wall have increased as a result of the retrofit.

A brief overview of selected seismic retrofitting techniques for URM structures is presented in the following text.

3.3. RC Jacketing—A Wall Retrofitting Technique

The RC jacketing technique (Figure 5) consists of constructing one- or two-sided RC jackets which need to be attached to the exterior and/or interior wall surfaces [30]. A jacket consists of a 3 to 5 cm thick concrete overlay with reinforcement in the form of a steel mesh (usually small size bars, 4 to 8 mm diameter). RC jackets are usually attached to an existing masonry wall via steel anchors inserted in pre-drilled holes, which are subsequently filled with cement- or epoxy-based grout. The required size and spacing of anchors depend on seismic demand and the required jacket thickness [41,42]. Either cast-in-place concrete or sprayed concrete (shotcrete) can be used for RC jackets.

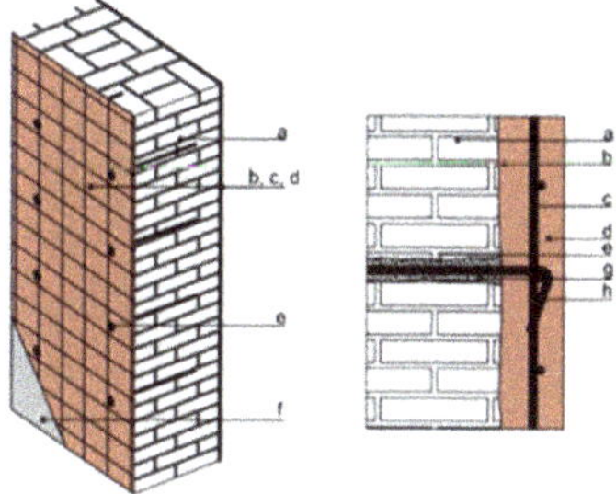

Figure 5. A schematic diagram of a retrofitted URM wall with RC jacketing (Legend: (a) existing masonry wall; (b) first layer of concrete; (c) steel wire mesh; (d) second layer of concrete; (e) steel anchors; (g) a grouted hole; (f) cement-based plaster; and (h) a steel anchor with a 90-degree hook).

The design of masonry walls retrofitted by RC jackets is performed by considering the stiffness of the retrofitted wall as the sum of the stiffnesses of the original masonry wall and the RC jackets. As a result, the internal shear force in the jacket is proportional to its stiffness relative to the total wall stiffness. It is worth noting that the stiffness of the jacket is influenced by the jacket thickness and mechanical properties of concrete (modulus of elasticity).

A significant increase in the shear capacity and stiffness of the masonry walls retrofitted using RC jacketing has been reported based on experimental studies on masonry wall specimens subjected to monotonic and/or reversed cyclic lateral loading [41,43–46]. The tests revealed that RC jacketing was able to increase the lateral capacity of the specimens by a factor of 2.0 to 3.0. Specimens with two-sided jacketing exhibited higher ductility and energy dissipation capacity. A few researchers performed experimental studies on the shear capacity of masonry wallets reinforced with ferrocement [47,48]. Shaking table testing of a four-story masonry building model retrofitted with RC jacketing was also performed [49].

In most cases, researchers used steel reinforcement (welded wire mesh) for RC jacketing; however, a few researchers tested specimens with jacketing consisting of fiber-reinforced concrete with glass or steel fibers, which exhibited superior performance compared to specimens retrofitted with RC jacketing [50].

3.4. A New RC Overlay/Wall Attached to the Existing One—A Wall Retrofitting Technique

When an existing URM wall has deficient gravity and lateral load-resisting capacity, it can be strengthened by constructing a new RC shear wall which is attached to the existing URM wall [51]. This is essentially a similar concept to RC jacketing, except that the thickness of the new wall is 10–15 cm, while the thickness of an RC jacket is usually on the order of 5 cm. Addition of a new RC wall results in a significant increase in lateral stiffness, shear and flexural capacity of the existing wall. A new RC wall is attached to the existing masonry wall in the same manner as previously explained for RC jacketing, except that the amount of wall reinforcement and anchors may be different. Retrofitting of wall foundations is usually required due to a significant increase in the shear and flexural capacity of a retrofitted wall.

This technique has proven to be effective for seismic retrofitting of damaged URM walls. Figure 6a shows a conceptual retrofitting scheme, which consists of constructing a new RC wall attached to the existing masonry wall by means of steel anchors. Figure 6b shows a possible location for a new RC wall attached to an existing exterior URM wall.

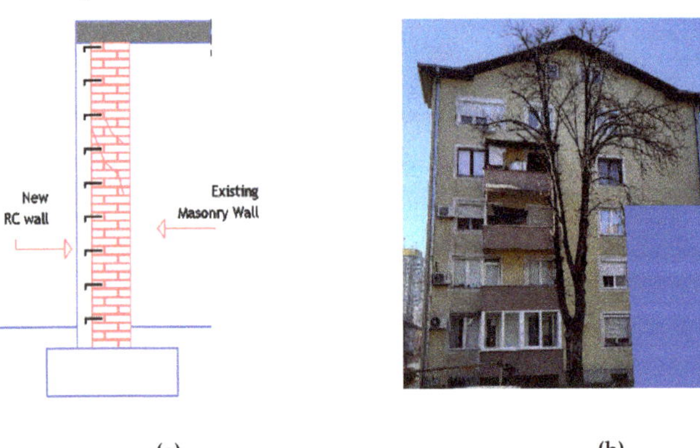

(a)   (b)

Figure 6. New RC walls: (**a**) a conceptual seismic retrofitting solution showing new RC wall and existing URM wall, and (**b**) a building elevation showing retrofit location.

3.5. FRP Overlays or Strips—A Wall Retrofitting Technique

Seismic retrofitting can also be achieved by applying fiber-reinforced polymer (FRP) overlays or strips on wall surfaces that were previously saturated by epoxy resin (or an alternative). FRP overlays may cover the entire wall surface (Figure 7a); alternatively, they could be applied in the form of strips aligned in horizontal, vertical, or diagonal directions (Figure 7b). FRP overlays and strips can be used either as one-sided or two-sided applications. To ensure an adequate anchorage, these FRP overlays/strips can be either wrapped (extended) at the wall ends, or custom-designed fiber anchors can be installed along the wall perimeter [30] (Figure 7a).

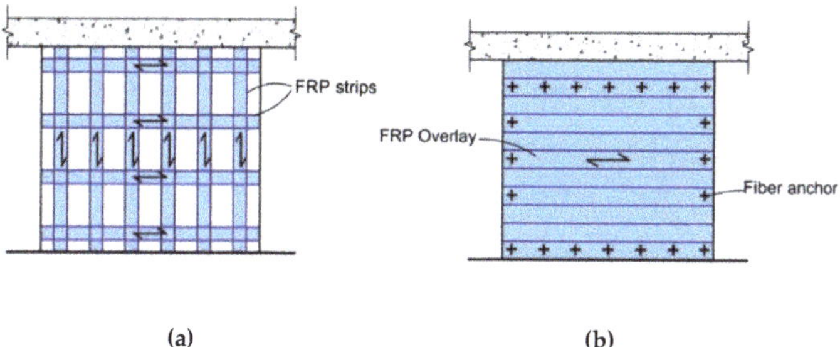

Figure 7. Seismic retrofitting of masonry walls using FRPs: (**a**) a FRP overlay applied over the entire wall surface, and (**b**) horizontal and vertical FRP strips [30].

FRP overlays act in a similar manner to RC jackets and are suitable for application to existing masonry walls with a deficient shear capacity. The fibers act as tension reinforcement for the wall and should be aligned in the direction of tensile stresses. The required effective area of fibers per unit width and the FRP contribution to shear capacity of the retrofitted wall are governed by the bond and anchorage strength at the FRP-to-wall interface. Design procedures for FRP-based retrofitting of masonry structures are well established [52].

The FRP technology has been used in the last 30 years, initially for structural rehabilitation purposes (e.g., existing bridges) and later on for seismic retrofitting of RC and masonry structures before and after earthquakes. Numerous experimental research studies on masonry specimens retrofitted with FRP overlays have been performed since the 1990s. Early experimental research studies were focused on testing small specimens, such as masonry triplets with bonded GFRP fabrics (with glass fibers) which are subjected to monotonic shear loading. The results showed that the mode of failure was governed by the FRP tensile strength [53].

One of the earliest experimental research studies involved the testing of a URM wall specimen retrofitted with a GFRP overlay on one side and vertical strips on the other side of the wall. The specimen was subjected to reversed cyclic lateral loading, and the results showed a capacity increase by a factor of 2.2 and a ductility increase by a factor of 2.5 compared to an otherwise similar URM specimen [54]. Delamination of the GFRP overlay was observed in the region of high tensile stresses in the middle portion of the wall. A study on URM piers retrofitted with horizontal and vertical GFRP strips showed a significant increase in the shear capacity by a factor of more than 2.0 for specimens subjected to reversed cyclic loading [33]. The behavior was governed by the delamination of vertical GFRP strips. A significant loss of shear capacity was observed after delamination (debonding) of the FRP from the wall surface. A study on seven scaled URM wall specimens with FRP overlays showed that one-sided retrofitting resulted in significantly enhanced lateral capacity, stiffness, and energy dissipation of the test specimens [55]. The results show that an increase in the lateral capacity was proportional to the amount of FRP axial stiffness and that high FRP axial stiffness led to a brittle failure. A comprehensive experimental study comprised testing 28 full-size brick masonry wall specimens with four different FRP layouts [56]. The results of the study emphasized the importance of a careful design of FRP retrofit scheme as well as material compatibility in developing seismic retrofitting solutions. In another study, a URM wall specimen with a CFRP overlay (with carbon fibers) and custom-designed fiber anchors installed along vertical and horizontal wall edges was subjected to reversed cyclic loading [57]. The results showed a 50% increase in the shear capacity compared to the URM wall specimen, and a significant drift capacity of the retrofitted wall (4.5%).

The effectiveness of vertical GFRP strips (with glass fibers) for enhancing the seismic performance of masonry piers was studied on a single-story masonry building model subjected to pseudo-dynamic loading [58]. The results show that GFRP strips were effective in increasing the pier lateral capacity and stiffness but did not cause a change in the original failure mechanism (pier rocking).

Shaking table tests on five half-scale masonry wall specimens retrofitted by means of single-sided FRP configurations with glass, aramid, and carbon fibers were also performed [59,60]. The results show that retrofitting with GFRP fabrics improved the shear capacity of masonry walls by a factor of about 2.5. In another study, shaking-table tests of five masonry walls retrofitted using GFRP strips in four different configurations (including a full-surface overlay and a combination of strips) showed that all retrofitted specimens performed well during the design-level shaking, and three out of four GFRP configurations also performed well during the extreme-level shaking [61]. The tests showed that the use of vertical GFRP strips alone is able to improve the in-plane performance of URM walls. A shaking-table testing on a scaled URM building model retrofitted with CFRP strips was performed at ZAG, Slovenia [8]. The model retrofitted using CFRP strips resisted to 3.5 times stronger shaking compared to the control model and did not experience a collapse.

Experience related to the application of FRP technology in Serbia and the region is limited; however, this technology has been recently used for the retrofitting of masonry buildings after the 2020 Zagreb, Croatia earthquake [62].

3.6. Replacement of Existing Masonry Walls

In some cases, it may be necessary to demolish an existing damaged wall and replace it with a new masonry or RC wall or frame, with enhanced gravity and lateral load capacity. The same procedure can be applied when it is required to enlarge an opening in an existing load-bearing wall. Figure 8 illustrates the reconstruction process. Initially, scaffolding and formwork are installed as preparation for the demolition of a lower portion of the wall (Figure 8a). Subsequently, supports and cross beams are provided to support the upper portion of the wall (Figure 8b). A side view of the arrangement for supporting the upper portion of the wall during the intervention on its lower portion is shown in Figure 8c).

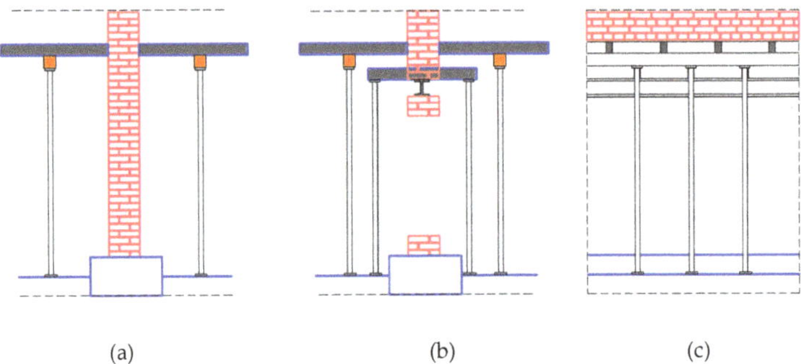

(a)   (b)   (c)

Figure 8. Demolition of a damaged URM wall and construction of a new wall: (**a**) scaffolding and formwork prepared for wall demolition; (**b**) setup for supporting upper portion of the wall, and (**c**) a side view of the arrangement for supporting upper portion of the wall.

This approach was used to increase the seismic resistance of the Diocese in Pančevo, Serbia, an existing URM building of cultural and historical importance which was constructed in 1832 (Figure 9) [63]. Loadbearing URM walls were 62 cm thick and were constructed in lime mortar. Due to earlier interventions, significant portions of load-bearing walls were removed and openings were formed in transverse walls (shown in solid red color on the floor plan, Figure 10). As a result of these interventions, the seismic integrity of the building in the transverse direction was jeopardized and a structural intervention was required.

Figure 9. The Diocese in Pančevo, Serbia—a view of the front façade.

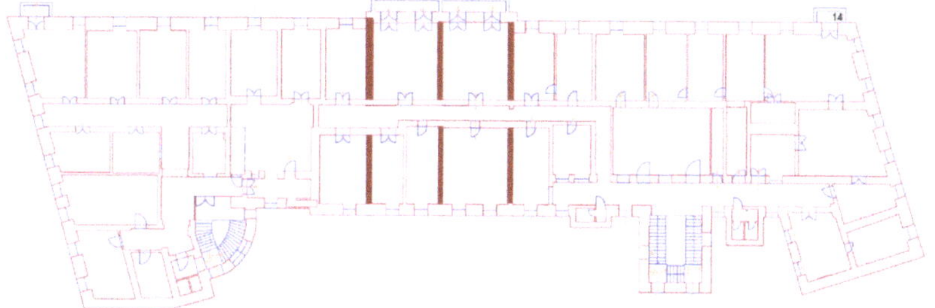

Figure 10. Floor plan of the Diocese in Pančevo.

In order to restore the lateral load-resisting capacity of the structure in the transverse direction, the width of new RC frames constructed at the locations of original walls was the same as wall thickness (Figure 10). RC beams were 72 cm high, while column dimension in the plane of the frame was 50 cm. Seismic analysis and design of new frames in the transverse direction at the ground floor level were carried out according to the PTN-S and PTN-R codes (and applicable codes for the design of RC structures). Figure 11a shows a scaffolding arrangement for transferring gravity loads from the upper floors to the ground, as well as reinforcement for the RC frame. Reinforcement detailing for a typical RC frame is shown in Figure 11b.

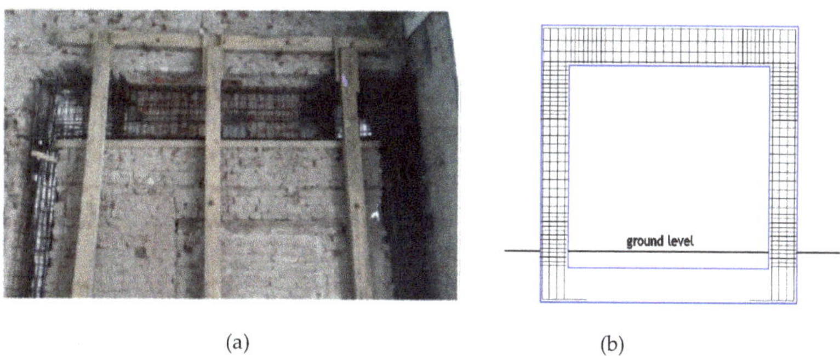

(a)   (b)

Figure 11. New RC frames were constructed as a part of the seismic retrofitting of the Diocese in Pančevo: (a) formwork and scaffolding arrangement and (b) frame reinforcement detailing.

3.7. Retrofitting of Flexible Floor Structures

The retrofitting of existing flexible floor structures in masonry buildings, usually made of timber, is often required, and it was prescribed by the PTN-R code. The following approaches were permitted by the code: (a) steel ties provided parallel with the walls on both sides (underneath the floors or roof); (b) diagonal bracing of existing timber floors in horizontal plane, and (c) replacement of an existing timber floor by a new RC floor [38]. Additionally, wall-to-floor connections may need to be strengthened. Figure 12 shows the application of the approach (b) for strengthening of the floors in a URM building after the 2010 Kraljevo earthquake [42]. Seismic assessment of the building showed that wall retrofitting would not be required provided that floor and roof structures acted as rigid diaphragms. A portion of the floor structure was a rigid RC floor, while another portion was a flexible timber floor (since the original building was extended in the horizontal direction). The design engineers decided to retrofit a flexible portion of the floor structure by means of a horizontal steel truss. The truss elements were anchored into RC ring beams which existed at each floor level.

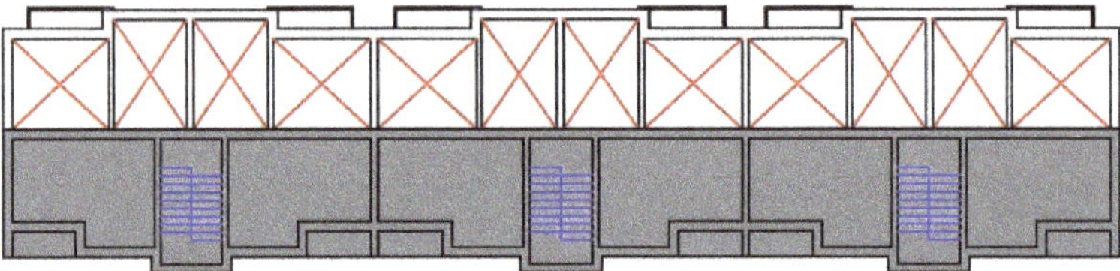

Figure 12. Retrofitting of an existing flexible floor structure by means of horizontal steel truss after the 2010 Kraljevo, Serbia earthquake.

3.8. Comparison of Seismic Retrofitting Techniques for Masonry Buildings

A comparison of seismic retrofitting techniques for masonry buildings is summarized in Table 1 [30]. The following criteria are deemed relevant for selecting the most suitable seismic retrofitting technique: (i) local availability of construction materials, (ii) required level of construction skills, (iii) construction cost, (iv) disruption to the occupants, and (v) required maintenance.

RC jacketing is often considered a more feasible solution compared to the retrofit performed using FRP technologies due to lower construction cost. Another advantage of RC jacketing is that advanced construction skills are not required for its implementation; however, disturbance to the occupants during the construction is significantly higher compared to the FRP retrofit. Advantages of FRPs include ease and speed of application, thus resulting in minimal disruption to the occupants. It should be noted that FRP technologies were not widely used for the retrofitting of buildings affected by the 2010 Kraljevo earthquake, but are currently used for various structural and seismic retrofitting projects in Serbia. Unit construction costs for different retrofitting techniques are included in Table 1 (column 6). These costs are based on the seismic retrofitting projects of URM buildings performed after the 2010 Kraljevo earthquake. Unit costs are expressed in EUR/m^2 and represent cost per m^2 of the wall area.

Table 1. A comparison of seismic retrofitting techniques for masonry walls.

Retrofitting Technique	Advantages	Disadvantages	Local Availability of Construction Materials	Required Level of Construction Skills	Construction Cost	Disruption to the Occupants	Required Maintenance
(1)	(2)	(3)	(4)	(5)	(6)	(7)	(8)
RC jacketing	(i) One of the most cost-effective retrofitting techniques. (ii) Able to enhance flexural and/or shear capacity and/or ductility.	(i) Adds mass/weight to the structure. (ii) Drilling holes through the existing walls may be required.	High	Low	Low (approximately 99 EUR/m^2 based on 5 cm thick RC jacket)	Moderate to high	Low
FRP overlays and strips	(i) Increases shear and/or flexural capacity of the existing wall. (ii) Lightweight. (iii) Rapid installation.	Requires fire and UV protection.	Low	Moderate to high	Moderate (approximately 110 EUR/m^2 for CFRP strips)	Low	Low
Replacement of an existing masonry wall with a new masonry or RC shear wall	One of the most effective retrofitting techniques.	(i) May increase seismic forces at the wall-to-floor slab interface. (ii) Requires new foundations. (iii) Need to drill holes through the existing RC floor slabs.	High	Medium	Moderate	High	Low

4. Seismic Retrofitting of Damaged URM Buildings after the 2010 Kraljevo Earthquake: A Case Study

4.1. The Earthquake and Its Consequences

The most damaging earthquake in Serbia since the beginning of the 21st century occurred on 3 November 2010, and had a magnitude (M_W) of 5.5 [64]. The epicenter was located close to the Sirča village, approximately 4 km north of Kraljevo (a town with a population of 68,000). The earthquake caused two fatalities and more than USD 100 million in damages [65]. Out of 16,000 buildings that experienced damage or collapse due to the earthquake, approximately 25% were found to be unsafe to occupy [66]. A large number of single-family dwellings, as well as some multi-family residential buildings, educational, and health facilities, were affected by the earthquake. Masonry buildings, which accounted for more than 90% of the building stock in the earthquake-affected area, were most severely affected by the earthquake [67].

Several multi-family URM buildings (three- to five-stories high) constructed after WWII (1945–1963) were damaged due to the earthquake and required repair and retrofit [42,67,68]. The masonry walls in these buildings were typically constructed using solid clay bricks, and their thickness ranged from 25 cm (interior walls) to 38 cm (exterior walls). Floor structures were in the form of ribbed RC slabs or semi-prefabricated clay and concrete floors, and RC tie-beams (ring beams) were provided at each floor level. In most cases, the walls experienced moderate damage in the form of cracks due to in-plane or out-of-plane seismic loads [14,42,67]. Some of the damaged buildings had vertical extensions (additional floors), and it was reported that the extensions which were not designed and constructed according to the existing technical regulations were damaged in many cases [69].

A URM residential building in Kraljevo was retrofitted before the 2010 earthquake due to the upcoming vertical extension, which prompted the need for seismic evaluation and retrofitting. The building was retrofitted using RC jacketing, and it did not experience any damage in the earthquake [69]. The building was located in Jug Bogdanova Street in Kraljevo, in the vicinity of a few other similar URM buildings which experienced moderate-to-severe structural damage and had to be repaired and retrofitted after the earthquake (see Figure 13).

This section discusses in detail a URM building in Kraljevo that was damaged due to the 2010 earthquake and was retrofitted after the earthquake by applying a seismic retrofitting approach, which was prescribed by the Serbian code PTN-R and was used in Serbia and the region since 1985.

4.2. Case Study Building: Description and Earthquake Damage

The case study building is located in Njegoševa Street No. 2 in Kraljevo, and was constructed around 1950 as a three-story residential building with a basement and a half-floor at the top (typical story height was 2.8 m), see Figure 14. The building had a rectangular plan shape with 22.2 m length and 16.0 m width for all floors, except for the top floor (extension) with smaller plan dimensions (11.0 m length and 10.9 m width). The walls at the lower three floors were constructed using 25 cm solid clay bricks in cement–lime mortar, while the walls at the top floor were constructed using modular (multi-perforated) clay blocks with 120 mm thickness. Floors and roofs were constructed using semi-prefabricated composite masonry and a concrete system which were considered to act as rigid diaphragms. The walls were constructed using URM construction, but RC tie-beams were provided at each floor level.

1-URM building retrofitted before the earthquake and
remained undamaged
2- Damaged URM building retrofitted after the earthquake
X- Other damaged URM buildings

Building 1

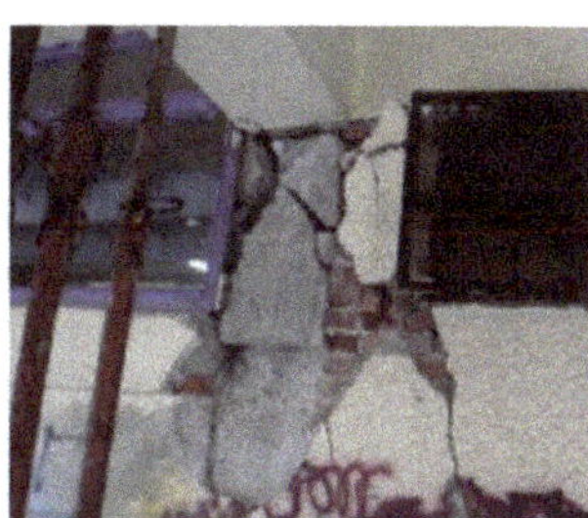

Building 2

Figure 13. Jug Bogdanova Street in Kraljevo, Serbia after the 2010 earthquake: examples of damaged and undamaged masonry buildings.

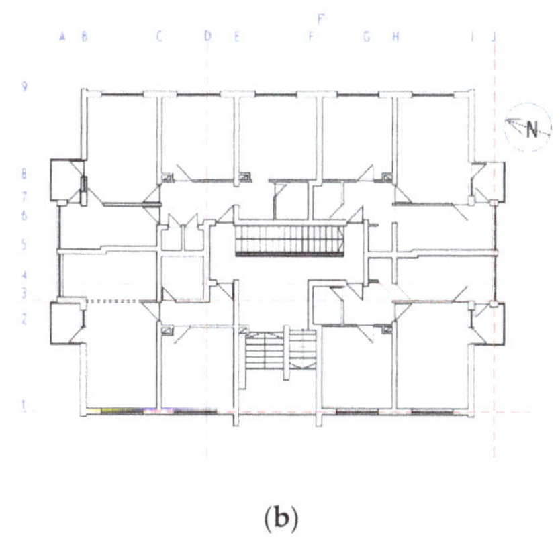

(a)

(b)

Figure 14. URM building located in Njegoševa Street No. 2, Kraljevo: (**a**) exterior views and (**b**) typical floor plan.

As the building was constructed around 1950, seismic effects were likely not considered in the original design. The building was damaged in the 2010 earthquake. Structural damage in the lower portion of the building was mostly in the form of inclined cracks due to in-plane seismic effects. Cracking was most prominent in longitudinal walls (N–S direction); for example, wide cracks were observed along the masonry-to-tie-beam interface in a longitudinal wall along gridline 5 at the second-floor level (Figure 15). The most extensive damage was observed in the extended portion of the building at the top floor level. Refer to [1] for more details related to the seismic performance of the building in the 2010 Kraljevo earthquake and a seismic evaluation of the damaged structure according to the PTN-S code and Eurocode 8.

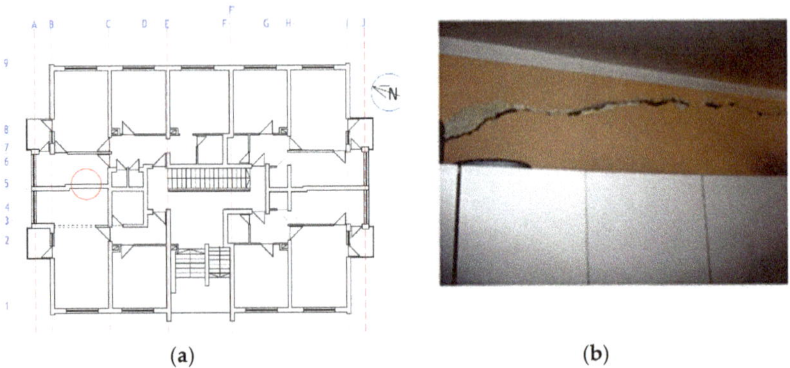

(a)   (b)

Figure 15. Example of earthquake damage in a case study building: (**a**) floor plan at the second-story level showing a damaged wall (red arrow) and (**b**) severe cracking in a longitudinal wall (gridline 5).

4.3. Seismic Retrofitting Approach

The building was retrofitted according to the PTN-R code. The main goal of seismic retrofitting was to enhance the overall structural integrity by constructing vertical RC jackets along the façade (embellished in navy blue color in Figure 16). The main reason for performing retrofitting at the exterior was to minimize the disruption to the building occupants.

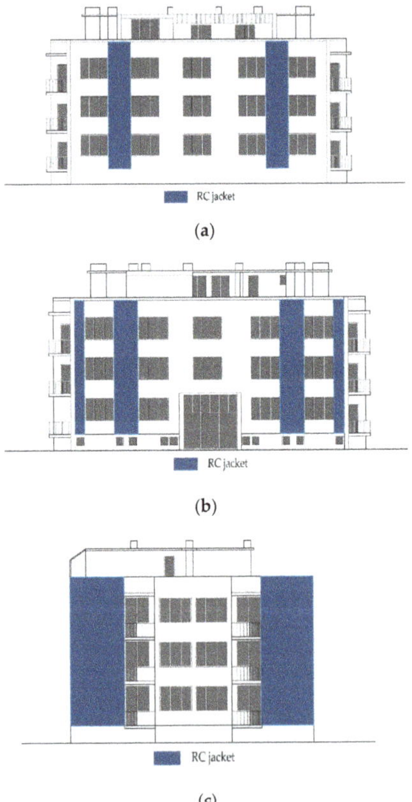

(a)

(b)

(c)

Figure 16. *Cont.*

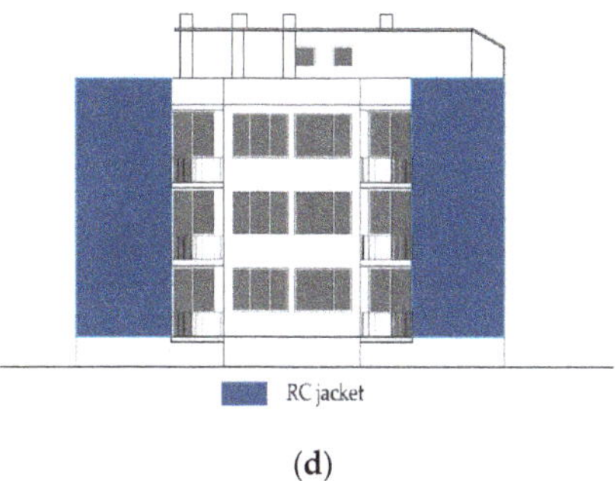

(**d**)

Figure 16. Elevations of the retrofitted building: (**a**) east façade; (**b**) west façade; (**c**) north façade, and (**d**) south façade.

RC jackets were constructed using 50 mm thick concrete reinforced by means of 6 mm diameter welded steel mesh at 100 mm spacing (Figure 17). The jackets were attached to the existing wall by means of 8 mm diameter steel anchors spaced at 300 mm horizontally and vertically. Prior to the retrofit, the wall surface was prepared by removing plaster and sandblasting. The structural cost of the retrofitting (based on m^2 of the built-up area) was approximately 8.4 EUR/m^2.

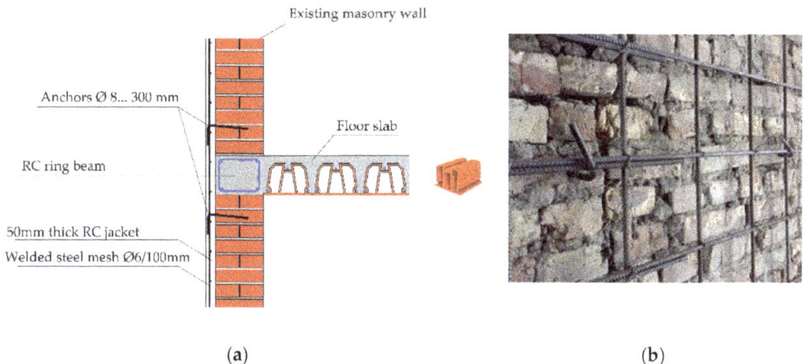

(**a**)   (**b**)

Figure 17. RC jackets: (**a**) vertical section of the wall and (**b**) an example of RC jacketing under construction [30].

4.4. Seismic Analysis and Numerical Models

Seismic evaluation and retrofit design of earthquake-damaged buildings in Kraljevo was performed in line with the technical regulations which were enforced in Serbia at the time of the 2010 earthquake, that is, the PTN-S code (which prescribed seismic design parameters and analysis procedure) and the PTN-R code (which contained provisions related to retrofit design). These codes prescribed linear elastic analysis for both the original and retrofitted structures. In situ testing of masonry materials was not performed after the earthquake, and material properties were assumed based on the original design specifications. As a result, the effect of non-linear material behavior on seismic response in the post-cracking stage of URM walls was not considered. It should be noted that the effect of non-linear seismic response of cracked URM walls for the case study building was considered in line with the EC8-3 provisions for masonry buildings (by reducing the wall stiffness).

Equivalent static seismic analysis according to the PTN-S code was performed using the following parameters: seismic intensity coefficient K_s of 0.05 (seismic intensity zone VIII according to the map shown in Figure 3a), building category coefficient K_o of 1.0

corresponding to Category I, dynamic response coefficient K_d of 1.0, and the ductility and damping coefficient K_p of 2.0 (corresponding to URM building). The soil was classified as Category II according to the PTN-S code. It should be noted that seismic hazard parameters for Kraljevo were revised after the earthquake; hence, the building site is currently located in seismic intensity zone IX.

Multi-modal seismic analysis was performed for both the original and retrofitted structure according to EC8-1. The design ground acceleration for soil type A was 0.2 g, while ground type B was considered for the site. Spectral accelerations for the elastic design spectrum $S_d(T)$ according to Eurocode 8 were divided by the behavior factor q of 1.5 for URM structures designed without seismic provisions (for original structure) and q = 2.5 (for retrofitted structure). Type 1 spectrum was deemed appropriate given the seismic hazard setting for the building site. Design response spectra for Kraljevo, Serbia, based on the PTN-S code and Eurocode 8 are presented in Figure 18.

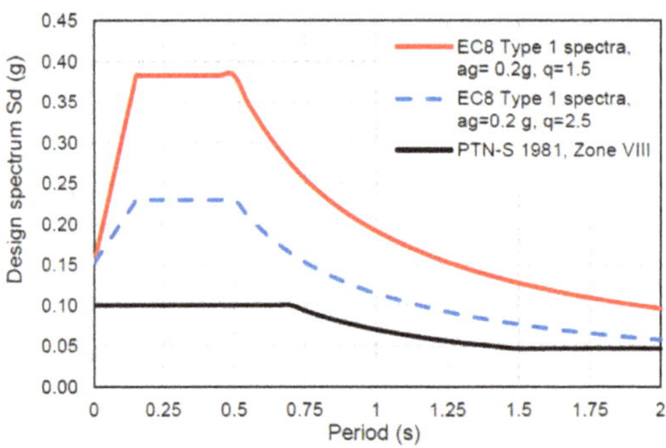

Figure 18. Design response spectra for Kraljevo, Serbia, according to Eurocode 8 (a_g = 0.2 g, ground type B, q = 1.5 and q = 2.5) and PTN-S code (seismic intensity VIII, soil Category II—valid at the time of the 2010 earthquake).

Due to time constraints and limited resources, it was not possible to perform detailed material testing in order to determine the mechanical properties of masonry. Hence, material properties were assumed to be equal to the values specified at the time of the original design. Similar values were also obtained by testing sample bricks extracted from buildings in Kraljevo after the earthquake. Masonry walls were constructed using solid clay bricks, with the estimated characteristic compressive strength of 2.5 MPa. Cement:lime:sand mortar with the mix proportions 1:3:9 was used, and its characteristic compressive strength was estimated at 2.0 MPa. Based on these characteristics, a masonry compressive strength of 2.41 MPa was obtained based on the PTN-Z code provisions, and the modulus of elasticity for masonry was taken as 2410.0 MPa. Since the masonry walls in the extended building portion (top floor) were constructed using modular clay blocks, their characteristic compressive strength was estimated at 10 MPa. The verification of load-bearing capacity for masonry structural components for the seismic retrofitting purposes was performed according to the PTN-Z code [25]. Eurocode 6 [26,70] requirements were not followed in the retrofit project; however, relevant provisions were used in this study.

RC jacketing was performed using low-strength concrete overlay (M20 grade with 20 MPa characteristic compressive strength based on the cube specimens), while GA240/360 steel grade (yield strength 240 MPa) and MAG500/560 steel grade (yield strength 500 MPa) were used for wall anchors and welded wire mesh, respectively.

A 3D numerical model of the building was created in the Tower finite element software package, which was developed in Serbia and has been widely used by academics and practicing engineers [71]. The walls were modelled as shell elements, while the slabs were modelled as plate elements. Floor and roof structures were modelled as rigid diaphragms. The foundations were simulated as fixed-base restraints. Two numerical models were developed for the original structure: (a) a model taking into account wall piers (no parapets) and (b) a model with parapets and spandrels, accounting for the stiffness of horizontal elements below and above the openings (Figure 19). The first model, in which wall piers are the main vertical elements of the lateral load-resisting system, may also be referred to as the "Cantilever model". Since the model ignores the effect of spandrels, the effect of the wall–slab interaction may not be accurately simulated in the seismic analysis. On the other hand, the second model is similar to the "Equivalent Frame Model" (EFM), which has also been used to model masonry structures for seismic analysis purposes [72]. A discussion on the pros and cons of both models is provided elsewhere [73].

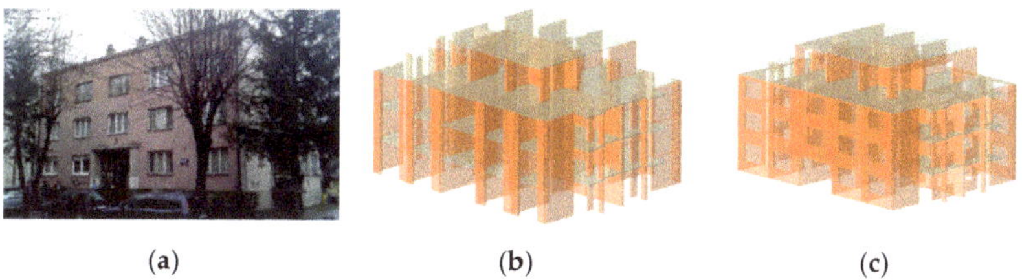

(a)   (b)   (c)

Figure 19. 3D numerical models: (**a**) actual building; (**b**) Cantilever model, and (**c**) Equivalent Frame Model (EFM).

The following three models were considered to account for the effect of cracking on the original and retrofitted structure: (a) Model 1, which considered uncracked (gross) properties of the original structure (referred to as Original 1 and Retrofitted 1); (b) Model 2, which considered the effect of cracking (20% stiffness reduction), which is referred to as Original 2 and Retrofitted 2, and (c) Model 3, which considers a 50% stiffness reduction (Original 3 and Retrofitted 3). Note that the PTN-S code did not include a provision related to the stiffness reduction (or an alternative provision to account for the effect of cracking in masonry and RC structures); hence, Model 1 is in line with the PTN-S requirements. On the other hand, Model 3 reflects the EC8-3 requirements, which prescribe a 50% stiffness reduction. Finally, Model 2 (20% stiffness reduction) reflects a situation wherein URM walls have experienced moderate cracking, which was true for many buildings affected by the 2010 Kraljevo earthquake.

Dynamic properties of various numerical models were obtained from the modal analysis. It should be noted that the total seismic weight calculated according to the PTN-S was 14,032 kN. Table 2 presents fundamental periods for the original structure in both directions (N–S and E–W). A comparison of the two models (Cantilever versus EFM) can be made based on the fundamental period values. It can be seen from the table that fundamental periods are consistently higher for the Cantilever model compared to the EFM model, and the difference is on the order of 15 to 20%.

Table 2. Fundamental periods for the original structure.

	Original 1		Original 2		Original 3	
	Cantilever Model	EFM	Cantilever Model	EFM	Cantilever Model	EFM
	(1)	(2)	(3)	(4)	(5)	(6)
N–S direction	0.304	0.240	0.332	0.266	0.401	0.335
E–W direction	0.288	0.237	0.316	0.264	0.381	0.332

The modelling of RC jacketing is an important aspect of the project. According to the PTN-R code, a retrofitted masonry wall with an RC jacket is modelled as an equivalent masonry wall with the thickness equal to the thickness of the original wall plus additional thickness (equal to four times the thickness of an RC jacket), as discussed in Section 2. This concept is illustrated in Figure 20. Figure 20a shows a horizontal section of an actual 25 cm thick masonry wall retrofitted by a 5 cm thick RC jacket, while Figure 20b shows a model recommended by PTN-R, where a 5 cm thick RC jacket is represented by an equivalent 20 cm thick masonry wall. As a result, a 45 cm thick masonry wall is used for seismic analysis purposes. Unfortunately, the PTN-R code did not have a commentary; hence, the basis for this provision is not provided as a part of the code. Based on the fundamental mechanic principles for a composite masonry and RC section such as the one shown in Figure 20a, it can be assumed that the stiffness is determined as a sum of the stiffnesses for the masonry and RC components, which are characterized by different mechanical properties (modulus of elasticity E and modulus of rigidity G), as well as different thicknesses. The stiffness of the composite section can be assumed to have the mechanical properties of masonry. The equivalent thickness for the composite section can be determined by estimating a ratio of E_c/E_m, corresponding to the moduli of elasticity for concrete E_c and masonry E_m. The equivalent thickness according to the PTN-R code can be obtained when the E_c/E_m ratio is approximately equal to 4.0, which is a reasonable assumption when concrete with low-to-moderate characteristic compressive strength is used for RC jackets, as is the case with the retrofit solution for the case study building. According to the EC8-3 code, the designer is expected to simulate the effect of an RC jacket by modelling it as a separate shell layer, or a part of a composite equivalent column section, where masonry and concrete materials would be simulated using appropriate mechanical and geometric properties (see Figure 20c).

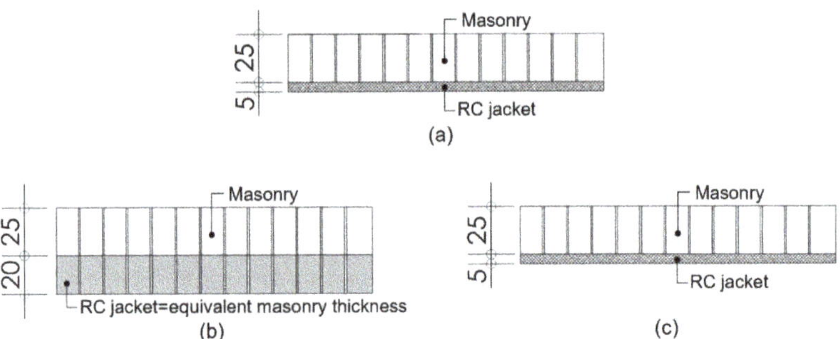

Figure 20. Numerical models for simulating RC jacketing: (**a**) actual wall section; (**b**) equivalent masonry section according to PTN-S, and (**c**) numerical model according to EC8-3.

Numerical models for the original and retrofitted structure are shown on Figure 21. It is worth noting that the vertical RC jackets are embellished in a darker color (Figure 21b).

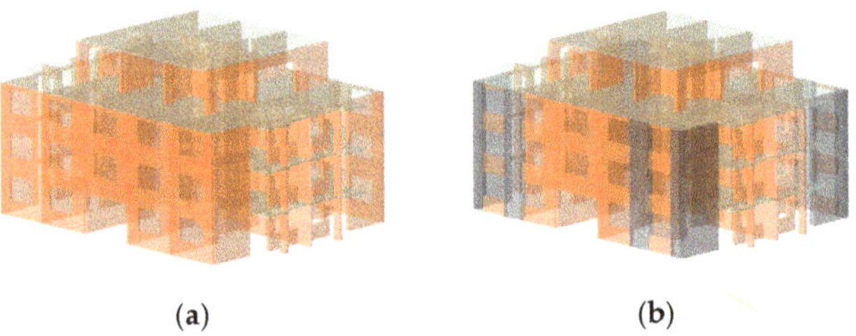

Figure 21. Numerical models: (**a**) original structure and (**b**) retrofitted structure.

Fundamental periods for the retrofitted structure, which were obtained using three different numerical models which take into account the extent of cracking and two different models for simulating masonry wall characteristics (Cantilever and EFM), are summarized in Table 3. It can be seen from the table that the periods for the Cantilever model are higher than for the EFM model; a similar phenomenon was observed for the original structure (Table 2). It can be noticed that the difference ranges from 18 to 23% for the N–S direction, while the difference for the E–W direction is smaller (on the order of 10%); this can be explained by a larger number of openings (windows) in the N–S direction (see Figure 14), which can be better simulated by the EFM model. It should be noted that the periods for both models are within the "plateau" range of the EC8-1 design spectra.

Table 3. Fundamental periods for the retrofitted structure.

	Retrofitted 1		Retrofitted 2		Retrofitted 3	
	Cantilever Model	EFM	Cantilever Model	EFM	Cantilever Model	EFM
	(1)	(2)	(3)	(4)	(5)	(6)
N–S direction	0.299 s	0.231 s	0.327 s	0.258 s	0.396 s	0.326 s
E–W direction	0.253 s	0.227 s	0.281 s	0.253 s	0.352 s	0.319 s

The effect of retrofitting and the extent of cracking on fundamental periods (based on the results presented in Tables 2 and 3) is illustrated in Figure 22. It can be seen from the chart that fundamental period values are consistently smaller for the retrofitted structure (for all models), but the difference is insignificant (less than 2%); this indicates a minor effect of RC jacketing on the stiffness increase, which can be attributed to one-sided jacketing applied at critical exterior locations in the building. An important observation is related to the effect of extent of cracking on the fundamental period values. It can be seen from the chart that the fundamental period of 0.401 s for Model 3 (50% stiffness reduction in line with the EC8-3 requirements) is by approximately 30% higher than the corresponding period of 0.304 s for Model 1 (no stiffness reduction—in line with the PTN-S requirements).

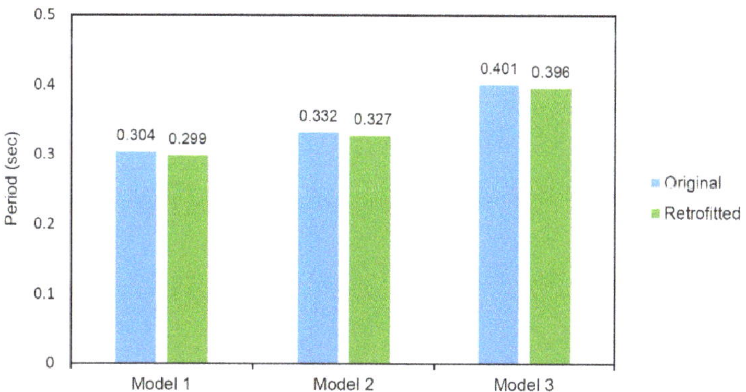

Figure 22. The effect of cracking on the fundamental periods in the N–S direction for the original and retrofitted structure (Cantilever model).

4.5. Results and Discussion

The results of the seismic analysis enabled a comparison between the seismic demand and capacity for individual walls, as well as the entire structure, according to the PTN-S code and Eurocode 8 (Parts 1 and 3). Note that majority of the results presented in this section were obtained using the Cantilever model, due to higher seismic forces and a more conservative design in this case.

The seismic base shear forces for the original building, according to the PTN-S code and Eurocode 8, Part 1, were previously reported by the authors [1]. It was observed that seismic forces corresponding to the PTN-S code were significantly smaller than the corresponding values obtained according to the Eurocode 8 requirements; this can also be seen from the design response spectra presented in Figure 18. A difference in the magnitude of seismic forces for the case study building according to the PTN-S and EC8-1 codes is illustrated in Figure 23. At the time of the 2010 earthquake, Kraljevo was located in seismic intensity zone VIII, but seismic hazard zonation has been subsequently revised to seismic intensity zone IX, thereby resulting in higher seismic forces. On the other hand, EC8-1 seismic forces were determined based on the latest seismic hazard map for Serbia shown in Figure 3b. A significant difference, both in magnitude and distribution of seismic force up the building height, can be seen from the chart. The difference is particularly notable at the third-story level (elevation 8.4 m), at which the applied seismic forces are 442.2 kN, 844.4 kN, and 1668.4 kN for PTN-S (intensity zones VIII and IX) and EC8-1, respectively. It can be concluded that the applied force according to EC8-1 (1668.4 kN) is almost 3.8 times higher than the corresponding force (442.2 kN) used for seismic evaluation and retrofitting of the building after the earthquake (PTN-S, zone VIII).

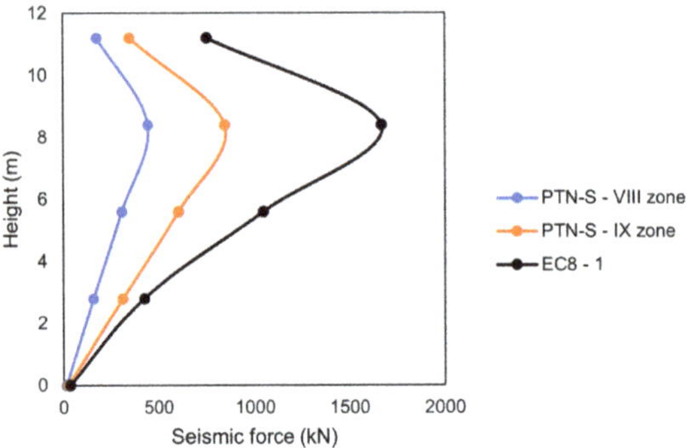

Figure 23. Applied seismic forces for the case study building at story levels determined according to the PTN-S code (seismic intensity VIII and IX) and EC8-1 (a_g = 0.2 g, ground type B, q = 1.5, and q = 2.5).

To illustrate the effectiveness of retrofitting, seismic base shear force V_{Ed} (kN) (seismic demand) was compared with the shear capacity at the ground floor level V_{Rd} (kN), which was taken as equal to the sum of capacities for all walls aligned in the same direction (N–S or E–W).

The results for the longitudinal (N–S) direction are summarized in Tables 4 and 5 and illustrated in Figure 24. It can be seen from the chart that the capacity of the building was satisfactory according to the PTN-S code, since the capacity (C) versus demand (D) ratio, C/D, is larger than 1.0 both for the Original 1 (uncracked) model which is in line with the PTN-S code and the Original 2 model (cracked, 20% stiffness reduction), but it is not satisfactory for the Original 3 model (cracked, 50% stiffness reduction—in line with EC8-3), as shown in Figure 25.

Table 4. Comparison of seismic capacity (C) and demand (D) at the ground floor level of the building in N–S direction (for the original structure).

	Original 1		Original 2		Original 3	
	PTN-S	EC8	PTN-S	EC8	PTN-S	EC8
	(1)	(2)	(3)	(4)	(5)	(6)
C: Shear Capacity $V_{Rd}(kN)$	1231.57	2052.61	985.25	1642.09	615.78	1026.30
D: Design Shear Force $V_{Ed}(kN)$	930.59	2301.70	930.59	2301.70	930.59	2301.70
$\frac{C}{D} = V_{Rd}/V_{Ed}$	1.32	0.89	1.06	0.71	0.66	0.44

Table 5. Comparison of seismic capacity (C) and demand (D) at the ground floor level of the building in N–S direction (for the retrofitted structure).

	Retrofitted 1		Retrofitted 2		Retrofitted 3	
	PTN-S	EC8	PTN-S	EC8	PTN-S	EC8
	(1)	(2)	(3)	(4)	(5)	(6)
C: Shear Capacity $V_{Rd}(kN)$	1477.88	2463.13	1180.08	1970.51	738.94	1231.56
D: Design Shear Force $V_{Ed}(kN)$	969.88	2398.88	969.88	2398.88	969.88	2398.88
$\frac{C}{D} = V_{Rd}/V_{Ed}$	1.52	1.03	1.22	0.82	0.76	0.51

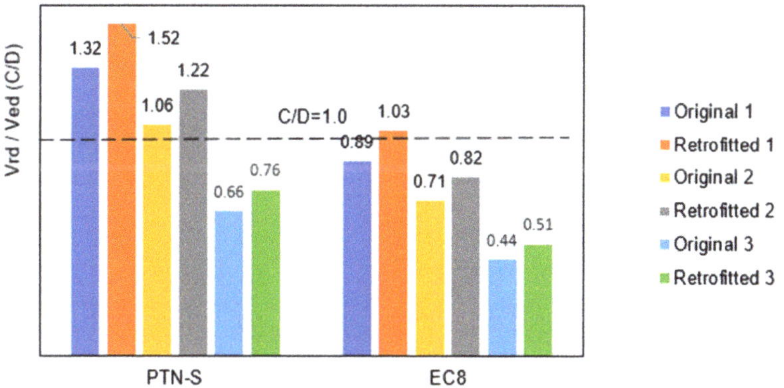

Figure 24. Seismic capacity versus demand (C/D) ratio for the ground floor of the building in N–S direction (for the original and retrofitted building).

The results also indicate that, according to the PTN-S code, the retrofit has resulted in an increased C/D ratio for the building to 1.52, 1.22, and 0.76 for Models 1, 2, and 3, respectively. The results shown in Table 5 indicate that the Retrofitted 3 model (which considers a 50% stiffness reduction) is not satisfactory, since the corresponding C/D value is less than 1.0. The results of the analysis performed according to the EC8-1 requirements have shown that the capacity of the structure is not satisfactory even after the retrofit for Models 2 and 3, since the corresponding C/D values are less than 1.0; however, the retrofit seems to be effective for Model 1, since the corresponding C/D ratio is 1.03 (in line with the PTN-S code).

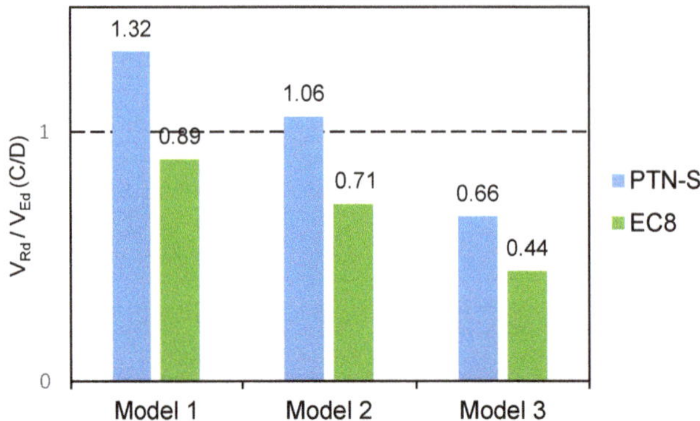

Figure 25. Seismic capacity versus demand (C/D) ratio for the ground floor of the building in N–S direction (for the original building).

A similar comparison is presented for the transverse (E–W) direction of the building, as shown in Tables 6 and 7 and Figure 26. The results (Table 6) have shown that the original building has satisfactory capacity (C/D ratio greater than 1.0), except for the Original 3 model, which is in line with the EC8-3 requirements (50% stiffness reduction); this is also illustrated in Figure 27. The results (Table 7) indicate that, according to the PTN-S code, the retrofit is effective in enhancing the seismic safety, expressed in terms of the C/D ratio (Retrofitted 1 model), but it was not deemed effective according to EC8-3 code requirements (Retrofitted 3 model) since the corresponding C/D ratio is 0.94.

Table 6. Comparison of seismic capacity (C) and demand (D) at the ground floor level of the building in E–W direction (for the original structure).

	Original 1		Original 2		Original 3	
	PTN-S	EC8	PTN-S	EC8	PTN-S	EC8
	(1)	(2)	(3)	(4)	(5)	(6)
C: Shear Capacity $V_{Rd}(kN)$	2088.35	3480.58	1670.68	2784.46	1044.18	1740.29
D: Design Shear Force $V_{Ed}(kN)$	930.59	2127.16	930.59	2127.16	930.59	2127.16
$\frac{C}{D} = V_{Rd}/V_{Ed}$	2.25	1.64	1.79	1.31	1.12	0.82

Table 7. Comparison of seismic capacity (C) and demand (D) at the ground floor level of the building in E–W direction (for the retrofitted structure).

	Retrofitted 1		Retrofitted 2		Retrofitted 3	
	PTN-S	EC8	PTN-S	EC8	PTN-S	EC8
	(1)	(2)	(3)	(4)	(5)	(6)
C: Shear Capacity $V_{Rd}(kN)$	2506.02	4176.69	2001.04	3341.36	1253.01	2088.34
D: Design Shear Force $V_{Ed}(kN)$	969.88	2216.97	969.88	2216.97	969.88	2216.97
$\frac{C}{D} = V_{Rd}/V_{Ed}$	2.58	1.88	2.06	1.51	1.29	0.94

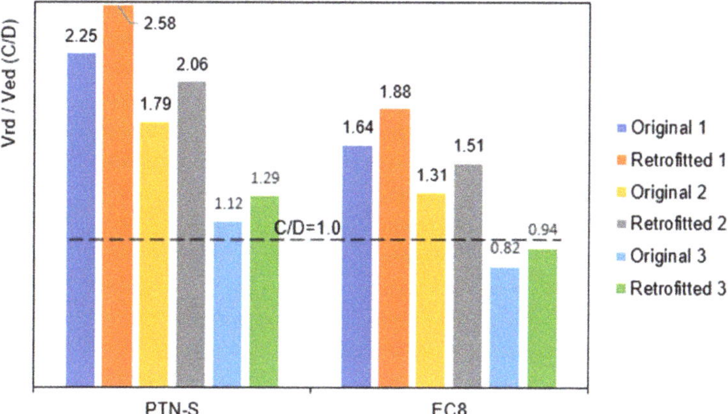

Figure 26. Seismic capacity versus demand (C/D) ratio for the ground floor of the building in E–W direction (for the original and retrofitted building).

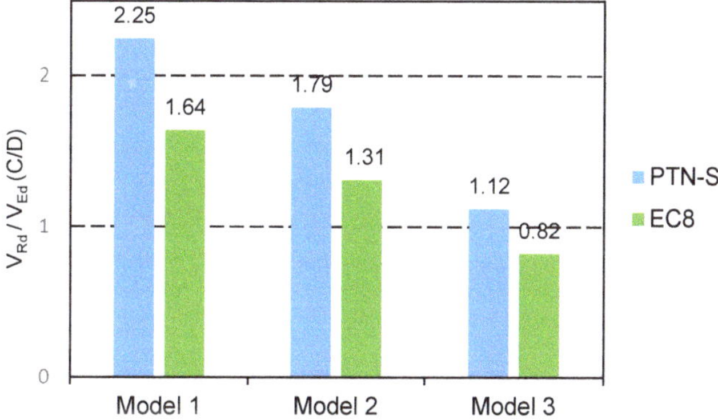

Figure 27. Seismic capacity versus demand (C/D) ratio for the ground floor of the building in E–W direction (for the original building).

The C/D ratio values for longitudinal (N–S) direction are generally lower than for the transverse (E–W) direction; this is in line with the findings of the previous study related to the same building [1], which was focused on examining the wall index (WI) as an indicator of seismic safety, based on the number of walls in each horizontal direction of a masonry building. The building under consideration is characterized by a significantly higher WI per floor ratio value for the transverse (E–W) direction (1.76%) compared to the longitudinal (N–S) direction (1.03%); this indicates that the lateral load-resisting capacity of the building for the N–S direction may not be adequate (but needs to be verified through design calculations).

The effect of retrofitting was also studied on an example of an interior wall in the transverse (E–W) direction (embellished in red color in Figure 28). Note that the wall was not retrofitted since only exterior walls were retrofitted at critical exterior locations (shown in blue color on the floor plan). The wall was located at the ground floor level of the building and was subjected to seismic shear demand V_{Ed} (kN), which was compared with the corresponding shear capacity V_{Rd} (kN). The procedure for determining the shear capacity for URM walls according to the PTN-S code and Eurocode 8 was explained in [1]. It should be noted that the masonry design code from former Yugoslavia, PTN-Z [25], as well as Eurocode 6 [26,70] were used for the wall capacity calculations.

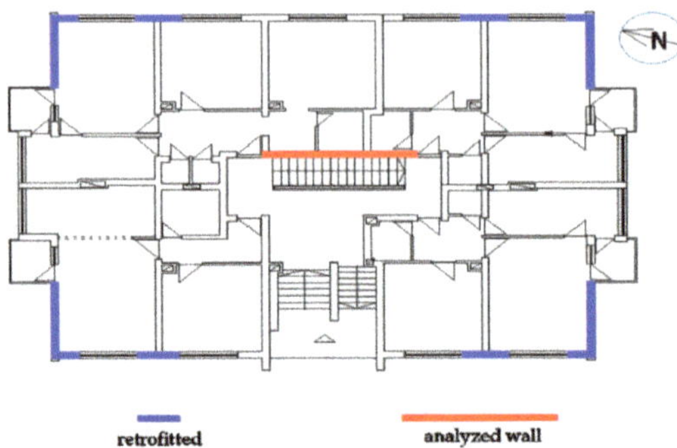

Figure 28. The floor plan showing retrofitted exterior walls (blue color) and the analyzed wall (red color).

The results are summarized in Tables 8 and 9 and illustrated in Figure 29. It can be seen from the chart that before retrofitting the capacity of the wall was inadequate since the corresponding C/D ratios were less than 1.0, according to both the PTN-S code as well as the EC8-1 code for all models. The analysis has shown that the retrofit has resulted in an increased C/D ratio for the wall according to the PTN-S code to 1.24, but only for the Retrofitted 1 model (uncracked), which is in line with that code. However, when the effect of cracking was considered, the corresponding C/D ratio was less than 1.0.

Table 8. Comparison of the seismic capacity (C) and demand (D) for the analyzed wall at the ground floor level (for the original structure).

	Original 1		Original 2		Original 3	
	PTN-S	EC8	PTN-S	EC8	PTN-S	EC8
	(1)	(2)	(3)	(4)	(5)	(6)
C: Shear Capacity $V_{Rd}(kN)$	177.00	295.00	141.6	236.00	88.50	147.50
D: Design Shear Force $V_{Ed}(kN)$	181.00	431.00	181.00	431.00	181.00	431.00
$\frac{C}{D} = V_{Rd}/V_{Ed}$	0.98	0.68	0.78	0.55	0.49	0.34

Table 9. Comparison of the seismic capacity (C) and demand (D) for the analyzed wall at the ground floor level (for the retrofitted structure).

	Retrofitted 1		Retrofitted 2		Retrofitted 3	
	PTN-S	EC8	PTN-S	EC8	PTN-S	EC8
	(1)	(2)	(3)	(4)	(5)	(6)
C: Shear Capacity $V_{Rd}(kN)$	212.40	354.00	169.6	283.20	106.20	177.54
D: Design Shear Force $V_{Ed}(kN)$	171.75	408.97	189.20	450.54	189.64	455.23
$\frac{C}{D} = V_{Rd}/V_{Ed}$	1.24	0.86	0.90	0.63	0.56	0.39

Based on the example of this wall and other similar walls in the building, it can be concluded that the presented seismic retrofit solution, which was developed based on the PTN-S and PTN-R codes from former Yugoslavia, does not meet the EC8-3 requirements for retrofitted masonry buildings.

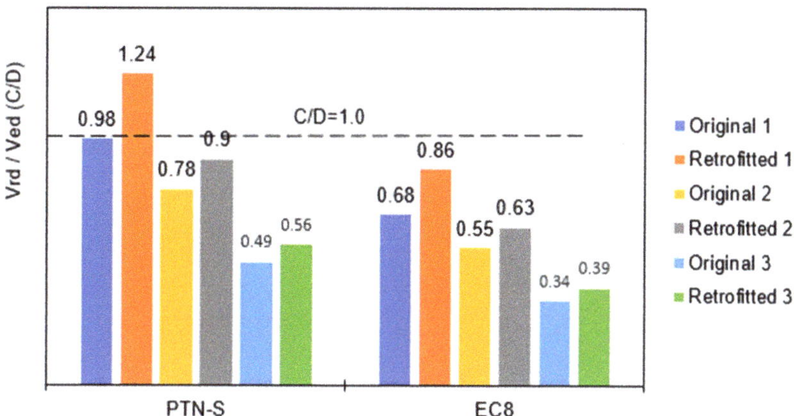

Figure 29. Seismic capacity versus demand (C/D) ratio in the analyzed wall for the original and retrofitted building.

Maximum lateral displacements and total drift ratio were determined for the original and retrofitted structure (all models) and presented in Tables 10 and 11. The results include elastic displacements Δ_e, design displacements Δ_d, and total drift ratio d, which is a ratio of the maximum design displacement Δ_d and the building height (11.2 m). It should be noted that the design displacements Δ_d were obtained by multiplying elastic displacement values by the behavior factor. In the case of EC8 calculations, the behavior factor (q) of 1.5 was used for the original structure (Table 10), while q = 2.5 was used for the retrofitted structure (Table 11). It can be seen from the tables that the displacement values are significantly higher for the analyses performed according to the EC8 code in comparison with the results according to the PTN-S code. For the Cantilever model, which was used in majority of the analyses presented in this section, drift ratio (d) ranges from 0.048% to 0.08% for the original structure and from 0.0079 to 0.130% for the retrofitted structure. It can be seen that drift values for the retrofitted structure are higher than for the original structure; this may not seem logical since the application of RC jackets results in the increased stiffness for the masonry structure. However, this result is not surprising when elastic displacements (Δ_e) are considered—displacements for the PTN-S model are similar for the original and retrofitted structure. For example, the Δ_e value for the Original 1 model is 3.59 mm (Table 10), while the corresponding displacement for the Retrofitted 1 model is 3.54 mm (Table 11); therefore, an increase in the drift value for retrofitted structure can be attributed to a higher q value. In the case of the results corresponding to analyses performed according to EC8, values of drift ratio are higher, ranging from 0.186 to 0.310% for the original structure (drift values for the retrofitted structure are almost the same). Elastic displacements (Δ_e) for the retrofitted structure are lesser compared to the original structure (as expected). For example, the Δ_e value for the Original 3 model is 23.17 mm (Table 10), while the corresponding displacement for the Retrofitted 3 model is 9.268 mm (Table 11); however, the drift ratios are based on design displacements (Δ_d) and reflect different q values used in these two cases (1.5 and 2.5 for the original and retrofitted structure, respectively). Overall, all lateral displacements and drift values are very low, as expected for a low- to mid-rise URM building.

4.6. Limitations of the Study: Seismic Analysis Procedure

It may be considered that a limitation of the study is associated with the use of a linear elastic seismic analysis procedure—as opposed to modern nonlinear analysis approaches. Note that application of linear elastic seismic analysis was justified at the time when retrofitting of the earthquake-damaged buildings in Kraljevo was performed (2010). Linear elastic seismic analysis was the default procedure for ordinary buildings, such as residential buildings, as per the Serbian technical regulations, including the PTN-S code for seismic design of new structures and the PTN-R code for seismic retrofitting of existing structures. It should be noted that linear elastic seismic analysis is an acceptable approach by the current European seismic code for existing buildings (EC8-3).

Table 10. Lateral displacements and total drift ratio for the original structure in N–S direction.

		Original 1		Original 2		Original 3	
		Cantilever Model	EFM	Cantilever Model	EFM	Cantilever Model	EFM
		(1)	(2)	(3)	(4)	(5)	(6)
PTN-S	Δ_e (mm)	3.590	2.270	4.240	2.690	6.010	3.900
	Δ_d (mm)	5.385	3.405	6.36	4.035	9.015	5.850
	d (%)	(0.048)	(0.030)	(0.056)	(0.036)	(0.080)	(0.052)
EC8	Δ_e (mm)	13.94	7.440	16.440	6.650	23.17	2.09
	Δ_d (mm)	20.91	11.16	24.66	9.975	34.755	3.135
	d (%)	(0.186)	(0.099)	(0.220)	(0.089)	(0.310)	(0.138)

Table 11. Lateral displacements and total drift ratio for the retrofitted structure in N–S direction.

		Retrofitted 1		Retrofitted 2		Retrofitted 3	
		Cantilever Model	EFM	Cantilever Model	EFM	Cantilever Model	EFM
		(1)	(2)	(3)	(4)	(5)	(6)
PTN-S	Δ_e (mm)	3.540	2.020	4.190	2.390	5.930	3.440
	Δ_d (mm)	8.850	5.05	10.475	5.975	14.825	8.600
	d (%)	(0.079)	(0.045)	(0.093)	(0.053)	(0.132)	(0.076)
EC8	Δ_e (mm)	8.330	1.090	9.840	1.290	9.268	1.050
	Δ_d (mm)	20.825	2.725	24.600	3.225	34.675	4.625
	d (%)	(0.186)	(0.024)	(0.219)	(0.029)	(0.309)	(0.041)

The nonlinear seismic response of masonry structures has been extensively studied in recent decades, and the corresponding numerical modelling and analysis approaches have been developed and incorporated into international seismic design codes. Non-linear static (pushover) analysis (NSA) is an internationally accepted analysis approach for simulating the nonlinear seismic response of building structures. NSA has been recommended for the seismic evaluation of existing buildings in the USA since the 1990s [74], and it has been incorporated into the current code for seismic evaluation of existing buildings ASCE/SEI 41-17 [27]. One of the main results of the NSA is a lateral force versus displacement curve (also known as pushover curve), which characterizes the response of a structure subjected to incrementally increasing monotonic lateral loading. The NSA results enable a designer to compare the lateral displacement/drift capacity for a structure under consideration and the corresponding displacement demand, which depends on the seismic hazard level and dynamic characteristics of the structural model. The displacement capacity can be determined for different performance limit states, which are established by the design codes, e.g., damage limitation and no collapse, according to EC8-1. NSA is one of the key components of performance-based earthquake engineering, and it can be used to help predict structural and non-structural damage and losses for buildings at different earthquake hazard levels.

Nonlinear analysis of masonry structures has been particularly challenging due to the composite nature of masonry structures, consisting of masonry elements, mortar, and reinforcement (in some cases). One of the early numerical models for nonlinear static analysis of masonry buildings was based on a simplified "story model", which considered the non-linear force-displacement behavior of individual walls idealized through a bilinear elastic–plastic model, and was developed in the 1970s by Tomaževič [75]. The non-linear modelling of URM structures at the micro level was studied by Lourenço [76], while the macro-modelling approach was extensively studied by several researchers due to its suitability for practical design applications. One of the most popular macro-models for simulating nonlinear seismic response of masonry structures is the equivalent frame model (EFM), proposed by Magenes [77], and further developed and implemented in the form of a TREMURI computer analysis program [72,78]. According to the EFM model, a masonry wall with openings is idealized as a moment frame consisting of the rigidly connected pier and spandrel elements. The TREMURI program enables the user to simulate the non-linear response of both piers and spandrels in a masonry wall subjected to in-plane seismic loading by means of a non-linear macro-element model [79]. The non-linear seismic response of URM walls obtained using the EFM model has been validated using the results of experimental studies [73]. The EFM approach for the non-linear analysis of masonry buildings is in line with the code provisions in the USA [27] and Europe (EC8-3) [21].

NSA of a typical six-story residential URM building from Sarajevo, Bosnia and Herzegovina was performed by Ademović et al. [9]. The building is representative of the typology which was the subject in this study. The NSA results were obtained for a micro-model (FEM) developed using the DIANA software package, while a macro-model was developed using the TREMURI software. Both models took into account material non-linearity characteristic for masonry walls. The results of the analysis (pushover curve for the transverse direction) show that the structure demonstrated a linear elastic behavior until the inter-story drift of approximately 0.07%. Under a further deformation increase, the structure experienced non-linear behavior until the failure took place. It should be noted that the case study building from Sarajevo is similar to the building considered in the current study, except for a very few walls provided in the longitudinal direction.

The application of NSA would be possible in the context of the current study; however, it may be argued whether the results of such an analysis would be useful in the context of the seismic retrofit project of the case study building. Since the case study building experienced damage in the 2010 Kraljevo earthquake, it can be expected that its seismic response to ground shaking was non-linear. Seismic evaluation of the building was performed according to the local codes, based on the given seismic hazard parameters, and the results indicated that the building was safe for seismic demand determined based on the PTN-S code. These results evidently did not reflect the actual seismic performance of the building; however, acceleration records for the Kraljevo earthquake were not available. As a result, seismic hazard parameters corresponding to the earthquake were not known, and it would be difficult to determine realistic displacement demand which is required for NSA. On the other hand, precise values for the mechanical properties of masonry materials were not available due to the lack of in situ material testing after the earthquake. Consequently, it was not possible to determine strength and deformation characteristics for nonlinear characterization (backbone curve) of masonry piers and spandrels, which are required as input for an NSA. It can be concluded that the results of an NSA would neither contribute toward an improved understanding of non-linear seismic behavior of the case study building nor be useful for optimizing the seismic retrofit solutions.

It is unfortunate that more precise data related to the material characteristics and seismic hazard for URM buildings affected by the 2010 earthquake were not available; however, limited information related to material characteristics and seismic hazard parameters is a reality in many post-earthquake situations. For that reason, the authors believe that simplified seismic analysis and numerical modelling approaches that account for material non-linearity characteristics will continue to be used in practice and are relevant for the engineering community.

5. Conclusions

This paper presents a study on seismic retrofitting of URM mid-rise residential buildings damaged due to the 2010 Kraljevo earthquake. Seismic evaluation and retrofitting of the buildings were performed according to the codes from the former Yugoslavia, which were enforced in Serbia until 2019, such as the seismic design code PTN-S and the code for seismic retrofitting of buildings PTN-R. A comparison of the results for the application of pertinent Yugoslav codes and Eurocode 8 was presented for a case study building in Kraljevo. The following relevant conclusions have been drawn based on this study:

(1) The case study building is a mid-rise URM building typical of residential construction in former Yugoslavia after WWII. Buildings of this type are vulnerable to earthquake effects and were exposed to a few damaging earthquakes, which caused moderate-to-severe structural damage in these buildings.

(2) The retrofit solution which was applied in the case study building consisted of RC jackets applied to exterior walls. The retrofit solution is acceptable by Eurocode 8, Part 3.

(3) The results of the seismic analysis and the design checks for the retrofitted structure, which were performed according to the pertinent Yugoslav codes, show that all walls in the retrofitted structure were satisfactory in terms of seismic safety.

(4) A comparison of the results of seismic analysis for the case study building performed according to the Yugoslav codes and Eurocode 8 has shown that the seismic demand according to Eurocode 8 is significantly higher compared to the seismic codes from former Yugoslavia, which were used for seismic evaluation and retrofitting design. The key finding is that the retrofit design performed according to the Yugoslav code does not meet the seismic safety requirements of Eurocode 8. It would be required to perform more extensive retrofit, likely for the interior walls (in addition to the exterior walls), in order to satisfy the seismic demand requirements according to Eurocode 8.

(5) There is a difference in stiffness estimation for the retrofitted walls for the numerical model developed according to the Yugoslav code PTN-R and Eurocode 8, Part 3. According to the PTN-R code, the thickness of a retrofitted wall needs to be increased to account for RC jackets by considering an equivalent masonry section with the RC jacket thickness increased four times. On the other hand, Eurocode 8, Part 3 prescribes a reduction in the wall stiffness to account for the effect of cracking.

The results of this study are relevant for Serbia, and other countries in the region which followed seismic design and retrofit codes developed in the former Yugoslavia in the past and are currently using Eurocode 8 as the main code for seismic design and retrofit of masonry structures.

The authors believe that this paper contributes to the knowledge base related to field applications of cost-effective and practical seismic retrofitting techniques for URM buildings which are suitable for application in countries with limited human and financial resources, compounded by limited experience related to seismic retrofitting.

6. Recommendations for Future Research Studies

The findings of studies related to seismic evaluation and retrofitting of earthquake-damaged buildings, such as the one presented in this paper, are useful for seismic risk and resilience studies on local and/or regional levels. The authors believe that future research studies related to the topic of this study should be integrated with seismic risk studies, which will provide useful information for predicting losses due to future earthquakes and enhancing the resilience of communities at risk. Unfortunately, seismic risk studies in Serbia are practically non-existent; however, an ongoing initiative to undertake a seismic risk study for Serbia is underway by members of the Serbian Association for Earthquake Engineering (SUZI-SAEE). Notable past initiatives at the regional level (which included Serbia) resulted in the development of the European seismic risk model in the framework of the SERA project [80]. A World Bank's study on the seismic risk associated with urban residential buildings in Eastern Europe and Central Asia [81] provided useful information related to Serbia's capital Belgrade.

There has been a limited effort to determine empirical fragility functions based on building damage data after past earthquakes in Serbia. Notable empirical seismic risk studies were performed on masonry buildings after the damaging earthquakes in Italy [82–84] and a similar methodology could be applied in Serbia and the neighboring countries. A unique study of this kind was performed after the 2010 Kraljevo earthquake on a sample of 1.193 damaged masonry buildings, and it resulted in the development of empirical fragility functions which will be useful for future seismic risk studies in Serbia and the region [66,85]. Expert-based fragility functions were recently developed for masonry buildings in Serbia [12].

Seismic risk studies on portfolios of existing urban masonry buildings have been performed in a few countries, including Italy [86]. A methodology for rapid seismic assessment of urban masonry and RC buildings in the countries surrounding the Adriatic and the Ionian Sea has been recently developed within framework of the Interreg's Adriseismic project [87]. The Serbian team participating in the project performed an initial pilot study, in which the proposed methodology was applied on typical masonry and RC buildings in the center of Serbia's capital Belgrade.

One of the important aspects of seismic risk studies is the estimation of costs associated with the recovery after possible future earthquakes. Methodology developed in recent Italian studies [88] can be applied to estimate retrofit cost for masonry buildings after future earthquakes in Serbia.

Author Contributions: Conceptualization, P.B. and S.B.; methodology, P.B. and S.B.; software, P.B; validation, P.B. and R.C.; formal analysis, P.B.; investigation, S.B., P.B. and R.C.; resources, P.B. and S.B.; data curation, R.C.; writing—original draft preparation, P.B. and S.B.; writing—review and editing, S.B. and R.C.; visualization, P.B. and R.C; supervision, P.B.; project administration, R.C.; funding acquisition, P.B. All authors have read and agreed to the published version of the manuscript.

Funding: This research received no external funding.

Data Availability Statement: Data related to this study is archived in the City of Kraljevo administration office, because the 2010 Kraljevo earthquake took place more than 10 years ago.

Acknowledgments: The authors acknowledge the City Administration of the City of Kraljevo for providing information related to the rehabilitation of buildings after the 2010 Kraljevo earthquake. The authors also acknowledge Nikola Blagojević, doctorand at the Swiss Federal Institute of Technology (ETH) at Zurich, Switzerland, for reviewing the manuscript and providing valuable comments.

Conflicts of Interest: The authors declare no conflict of interest.

References

1. Brzev, S.; Blagojević, P.; Cvetković, R. Simplified seismic assessment of unreinforced masonry residential buildings in the Balkans: The case of Serbia. *Buildings* **2021**, *11*, 392.
2. Statistical Office of the Republic of Serbia (SORS). Number and the floor space of housing units and dwellings according to the occupancy status, by settlements. In The Census of Population, Households and Dwellings; Statistical Office of the Republic of Serbia, Belgrade, Serbia: 2011. Available online: https://www.stat.gov.rs/sr-latn/oblasti/popis/popis-2011/popisni-podaci-eksel-tabele/ (accessed on 10 December 2022).
3. Atalić, J.; Uroš, M.; Šavor Novak, M.; Demšić, M.; Nastev, M. The Mw5.4 Zagreb (Croatia) earthquake of March 22, 2020: Impacts and response. *Bull. Earthq. Eng.* **2021**, *19*, 3461–3489. [CrossRef] [PubMed]
4. Stepinac, M.; Lourenço, P.B.; Atalić, J.; Kišiček, T.; Uroš, M.; Baniček, M.; Šavor Novak, M. Damage classification of residential buildings in historical downtown after the M_L5.5 earthquake in Zagreb, Croatia in 2020. *Int. J. Disaster Risk Reduct.* **2021**, *56*, 102140. [CrossRef]
5. Miranda, E.; Brzev, S.; Bijelić, N. Petrinja, Croatia December 29, 2020, Mw 6.4 Earthquake. StEER-EERI Joint Reconnaissance Report, 2021. Available online: https://www.designsafe-ci.org/data/browser/public/designsafe.storage.published/PRJ-2959 (accessed on 15 December 2022).
6. Tomaževič, M.; Weiss, P. Seismic behavior of plain- and reinforced-masonry buildings. *J. Struct. Eng.* **1994**, *120*, 323–338. [CrossRef]

7.	Andonov, A.; Baballëku, M.; Baltzopoulos, G.; Blagojević, N.; Bothara, J.; Brûlé, S.; Brzev, S.; Veliu, E. *EERI Earthquake Reconnaissance Team Report: M6.4 Albania Earthquake on 26 November 2019*; Earthquake Engineering Research Institute: Oakland, CA, USA, 2022. Available online: http://www.learningfromearthquakes.org/images/earthquakes/2019_Albania_Earthquake/EERI_Earthquake_Reconnaissance_Report_-_M6.4_Albania_Earthquake_on_November_26_2019.pdf (accessed on 20 December 2022).
8.	Tomaževič, M.; Klemenc, I.; Weiss, P. Seismic upgrading of old masonry buildings by seismic isolation and CFRP laminates: A shaking-table study of reduced scale models. *Bull. Earthq. Eng.* **2009**, *7*, 293–321. [CrossRef]
9.	Ademovic, N.; Hrasnica, M.; Oliveira, D.V. Pushover analysis and failure pattern of a typical masonry residential building in Bosnia and Herzegovina. *Eng. Struct.* **2013**, *50*, 13–29. [CrossRef]
10.	Ademović, N.; Oliveira, D.V.; Lourenço, P.B. Seismic evaluation and strengthening of an existing masonry building in Sarajevo, B&H. *Buildings* **2019**, *9*, 30.
11.	Grünthal, G. *European Macroseismic Scale 1998*; Centre Europèen de Géodynamique et de Séismologie: Luxembourg, 1998; Volume 15, ISBN 2879770084.
12.	Blagojević, N.; Brzev, S.; Petrović, M.; Borozan, J.; Bulajić, B.; Marinković, M.; Hadzima-Nyarko, M.; Koković, V.; Stojadinović, B. Residential building stock in Serbia: Classification and vulnerability for seismic risk studies. *Bull. Earthq. Eng.* **2023**, *under review*.
13.	Berg, G.V. *The Skopje, Yugoslavia Earthquake July 26, 1963*; American Iron & Steel Institute: Washington, DC, USA, 1964.
14.	Manić, M.; Bulajić, B. Ponašanje zidanih zgrada u kraljevačkom zemljotresu od 03. novembra 2010. godine—Iskustva i pouke (Behaviour of masonry buildings during the November 03, 2010 Kraljevo earthquake). *Izgradnja* **2013**, *67*, 235–246. (In Serbian)
15.	PTP-12. *Privremeni Tehnički Propisi za Građenje u Seizmičkim Područjima (Provisional Technical Regulations for Construction in Seismic Regions)*; Official Gazette of SFRY No. 39/64; Yugoslav Institute for Standardization: Belgrade, Yugoslavia, 1964. (In Serbian)
16.	PTN-S. *Pravilnik o Tehničkim Normativima za Izgradnju Objekata Visokogradnje u Seizmičkim Područjima (Technical Regulations for the Design and Construction of Buildings in Seismic Regions)*; Official Gazette of SFRY No. 31/81 (Amendments 49/82, 29/83, 21/88, 52/90); Yugoslav Institute for Standardization: Belgrade, Yugoslavia, 1981; Available online: https://iisee.kenken.go.jp/worldlist/64_Serbia/64_Serbia_Code.pdf (accessed on 13 December 2022).
17.	Fajfar, P. Analysis in seismic provisions for buildings: Past, present and future. The fifth Prof. Nicholas Ambraseys lecture. *Bull. Earthq. Eng.* **2018**, *16*, 2567–2608. [CrossRef]
18.	PGK. *Pravilnik za Građevinske Konstrukcije (Building regulations)*; Official Gazette of Republic of Serbia No. 89/2019, 52/2020, 122/2020; Institute for Standardization of Serbia: Belgrade, Serbia, 2019. (In Serbian)
19.	EN 1998-1:2005. Eurocode 8—Design of Structures for Earthquake Resistance-Part 1: General Rules, Seismic Actions and Rules for Buildings. European Committee for Standardization: Bruxelles, Belgium, 2005.
20.	SRPS EN 1998-1/NA:2018. Evrokod 8—Projektovanje Seizmički Otpornih Konstrukcija Deo 1: Opsta Pravila, Seizmicka Dejstva i Pravila za Zgrade (Eurocode 8—Design of Structures for Earthquake Resistance-Part 1: General Rules, Seismic Actions and Rules for Buildings). Institute for Standardization of Serbia: Belgrade, Serbia, 2018. (In Serbian)
21.	EN 1998-3:2005. Eurocode 8—Design of Structures for Earthquake Resistance-Part 3: Assessment and Retrofitting of Buildings. European Committee for Standardization: Bruxelles, Belgium, 2005.
22.	SRPS EN 1998-3/NA:2018. Evrokod 8—Projektovanje seizmički otpornih konstrukcija Deo 3: Evaluacija i ojačanje konstrukcija (Eurocode 8—Design of structures for earthquake resistance-Part 3: Assessment and retrofitting of buildings, EN 1998-3:2005). Institute for Standardization of Serbia: Belgrade, Serbia, 2018. (In Serbian)
23.	Seismological Survey of Serbia (SSS). *Seismic Hazard Maps for Serbia*; Seismological Survey of Serbia: Belgrade, Serbia, 2018. Available online: http://www.seismo.gov.rs/Seizmicnost/Karte_hazarda_e.htm (accessed on 10 December 2022).
24.	PTN-R. *Pravilnik o Tehničkim Normativima za Sanaciju, Ojačanje i Rekonstrukciju Objekata Visokogradnje Oštećenih Zemljotresom i za Rekonstrukciju i Revitalizaciju Objekata Visokogradnje (Technical Regulations for Repair, Strengthening and Reconstruction of Building Structures Damaged by Earthquakes and for Reconstruction and Rehabilitation of Buildings)*; Službeni list SFRJ No. 52/85; Yugoslavia, 1985. (In Serbian)
25.	PTN-Z. *Pravilnik o Tehničkim Normativima za Zidane Zidove (Technical Norms Regulation for Masonry Walls)*; Official Gazette of SFRY No. 87/91; Yugoslav Institute for Standardization: Belgrade, Yugoslavia, 1991. (In Serbian)
26.	EN 1996-1-1:2004. Eurocode 6—Design of Masonry Structures. Part 1-1: General rules for reinforced and unreinforced masonry structures. European Committee for Standardization: Bruxelles, Belgium, 2004.
27.	ASCE/SEI. *Seismic Evaluation and Retrofit of Existing Buildings, Standard (ASCE/SEI 41-17)*; American Society of Civil Engineers/Structural Engineering Institute: Reston, VA, USA, 2017.
28.	Thermou, G.E.; Elnashai, A.S. Seismic retrofit schemes for rc structures and local–global consequences. *Prog. Struct. Eng. Mater.* **2006**, *8*, 1–15. [CrossRef]
29.	Thermou, G.E.; Pantazopoulou, S.J.; Elnashai, A.S. Upgrading of RC Structures for a target response shape. In Proceedings of the 13th World Conference on Earthquake Engineering, Vancouver, BC, Canada, 1–6 June 2004.
30.	Brzev, S.; Begaliev, U. *Practical Seismic Design and Construction Manual for Retrofitting Schools in the Kyrgyz Republic*; World Bank group: Washington, DC, USA, 2018; pp. 1–251. Available online: http://documents.worldbank.org/curated/en/505451593451983935/Practical-Seismic-Design-and-Construction-Manual-for-Retrofitting-Schools-in-The-Kyrgyz-Republic (accessed on 15 December 2022).
31.	Salaman, A.; Stepinac, M.; Matorić, I.; Klasić, M. Post-earthquake condition assessment and seismic upgrading strategies for a heritage-protected school in Petrinja, Croatia. *Buildings* **2022**, *12*, 2263. [CrossRef]

32. Triller, P.; Tomaževič, M.; Lutman, M.; Gams, M. Seismic behavior of strengthened URM masonry—An overview of research at ZAG. International Conference on Analytical Models and New Concepts in Concrete and Masonry Structures AMCM'2017. *Procedia Eng.* **2017**, *193*, 66–73. [CrossRef]
33. Abrams, D.; Smith, T.; Lynch, J.; Franklin, S. Effectiveness of rehabilitation on seismic behavior of masonry piers. *J. Struct. Eng.* **2007**, *133*, 32–43. [CrossRef]
34. Marinković, M.; Baballëku, M.; Isufi, B.; Blagojević, N.; Milićević, I.; Brzev, S. Performance of RC cast-in-place buildings during the November 26, 2019 Albania earthquake. *Bull. Earthq. Eng.* **2022**, *20*, 5427–5480. [CrossRef]
35. Gkournelos, P.D.; Triantafillou, T.C.; Bournas, D.A. Seismic upgrading of existing reinforced concrete buildings: A state-of-the-art review. *Eng. Struct.* **2021**, *240*, 112273. [CrossRef]
36. Wararuksajja, W.; Leelataviwat, S.; Warnitchai, P.; Bing, L.; Tariq, H.; Naiyana, N. Seismic strengthening of soft-story RC moment frames. In Proceedings of the 25th National Convention on Civil Engineering, Chonburi, Thailand, 9 July 2020.
37. UNIDO. *Building Construction Under Seismic Conditions in the Balkan Region: Repair and Strengthening of Reinforced Concrete, Stone and Brick Masonry Buildings*, 1st ed.; United Nations Industrial Development Programme: Vienna, Austria, 1983; Volume 5, 231p.
38. Tomaževič, M. *Earthquake-Resistant Design of Masonry Buildings*; Imperial College Press: London, UK, 1999; 268p.
39. Uroš, M.; Todorić, M.; Crnogorac, M.; Atalić, J.; Šavor Novak, M.; Lakušić, S. (Eds.) *Potresno Inženjerstvo—Obnova Zidanih Zgrada*; Građevinski fakultet, Sveučilišta u Zagrebu: Zagreb, Croatia, 2021; ISBN 978-953-8163-43-7.
40. Muravljov, M.; Stevanović, B.; Ostojić, D. *Sanacije Građevinskih Konstrukcija i Objekata*, 1st ed.; Faculty of Civil Engineering, University of Belgrade: Belgrade, Serbia, 2022; pp. 1–271. (In Serbian)
41. Churilov, S.; Dumova-Jovanoska, E. Analysis of masonry walls strengthened with RC jackets. In Proceedings of the 15th World Conference on Earthquake Engineering, Lisbon, Portugal, 24–28 September 2012.
42. Ostojić, D.; Stevanović, B.; Muravljov, M.; Glišović, I. Sanacija i ojačanje zidanih objekata oštećenih zemljotresom u Kraljevu. In Proceedings of the 4th International Conference, Civil engineering—Science and practice, Žabljak, Montenegro, 20–24 February 2012. (In Serbian).
43. Medić, S.; Hrasnica, M. In-plane seismic response of unreinforced and jacketed masonry walls. *Buildings* **2021**, *11*, 472. [CrossRef]
44. El Gawady, M.; Lestuzzi, P.; Badoux, M. Retrofitting of masonry walls using shotcrete. In Proceedings of the 2006 NZSEE Conference, Napier, New Zealand, 10–12 March 2006.
45. Sheppard, P.; Terčelj, S. The effect of repair and strengthening methods for masonry walls. In Proceedings of the 7th World Conference on Earthquake Engineering, Istanbul, Turkey, 8–13 September 1980.
46. Proença, J.M.; Gago, A.S.; Costa, A.V.; André, A.M. Strengthening of masonry wall load bearing structures with reinforced plastering mortar solution. In Proceedings of the 15th World Conference on Earthquake Engineering, Lisbon, Portugal, 24–28 September 2012.
47. Kadam, S.B.; Singh, Y.; Li, B. Strengthening of unreinforced masonry using welded wire mesh and micro-concrete—behaviour under in-plane action. *Constr. Build. Mater.* **2014**, *54*, 247–257. [CrossRef]
48. Lin, Y.; Biggs, D.; Wotherspoon, L.; Ingham, J.M. In-plane strengthening of unreinforced concrete masonry wallettes using ECC shotcrete. *J. Struct. Eng.* **2014**, *140*, 04014081. [CrossRef]
49. Jurukovski, D.; Krstevska, L.; Alessi, R.; Diotallevi, P.P.; Merli, M.; Zarri, F. Shaking table tests of three four-storey brick masonry models: Original and strengthened by RC core and RC jackets. In Proceedings of the 10th World Conference on Earthquake Engineering, Madrid, Spain, 19–24 July 1992.
50. Hutchison, D.; Yong, P.; McKenzie, G. Laboratory testing of a variety of strengthening solutions for brick masonry wall panels. In Proceedings of the 8th World Conference on Earthquake Engineering, San Francisco, CA, USA, 21–28 July 1984.
51. FEMA. *Techniques for the Seismic Rehabilitation of Existing Buildings*; FEMA 547; Federal Emergency Management Agency: Washington, DC, USA, 2006; 571p.
52. INRC. *Guide for the Design and Construction of Externally Bonded FRP Systems for Strengthening Existing Structures: Materials, RC and PC structures, Masonry Structures*; CNR-DT 200 R1/2013; Italian National Research Council, Advisory Committee on Technical Recommendations for Construction: Rome, Italy, 2014; 154p.
53. Ehsani, M.R.; Saadatmanesh, H.; Al-Saidy, A. Shear behavior of URM retrofitted with FRP overlays. *J. Compos. Constr. ASCE* **1997**, *1*, 17–25. [CrossRef]
54. Reinhorn, A.M.; Madan, A. *Evaluation of Tyfo-W Fiber Wrap System for In-Plane Strengthening of Masonry Walls*; Report No. 95-0002; Department of Civil Engineering, State University of New York at Buffalo: Buffalo, NY, USA, 1995.
55. ElGawady, M.; Lestuzzi, P.; Badoux, M. Static cyclic response of masonry walls retrofitted with fiber-reinforced polymers. *J. Compos. Constr.* **2007**, *11*, 50–61. [CrossRef]
56. Tomaževič, M.; Gams, M.; Berset, T. Seismic strengthening of brick masonry walls with composites: An experimental study. In Proceedings of the Structural Engineers World Congress, Villa Erba, Como, Italy, 4–6 April 2011.
57. Arifuzzaman, S.; Saatcioglu, M. Seismic retrofit of load bearing masonry walls by FRP sheets and anchors. In Proceedings of the 15th World Conference on Earthquake Engineering, Lisbon, Portugal, 24–28 September 2012.
58. Paquette, J.; Bruneau, M.; Brzev, S. Seismic testing of repaired unreinforced masonry building having flexible diaphragm. *J. Struct. Eng. ASCE* **2004**, *130*, 1187–1196. [CrossRef]
59. ElGawady, M.; Lestuzzi, P.; Badoux, M. In-plane seismic response of unreinforced masonry walls upgraded with fiber reinforced polymer. *J. Comp. for Constr. ASCE* **2005**, *9*, 524–535. [CrossRef]

60. ElGawady, M.; Lestuzzi, P.; Badoux, M. Aseismic retrofitting of unreinforced masonry walls using FRP. *Compos. Part B* **2006**, *37*, 148–162. [CrossRef]
61. Turek, M.; Ventura, C.E.; Kuan, S. In-plane shake-table testing of GFRP strengthened concrete masonry walls. *Earthq. Spectra* **2007**, *23*, 223–237. [CrossRef]
62. Kišiček, T.; Stepinac, M.; Renić, T.; Hafner, I.; Lulić, L. Strengthening of masonry walls with FRP or TRM. *Građevinar* **2020**, *72*, 937–953.
63. Cvetković, R.; Stojić, D.; Marković, N.; Conić, S. Analiza seizmičke otpornosti i ojačanje zidane konstrukcije Biskupije u Pančevu. Zbornik radova desetog međunarodnog naučno–stručnog savetovanja "Ocena stanja, održavanje i sanacija građevinskih objekata i naselja"; Vršac, Serbia: 14-16 June 2017 (In Serbian).
64. Seismological Survey of Serbia. *Izveštaj o rezultatima i aktivnostima Republičkog seiz-mološkog zavoda posle zemljotresa kod Kraljeva 03.11.2010 u 01:56 (Report on the results and activites of the Seismological Survey of Serbia after the earthquake in Kraljevo 03.11.2010 at 01:56)*; Seismological Survey of Serbia: Belgrade, Serbia, 2010. (In Serbian)
65. World Bank. *Europe and Central Asia (ECA) Risk Profiles: Serbia*; World Bank Group: Washington, DC, USA, 2017. Available online: https://www.gfdrr.org/en/publication/disaster-risk-profile-serbia (accessed on 10 December 2022).
66. Marinković, D.; Stojadinović, Z.; Kovačević, M.; Stojadinović, B. 2010 Kraljevo earthquake recovery process metrics derived from recorded reconstruction data. In Proceedings of the 16th European Conference on Earthquake Engineering, Thessaloniki, Greece, 8–21 June 2018.
67. Ostojić, D.; Muravljov, M.; Stevanović, B. Primeri sanacije višespratnih stambenih zidanih zgrada oštećenih zemljotresom u Kraljevu (Examples of housing rehabilitation multistory masonry buildings damaged in the earthquake in Kraljevo). *Izgradnja* **2011**, *5–6*, 315–325. (In Serbian)
68. Nikić, M. Sanacija stambene zgrade u Kraljevu oštećene u zemljotresu 2010. godine. Zbornik radova devetog međunarodnog naučno–stručnog savetovanja "Ocena stanja, održavanje i sanacija građevinskih objekata i naselja", Zlatibor, Serbia, 25-29 May 2015. (In Serbian).
69. Manić, M.; Bulajić, B. Zašto procena šteta na građevinskim objektima u kraljevačkom regionu nije izvršena ni godinu dana nakon zemljotresa od 03.11.2010. godine? *Izgradnja* **2012**, *66*, 269–308. (In Serbian)
70. SRPS EN 1996-1-1:2016. Evrokod 6—Projektovanje Zidanih Konstrukcija Deo 1-1: Opsta Pravila za Armirane i Nearmirane zidane Konstrukcije (Eurocode 6—Design of Masonry Structures. Part 1-1: General Rules for Reinforced and Unreinforced Masonry Structures). Serbian Institute for Standardization: Belgrade, Serbia, 2016. (In Serbian)
71. Radimpex. *Tower—3D Model Builder 8.4-x64 Edition*; Radimpex Software d.o.o: Belgrade, Serbia, 2022.
72. Lagomarsino, S.; Penna, A.; Galasco, A.; Cattari, S. TREMURI program: An equivalent frame model for the nonlinear seismic analysis of masonry buildings. *Eng. Struct.* **2013**, *56*, 1787–1799. [CrossRef]
73. Cattari, S.; Lagomarsino, S. Modelling the seismic response of unreinforced existing masonry buildings: A critical review of some models proposed by codes. In Proceedings of the 11th Canadian Masonry Symposium, Toronto, Canada, 31 May–3 June 2009.
74. Federal Emergency Management Agency. *Pre-Standard and Commentary for the Seismic Rehabilitation of Buildings*; FEMA 356: Washington, DC, USA, 2000.
75. Tomaževič, M. Dynamic modelling of masonry buildings: Storey mechanism model as a simple alternative. *Earthq. Eng. Struct. Dyn.* **1987**, *15*, 731–749. [CrossRef]
76. Lourenço, P.B. Computational Strategy for Masonry Structures. Ph.D. Thesis, Delft University of Technology, Delft, The Netherlands, 1996.
77. Magenes, G. A method for pushover analysis in seismic assessment of masonry buildings. In Proceedings of the 12th WCEE, Auckland, New Zealand, 30 January–4 February 2000.
78. Galasco, A.; Lagomarsino, S.; Penna, A.; Resemini, S. Non-linear seismic analysis of masonry structures. In Proceedings of the 13th World Conference on Earthquake Engineering, Vancouver, Canada, 1–6 August 2004.
79. Penna, A.; Lagomarsino, S.; Galasco, A. A nonlinear macro-element model for the seismic analysis of masonry buildings. *Earthq. Eng. Struct. Dyn.* **2013**, *43*, 159–179. [CrossRef]
80. Crowley, H.; Dabbeek, J.; Despotaki, V.; Rodrigues, D.; Martins, L.; Silva, V.; Danciu, L. *European Seismic Risk Model (ESRM20). EFEHR Technical Report 002, V1.0.1.*; EUCENTRE Foundation: Pavia, Italy, 2021. [CrossRef]
81. World Bank Group. *Earthquake Risk in Multifamily Residential Buildings, Europe and Central Asia Region*; World Bank Group: Washington, DC, USA, 2020.
82. Rota, M.; Penna, A.; Strobbia, C.L. Processing Italian damage data to derive typological fragility curves. *Soil Dyn. Earthq. Eng.* **2008**, *28*, 933–947. [CrossRef]
83. Del Gaudio, C.; De Martino, G.; Di Ludovico, M.; Manfredi, G.; Prota, A.; Ricci, P.; Verderame, G.M. Empirical fragility curves for masonry buildings after the 2009 L'Aquila, Italy, earthquake. *Bull. Earthq. Eng.* **2019**, *17*, 6301–6330. [CrossRef]
84. Chieffo, N.; Clementi, F.; Formisano, A.; Lenci, S. Comparative fragility methods for seismic assessment of masonry buildings located in Muccia (Italy). *J. Build. Eng.* **2019**, *25*, 100813. [CrossRef]
85. Stojadinović, Z.; Stojadinović, B.; Kovačević, M.; Marinković, D. Data-Driven Housing Damage and Repair Cost Prediction Framework Based on the 2010 Kraljevo Earthquake Data. In Proceedings of the 16th World Conference on Earthquake Engineering, Santiago, Chile, 9–13 January 2017.

86. Ruggieri, S.; Calò, M.; Cardellicchio, A.; Uva, G. Analytical-mechanical based framework for seismic overall fragility analysis of existing RC buildings in town compartments. *Bull. Earthq. Eng.* **2022**, *20*, 8179–8216. [CrossRef]
87. Predari, G.; Stefanini, L.; Marinković, M.; Stepinac, M.; Brzev, S. Adriseismic methodology for expeditious seismic assessment of unreinforced masonry buildings. *Buildings* **2023**, *accepted for publication*. [CrossRef]
88. Di Ludovico, M.; De Martino, G.; Prota, A.; Manfredi, G.; Dolce, M. Relationships between empirical damage and direct/indirect costs for the assessment of seismic loss scenarios. *Bull. Earthq. Eng.* **2022**, *20*, 229–254. [CrossRef]

Article

Adriseismic Methodology for Expeditious Seismic Assessment of Unreinforced Masonry Buildings

Giorgia Predari [1], Lorenzo Stefanini [1,*], Marko Marinković [2], Mislav Stepinac [3] and Svetlana Brzev [4]

1   Department of Architecture, University of Bologna, 40136 Bologna, Italy
2   Faculty of Civil Engineering, University of Belgrade, 11000 Belgrade, Serbia
3   Faculty of Civil Engineering, University of Zagreb, 10000 Zagreb, Croatia
4   Department of Civil Engineering, University of British Columbia, Vancouver, BC V6T 1Z4, Canada
*   Correspondence: lorenzo.stefanini2@unibo.it

Abstract: The paper describes a novel Adriseismic method for expeditious assessment of seismic risk associated with unreinforced masonry buildings. The methodology was developed for the Adriseismic project of the Interreg ADRION programme, with the aim to develop and share tools for increasing cooperation and reducing seismic risk for six participating countries within the region surrounding the Adriatic and the Ionian Seas. The method is applicable to unreinforced masonry buildings characterised by three main seismic failure mechanisms, namely masonry disintegration, out-of-plane failure, and in-plane damage/failure. Depending on the input parameters for a specific structure, the assessment yields a qualitative output that consists of the masonry quality index, the index of structural response, the level of seismic risk, and the most probable collapse mechanism. Both input and output of the method are applied in the spreadsheet form. The method has so far been applied in urban areas of participating countries in the project, including Mirandola, Italy; Kaštela, Croatia; Belgrade, Serbia. In parallel, the methodology has been validated by performing a detailed seismic assessment of more than 25 buildings, and the results have been compared with the results of the proposed expeditious method. The results show a good correlation between the two methods, for example, the structural response index obtained from the expeditious method and the capacity/demand ratio obtained from the conventional assessment method.

Keywords: unreinforced masonry structures; Adriseismic project; seismic vulnerability assessment; risk prediction; seismic failure mechanisms; existing buildings

Citation: Predari, G.; Stefanini, L.; Marinković, M.; Stepinac, M.; Brzev, S. Adriseismic Methodology for Expeditious Seismic Assessment of Unreinforced Masonry Buildings. *Buildings* **2023**, *13*, 344. https://doi.org/10.3390/buildings13020344

Academic Editor: Rita Bento

Received: 23 December 2022
Revised: 19 January 2023
Accepted: 23 January 2023
Published: 26 January 2023

1. Introduction

The countries surrounding the Adriatic and Ionian Seas, including Italy, Slovenia, Croatia, Bosnia and Herzegovina, Serbia, Montenegro, Albania, and Greece, have been historically exposed to major seismic events, both in terms of intensity and frequency [1,2]. These countries are characterised by the high seismic exposure and significant stock of vulnerable unreinforced masonry (URM) buildings, which were designed to withstand only gravity loads and were unable to sustain the effects of moderate to strong earthquakes without substantial damage. Historical URM buildings located in urban centres appear to be most vulnerable due to numerous transformations which they have undergone over time, and the challenges encountered while carrying out structural interventions, as demonstrated by the many seismic retrofitting techniques introduced since the 1970s [3]. At the same time, these urban areas are also characterised by the highest population density and the unique identity of a specific place. The combination of high seismic hazard, high vulnerability, and high exposure makes seismic risk reduction an issue of fundamental importance for these urban areas. Recent earthquakes in the region, including Albania [4], Greece [5,6], and Croatia [7,8], once again confirmed the high seismic vulnerability of URM buildings, which experienced substantial damage or collapse due to these seismic events.

It was estimated that URM buildings account for 45% to 61% of the existing building stock within the region surrounding the Adriatic and Ionian Seas [9]. Seismic assessment of URM buildings is often challenging because an in-depth seismic analysis is both time-consuming and expensive. For this reason, methods for expeditious seismic assessment have been considered as an alternative to detailed approaches. They are less time-consuming, but accurate enough to guide informed planning at the urban or regional level.

It is possible to classify seismic vulnerability assessment methods into analytical and empirical ones [10], although hybrid assessment procedures are also available [11]. Empirical methods are mainly based on rapid post-earthquake damage observations, while analytical methods are usually based on performance limit states, mechanical characteristics of construction materials, and include detailed vulnerability assessment algorithms [12]. Hybrid methods are a combination of empirical and analytical ones—they are derived from statistical approaches and consider the actual effects of past earthquakes on different types of structures, as well as the results from analytical methods [13,14].

Probable damage matrices are classic examples of empirical methods, which attempt to predict the effect of an earthquake on structures with known characteristics [15]. Empirical methods of this type have evolved over time. For example, the study by Braga [16] used the macroseismic intensity scale MSK-76 [17] to define damage matrices based on the observed damage after the Irpinia earthquake (Campania, Italy). The same method, implemented by Giovinazzi and Lagomarsino [18], proved particularly effective for application to large urban areas, such as the cities of Faro, Portugal [19], Barcelona, Spain [20], and Lisbon, Portugal [21]. Another widespread empirical method is the Rapid Visual Screening Method, presented in FEMA154 [22], which assigns a score for the building on the basis of visual sidewalk screening, providing results to the user in about 30 min. Approaches based on general considerations, inspired by field experience, have been developed in several countries, including Japan [23], Turkey [24], and Canada [25].

Analytical methods can generally rely on more accurate results but are more time-consuming. They may include the use of collapse multipliers, e.g., the Vulnus System, developed in Italy [26], or the FaMIVE method [27]. Calvi [28] proposed a displacement-based vulnerability assessment approach applicable to both reinforced concrete and URM buildings, and served as the basis for other methods for URM buildings [29–31].

The expeditious method presented in this paper was developed for application in several countries within the region surrounding the Adriatic and Ionian Seas. Similar to the EMS-98 scale [32] which is applicable in Europe, the proposed method aims at performing seismic risk assessment at the international (regional) level. It can be used to expeditiously evaluate individual buildings, and potentially large areas within a short timeframe, without the need to define complex frameworks or acquire empirical data from previous seismic events [33]. Building-specific vulnerabilities depend on regional construction characteristics [9] and are different from seismic risk, which is the product of vulnerability, exposure, and hazard. The procedure yields the following output indicators: (i) index of structural response, defined as an inverse value to the vulnerability; (ii) masonry quality index (MQI) [34,35]; (iii) the most probable collapse mechanism; (iv) seismic risk index. Both input and output values were simplified as much as possible, often using general categories (I, II, III, etc.), with the values that were attributed *a priori*. The method enables fast data input in the spreadsheet form, easy interpretation of the results, and was designed to be implemented on Geographical Information System (GIS) platforms to facilitate large-scale applications.

During the Adriseismic project [36], three pilot cases were performed to assess the real-life applicability of the procedure. The selected sites were located in Bologna (Italy) [37], Kaštela (Croatia), and Rethymno (Greece). This paper presents instead the results obtained for three urban sites located in the cities of Mirandola (Italy), Kaštela (Croatia), and Belgrade (Serbia), which were considered relevant for the case study purposes, too. These case studies enabled evaluation and testing of the procedure, in terms of its applicability to sites characterised by different seismic hazard levels and masonry buildings constructed using different techniques. Furthermore, the expeditious method was verified on a pilot sample of 25 buildings, for which the results of seismic assessment were previously determined using conventional seismic assessment methods. Some of the buildings were previously assessed using the LV1 expeditious method [38], while others were assessed by means of linear elastic dynamic analysis.

The features such as expeditiousness, applicability at different localities, and qualitative, easily interpretable output distinguish the proposed method from existing approaches and facilitate its large-scale application, thereby increasing knowledge of the state of the existing heritage, therefore targeting more in-depth analyses on the most critical buildings.

2. The Adriseismic Method

2.1. An Overview of the Method

The Adriseismic method was developed based on the following criteria:

1. **Input information is easily accessible**: to be effective, the method must be applicable on a large scale. For that reason, assessment of an individual building should not require in-depth investigations, such as detailed condition surveys or historical analysis, instead, an assessment should be based on general data, such as cadastral plans and other types of building information.
2. **The assessment is performed quickly**: ideally, it should take no more than a few minutes to evaluate an individual building once all the information is available.
3. **The output can be easily understood by non-experts**: a long-term goal of the project is rapid dissemination of the method; hence, it is expected that the results can be incorporated in urban planning tools (through municipal maps), as well as property evaluations in the insurance sector.
4. **The method is internationally applicable**: it is essential for the procedure to include features unrelated to a specific country or region.

These four criteria guided the method development. On one hand, it was important to maintain the consistency with the initial objectives, while on the other hand it was important to impose certain constraints, such as international applicability and a balance between the accuracy of the results and the speed of processing input data.

From the operational point of view, the method comprises the following four main phases (see Figure 1):

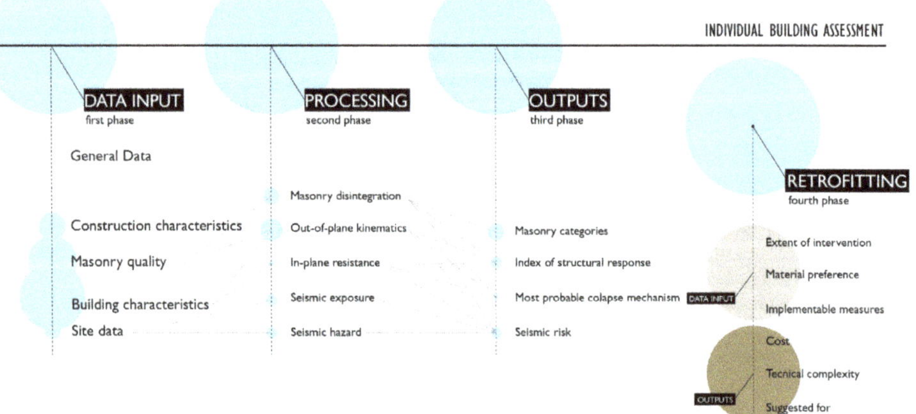

Figure 1. General structure of the proposed Adriseismic method, showing macro-phases and intermediate steps.

1. **Data input:** the user needs to enter input data related to specific building.
2. **Processing:** the input data are processed according to the algorithm.
3. **Output:** the system provides output (results).
4. **Retrofitting:** this phase is currently not directly linked to the seismic assessment of a building, but a structure intervention strategy can be suggested for enhancing seismic resistance of a specific building.

The expeditious method can be applied using a Microsoft Excel spreadsheet. Input data are organised into the following sections:

1. **General data**—this section provides general information related to a specific building and does not affect the assessment results.
2. **Construction characteristics**—information related to the prevalent construction techniques for foundations, vertical structural elements, floors, and roofs is selected from an existing database via a drop-down menu [9,39]. The provided information does not affect the assessment results.
3. **Masonry quality**—masonry assessment is performed using the Masonry QualityI method, M.Q.I. [34,35]. The user is requested to input nine masonry characteristics for the building, which strongly influence the seismic vulnerability assessment.
4. **Building characteristics**—this section requires the user to provide 12 input data related to the intended use of the building and its morphological and structural configuration. Whenever possible, the weights attributed to each parameter were defined using normative values or considerations based on simplified schemes. Input data, such as intended use, irregularities, and expected ductility, were derived using indirect considerations from Eurocode 8 [40,41]. (See Appendix B.) Other parameters were set using equilibrium-based considerations, as explained in the next section. Input data influence the key results, namely the index of structural response, the seismic risk, and the most probable collapse mechanism.
5. **Site data**—this section consists of three input parameters referring to the seismic zone of the building site according to the Eurocode 8 requirements [40] and is used to determine the seismic hazard level.

Detailed information related to each input section is contained in Appendix A.

The following output indicators are obtained as a result of the assessment:

1. **Masonry category**—it depends exclusively on the masonry quality. According to the M.Q.I. method, which serves as the basis for determining the masonry category, there are three possible categories (A to C), depending on the capacity of a masonry structure to resist vertical, out-of-plane, and in-plane load actions.
2. **Index of structural response**—associates the presumed building capacity to a numerical value (in the range from 0 to 1) and the corresponding category (from I to VI). A specific value is determined by analysing the following three main masonry failure mechanisms (in a decreasing extent of impact): wall disintegration, out-of-plane failure, and in-plane failure. The masonry quality and the building characteristics input data also influence the failure mechanism.
3. **Probable collapse mechanism**—a hypothesis regarding the most probable collapse mechanism (disintegration of masonry, out-of-plane kinematic mechanism, or in-plane failure) is formulated for the building based on the input data;
4. **Seismic risk**—it is calculated based on the index of structural response and the required site data (the higher the number, the greater the risk); the risk is also presented as a category (ranging from "none" to "very high").
5. **Retrofitting**—when specific structural deficiencies are noted, the user may wish to suggest specific actions to mitigate the risk. According to the possible choices in terms of the type of the structural intervention, simple qualitative information is provided to indicate its feasibility for a specific building.

Both input and output were simplified as much as possible. Instead of providing a numerical input to describe the building characteristics, general categories (I, II, and

III) were used. The use of qualitative indicators instead of numerical values also appears to be useful since it facilitates the application in different contexts by partially removing a language barrier. For example, "I" indicates a low quantity in an absolute sense, e.g., a regular building is assigned category "I" to indicate the absence of irregularity, "III" indicates a very high amount/abundance (e.g., high energy dissipation capacity), while "II" is used for intermediate values.

Figure 2 compares the assessment form of the Adriseismic method to the general diagram.

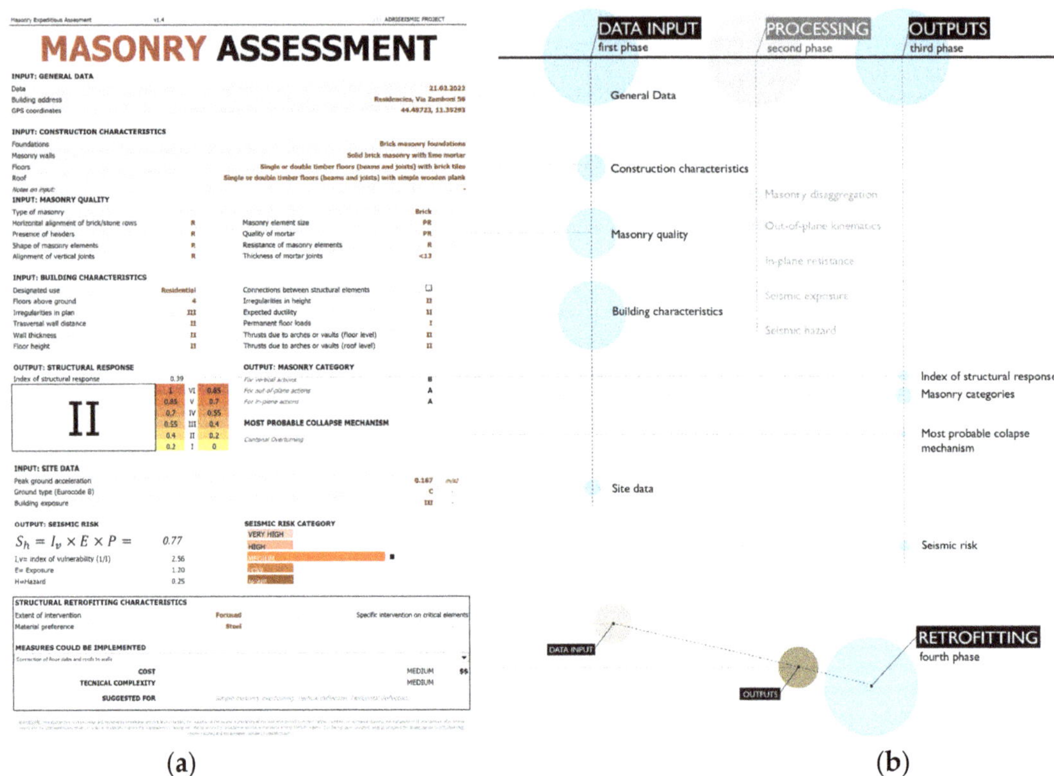

Figure 2. Adriseismic method: (**a**) assessment form, generated via a Microsoft Excel spreadsheet, showing input data and output (results), and (**b**) a diagram showing the main input categories (note that the processing phase is shown in grey).

Input data related to the masonry quality have been presented as proposed by the M.Q.I. method [34,35]. Each input characteristic is assigned a qualitative indicator, e.g., F. (fulfilled), P.F. (partially fulfilled), or N.F. (not fulfilled), as seen in Table 1. These qualitative input categories (I, II, III, F., P.F., N.F.) are associated with numerical data, which are essential for obtaining the output.

Table 1. Input parameters: main values.

Building Characteristics		MQI	
I	Low Value	N.F.	Not fulfilled
II	Medium Value	P.F.	Partially fulfilled
III	High Value	F.	Fulfilled

The output (results) was presented using the same quantitative indicators as the input. For example, the parameter "index of structural response" is defined by six different classes (I to VI), where I indicates a highly vulnerable building and VI indicates a building without obvious structural deficiencies. The class of a building was assigned based on the numerical value (ranging from 0 to 1), which describes the structural behaviour and directly depends on the inputs. The ranges shown in Table 2 were assigned considering wider ranges for the lower classes and narrower ranges for the higher ones; this increases the probability of assigning precautionary values.

Table 2. Index of structural response: categories and values.

Class	Minimum Value	Maximum Value
I	0.00	0.20
II	0.21	0.40
III	0.41	0.55
IV	0.56	0.70
V	0.71	0.85
VI	0.86	1.00

The use of building class is similar to the decree "Sismabonus" [42], which is used to evaluate the seismic capacity of existing buildings by means of letters (A to F). The use of defined classes and numerical ranges facilitates ease of application and interpretation, and the results can be transferred to urban planning maps or GIS for use by a wider community (non-experts).

2.2. Processing of Input Data

After the input data have been entered in the assessment form, the processing takes place in the background (it is not visible to the user). One of the key aspects of the processing stage is determination of the most critical masonry failure mechanism for a specific building. The Adriseismic method takes into account realistic seismic behaviour and failure mechanisms for URM structures, including (i) masonry disintegration, (ii) out-of-plane kinematic mechanisms, (iii) and in-plane failure. Figure 3 shows a schematic representation of these failure mechanisms.

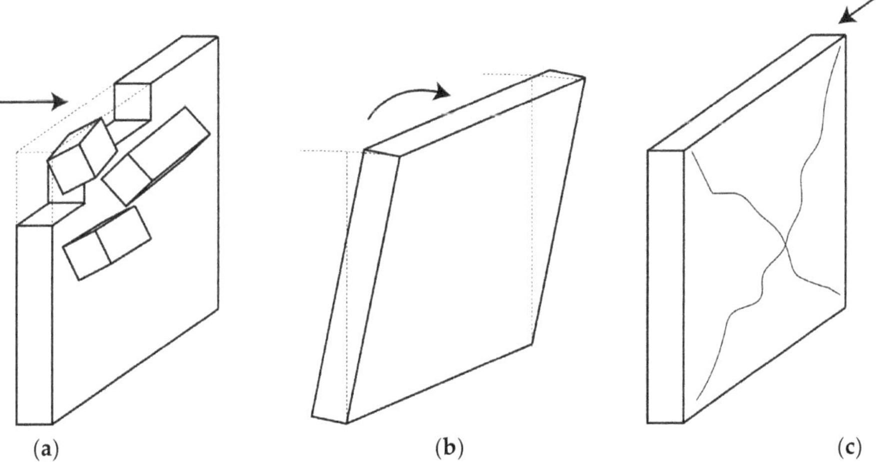

(a)   (b)   (c)

Figure 3. Masonry failure mechanisms—(a) masonry disintegration, (b) out-of-plane kinematic mechanism, and (c) in-plane failure.

Disintegration concerns the loss of cohesion of the wall and is typical of loose (poorly cohesive) and weak (low strength) masonry, (Figure 3a). Out-of-plane kinematic mechanism, such as overturning (Figure 3b), can be simulated by a rigid macro-element, which does not depend on material strength characteristics. Collapse due to reaching the ultimate resistance in plane, also known as in-plane failure (Figure 3c), usually requires the greatest seismic energy and can be expected in cohesive masonry which is found in structures with well-connected load-bearing elements [43].

Parameters related to building characteristics directly influence the activation of four out-of-plane kinematic mechanisms, including simple masonry overturning, vertical bending, horizontal bending, and corner overturning, [44] which were considered in this study. Each parameter was assigned a 0 value if it does not influence the activation of a specific out-of-plane mechanism, 0.5 for the case of a small effect, and 1 in the case of a large effect (see Table 3).

Table 3. Building characteristics related to out-of-plane mechanisms and the corresponding weights attributed.

Name and Description	Input	Weights Attributed to Kinematic Mechanism			
		Simple Masonry Overturning	Vertical Bending	Horizontal Bending	Corner Overturning
Transversal wall distance	I	0.00	0.00	1.00	0.00
	II	0.00	0.00	0.50	0.00
	III	0.00	0.00	0.00	0.00
Wall thickness	I	0.00	0.00	0.00	0.00
	II	0.50	0.50	0.50	0.50
	III	1.00	1.00	1.00	1.00
Floor height	I	1.00	1.00	1.00	1.00
	II	0.50	0.50	0.50	0.50
	III	0.00	0.00	0.00	0.00
Permanent floor weights	I	0.00	1.00	0.00	0.00
	II	0.00	0.50	0.00	0.50
	III	0.00	0.00	0.00	1.00
Thrusts due to arches and vaults	I	1.00	1.00	1.00	0.00
	II	0.50	0.50	0.50	0.00
	III	0.00	0.00	0.00	0.00
Thrusts due to the roof	I	1.00	0.00	0.00	1.00
	II	0.50	0.00	0.00	0.50
	III	0.00	0.00	0.00	0.00

The influence of these characteristics was assessed by monitoring the variation in spectral acceleration for the activation of each kinematic mechanism while the other parameters remained unchanged. C.I.N.E. spreadsheets developed in Italy [45] were used for this purpose. The impact was defined as irrelevant when the variation in a parameter changed the original spectral acceleration by less than 25%, moderate when changes in spectral acceleration ranged from 25 to 75%, and large when changes were greater than 75%.

As for the other input parameters, the categories "I", "II", and "III" indicate poor or abundant values. For example, in the case of small transversal wall distances, category 'I' is used, while for very thick walls the category 'III' should be selected. In order to minimise the complexity of survey and data acquisition, the parameters "thrust due to arches and vaults" and "thrusts due to the roof" were set to depend on the floor weight. For example, in case of a construction solution that generates a thrust on the vertical masonry, a higher floor weight corresponds to a higher thrust.

2.3. Results of the Assessment

2.3.1. Masonry Categories

The M.Q.I. method assigns masonry categories ranging from A to C, where A is the least vulnerable, B is intermediate, and C is the most vulnerable. The rating is assigned for all possible load actions, namely vertical actions (gravity) (MQIv), horizontal out-of-plane actions (MQIop), and horizontal in-plane actions (MQIip).

The M.Q.I. ratings are formulated for each load action based on Equations (1)–(3). The symbols in the equations are explained in the nomenclature section. The results are associated with the corresponding masonry category (A, B, or C).

$$MQI_v = r_v \cdot g \cdot m \cdot S_s \cdot \left(H_j + W_c + S_r + V_j + S_d + M_m\right) \tag{1}$$

$$MQI_{op} = r_{op} \cdot g \cdot m \cdot S_s \cdot \left(H_j + W_c + S_r + V_j + S_d + M_m\right) \tag{2}$$

$$MQI_{ip} = r_{ip} \cdot g \cdot m \cdot S_s \cdot \left(H_j + W_c + S_r + V_j + S_d + M_m\right) \tag{3}$$

Table 4 shows the ranges of values associated with each category. The numerical values obtained from equations are divided by a factor of 10 for the sake of convenience.

Table 4. Masonry categories and M.Q.I. values.

Load Action	Masonry Categories	Min	Max
	A	0.50	1.00
MQI$_v$—Vertical action	B	0.25	0.499
	C	0.00	0.249
	A	0.70	1.00
MQI$_{op}$—Horizontal out-of-plane action	B	0.40	0.699
	C	1.00	0.399
	A	0.50	1.00
MQI$_{ip}$—Horizontal in-plane action	B	0.30	0.499
	C	0.00	0.299

2.3.2. Index of Structural Response

Index of structural response is the key parameter used for assessing the seismic behaviour of the construction. It is defined considering an M.Q.I. rating and building characteristics. In order to evaluate all masonry damage mechanisms, the following equations have been proposed:

$$P_{sd} = D_u + I_p + I_h + E_d + F_n \tag{4}$$

$$S_{mo} = S_w + W_t + I_{nh} + P_{fw} + P_a + P_r \tag{5}$$

$$V_b = S_w + W_t + I_{nh} + P_{fw} + P_a + P_r \tag{6}$$

$$H_b = S_w + W_t + I_{nh} + P_{fw} + P_a + P_r \tag{7}$$

$$C_o = S_w + W_t + I_{nh} + P_{fw} + P_a + P_r \tag{8}$$

The equations assign a numerical value to parameters linked to seismic demand P_{sd} (Equation (4)), simple masonry overturning S_{mo} (Equation (5)), vertical bending V_b (Equation (6)), horizontal bending H_b (Equation (7)), and corner overturning C_o (Equation (8)). The meaning of the symbols can be found in the Nomenclature section at the end of the paper. The resulting values for these parameters have been normalised using the traditional max–min formulation (Equation (9)) to make them comparable. The maximum value was set equal to 1, and the minimum value was 0 (the same applies to the index of structural response and the M.Q.I.).

$$norm = \frac{v_i - v_{min}}{v_{max} - v_{min}} \tag{9}$$

Once the five values have been defined according to Equations (4)–(8), it is possible to calculate the index of structural response, I_{sr}. Two alternative formulations are used: the first considers the main structural elements disconnected from each other (input n.25 set "off", see Appendix A, Table A4), and the second assumes them as linked (input n.25 set "on"):

$$I_{sr,1} = \frac{P_{sd,n} + \min\left(S_{mo,n}; V_{b,n}; H_{b,n}; C_{o,n}; MQI_v; MQI_{op}; MQI_{ip}\right)}{2} \tag{10}$$

$$I_{sr,2} = \frac{P_{sd,n} + \min\left(MQIv; MQIo; MQIi\right)}{2} \tag{11}$$

Equation (10) offers the index of structural response as the mathematical average of the seismic demand parameters (normalised) and the minimum value for the four kinematic mechanisms and the M.Q.I. results. In this way, the most probable damage mechanism of the analysed structure is evaluated, choosing the one with the lowest numerical value on a 0–1 scale. For example, a building constructed with poor masonry quality but with good construction characteristics will still have a low index of structural response.

Equation (11) is related to the hypothesis of well-connected structural elements. The four out-of-plane kinematic mechanisms are not included in the equation. In this case, the index only depends on the parameters that directly influence the seismic demand (designated use, floors above ground, irregularity in plan, irregularity in height, expected ductility) and the quality of masonry.

A numerical value, ranging from 0 (very vulnerable building) to 1 (building without vulnerability), is used to assign a class (from I to VI), according to the range shown in Table 2. The assigned class allows for comparison with other buildings and can be used for urban planning purposes.

2.3.3. Most Probable Collapse Mechanism

The previously presented Equations (5)–(8), together with the M.Q.I. results and the data provided by the user about the "connections between structural elements", allow hypotheses to be formulated regarding the most probable collapse mechanism for the building. The output is provided based on the following criteria:

- When the masonry quality is class C, for one or all three actions, the masonry may disaggregate for very low seismic values.
- When the masonry quality is average or good class A or B, but the connections between structural elements are good, the evaluation system indicates the out-of-plane kinematic as the most probable collapse mechanism with the lowest value among the four investigated in the method, Equations (5)–(8).
- When neither of the two conditions presented is verified, the building may develop global behaviour, and collapse may occur due to reaching ultimate strength in vertical plane.

2.3.4. Seismic Risk

The final output is related to seismic risk, S_h. Using the same concept as before, a numerical value is assigned to a qualitative risk category for the building. The risk S_h is calculated using the widely used formulation [46]:

$$S_h = I_v \cdot E \cdot H \tag{12}$$

Unlike the index of structural response, the risk is not determined as a numerical value because its magnitude depends on the ground acceleration provided by the user. However, as the number describing the risk increases, the risk category increases, as shown in Table 5.

Table 5. Seismic risk categories and values.

Categories	Min	Max
Very high	1.40	
High	0.90	1.39
Medium	0.45	0.89
Low	0.10	0.44
None	0.00	0.09

The ranges in the table were calibrated for different conditions within the entire Adriatic–Ionian Sea region. For example, the risk was differentiated into high and very high for buildings located in areas subject to significant acceleration, but characterised by different exposures.

2.3.5. Retrofitting

Finally, the method provides recommendations related to the possible seismic intervention (retrofitting). The output is intended to be qualitative and identifies the intervention based on deficiencies that may have emerged during the investigation phase and confirmed by the results. The section is also organised into input and output, but there is no intermediate processing phase for the input information and the output. The intervention can be selected from a defined list. As shown in Table 6, it is possible to select the extent of intervention and the main material that should be used. Based on these preferences, compatible interventions are offered by the method.

Table 6. Seismic retrofitting intervention.

Name and Description	Possible Values
Extent of intervention (How extensive the intervention is)	Extensive, localised, none
Material preference (Main material used for the retrofitting)	Concrete, composite (fibre-reinforced polymers), masonry, wood, steel

Currently, 56 possible interventions have been included in the database; out of these, 28 interventions are localised, while the remaining ones are extensive.

Figure 4 shows five main types of materials that can be used for interventions. The blue colour shows the total number of interventions for each material, while a beige bar gives an indication of the extent (extensive or localised). For each intervention, a few qualitative indications are available to help guide the choice. Based on the Adriseismic project deliverable D.T.2.1.2 [9], the cost, technical complexity, and critical issues are indicated, as seen in Table 7. For the Croatian scenario, the retrofitting prices were based on World Bank reports and Croatian methodology [47–50].

Table 7. Qualitative indicators for seismic retrofitting.

Indications Provided	Range of Values
Cost	Low/medium/high
Technical complexity	Low/medium/high
Explanation	A brief summary of seismic deficiencies addressed by the intervention

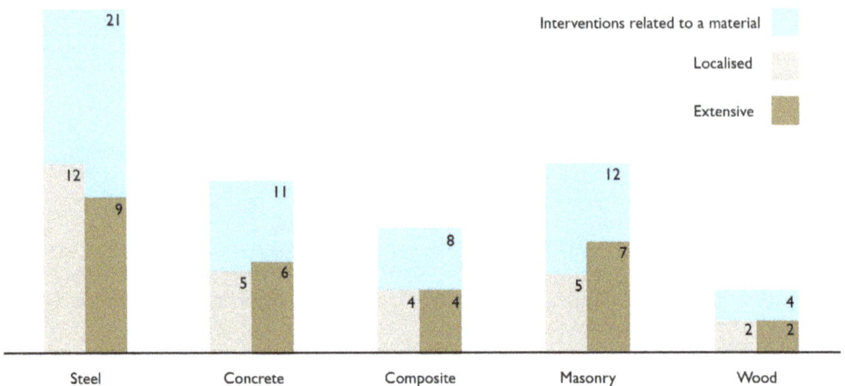

Figure 4. Interventions classified based on the main material and type (localised and extensive).

3. Application of the Adriseismic Method on Urban Case Studies

At the end of the development phase, the method was initially applied on real buildings. The aim was to verify the adaptability of the procedure in varied contexts and its compliance with the four guiding criteria illustrated in Section 2 (easily accessible information, quick assessment, outputs easily understandable, and international use). For this purpose, buildings in three different areas within three cities were studied: Mirandola (Italy), Kaštela (Croatia), and Belgrade (Serbia). The areas were selected according to Adriseismic project objectives; hence, URM buildings facing squares and having a historical identity for the place were considered.

3.1. The Mirandola Case Study

The city of Mirandola is a small town (approx. 23,000 inhabitants), located in northern Italy, in the region of Emilia-Romagna. It has an ancient history and dates back to the year 1000. Like many other settlements in the area, it is organised on an initial fortified structure and successive stratifications added over the centuries. The site is highly active from a seismic point of view, although the acceleration (PGA) expected for the area, at the Life Safety limit state, is 0.14 g, which is about half of that actually recorded in the last major earthquake that occurred in 2012. The combination of these two elements (settlement with a long history and high seismic hazard) made the area very relevant for applying the Adriseismic method. Figure 5 shows a satellite view of the case study location.

Figure 5. A view of the Mirandola case study site and the three buildings analysed.

Three buildings facing the Conciliation Square were selected for the study, as seen in Figure 6. All three buildings have masonry load-bearing structures and are a part of larger aggregates.

(a)   (b)   (c)

Figure 6. Selected buildings in Mirandola, Italy: (**a**) Building 1, (**b**) Building 2, and (**c**) Building 3; © Google.

The information to assess the seismic response was obtained through cadastral plans and site surveys. In parallel, historical analyses were carried out to reconstruct the main events that affected the buildings over time. Following the project objectives, no specific material investigations were carried out.

Building (a) (Figure 6) has three floors above ground and an attic. The lowest level is intended for commercial use, while the others are for residential use. The original structure dates back to the 16th century and has been modified over time. A central block has undergone numerous extensions until it was connected to the other units, generating a complex aggregate. Building (b) is of the most recent construction, dating back to the 1930s. As in the previous case, the structure is part of a large aggregate, characterised by an internal courtyard and a highly irregular planimetric configuration. The assessment was carried out on the part of the building with homogeneous construction characteristics. The floors are composite, made of hollow bricks and reinforced concrete, while the walls are made of bricks that are not always interlocked well. The use of the building is public.

Building (c) dates back to the 16th century. It is the largest in plan and has probably undergone the largest number of structural interventions over the centuries. The current configuration is certainly different from the original intention and is the result of numerous extensions over time. The roof and floors are made of wood, although, given the many transformations that have taken place, they are made using different construction techniques.

Table 8 shows the main results of the expeditious assessment: index of structural response, masonry category, most probable collapse mechanism, seismic risk, and seismic risk category.

Table 8. Result of the evaluation of buildings in Mirandola.

Building Code	Index of Structural Response	Masonry Categories	Most Probable Collapse Mechanism	Seismic Risk	Seismic Risk Category
Building a	II (0.35)	B, B, A	Cantonal Overturning	0.60	Medium
Building b	III (0.54)	A, A, C	Disintegration	0.39	Low
Building c	II (0.32)	A, A, A	Horizontal deflection	0.66	Medium

3.2. The Kaštela Case Study

Similar to Italy, Croatia is affected by a high seismic hazard. A few disastrous earthquakes that have occurred in recent years, such as the 2020 one in Petrinja with a magnitude 6.4 on the Richter scale, have raised awareness of the scientific and political community who started to evaluate innovative measures to reduce seismic risk. The chosen area, Podvorje Square, is located in the south of Kaštel Sućurac, the administrative area of the city of Kaštela (Croatia), as seen in Figure 7.

Figure 7. A view of Kaštela case study site and the five buildings analysed.

The city, located just north of Split, is a small town (around 40,000 inhabitants), characterised by a very ancient history dating back to prehistoric times, which has resulted in a building heritage rich in dated constructions. The presence of potentially vulnerable buildings and a high seismic hazard, the expected acceleration (PGA) at the site at the Life Safety limit state is equal to 0.22 g, made the square relevant for the application of the Adriseismic method.

The area dates back to the 15th century and was built at the behest of Archbishop Bartul Averaldo, who intended to create fortifications at the centre of the present Archbishop's palace. The assessment was carried out on five buildings, characterised by different construction techniques, construction periods, and functions; see Figure 8.

Figure 8. Selected five buildings for the Kaštela study area. Elevations photographed or reconstructed by photogrammetry.

The input information was obtained using: (i) historical surveys which were used to find information to demonstrate the building modifications that have occurred over time or the presence of structural interventions; (ii) physical surveys such as on-site measurements and aerial-photogrammetric surveys. The full laser scanning of the area from the ground was conducted to create a 3D cloud of points (Figure 9). The laser scanner was a compact Leica BLK360 3D imaging laser scanner with an integrated spherical imaging system and thermography panorama sensor system. Laser scanning facilitates and speeds up, documenting and reviewing geometry, especially for large and complex buildings [51]. The point cloud was used to create an accurate 3D model of the area (Figure 8). Similarly, full 3D photogrammetry scanning was performed by unmanned aerial devices (drones), which produced several 360° images that were used in the post-analysis stage [52].

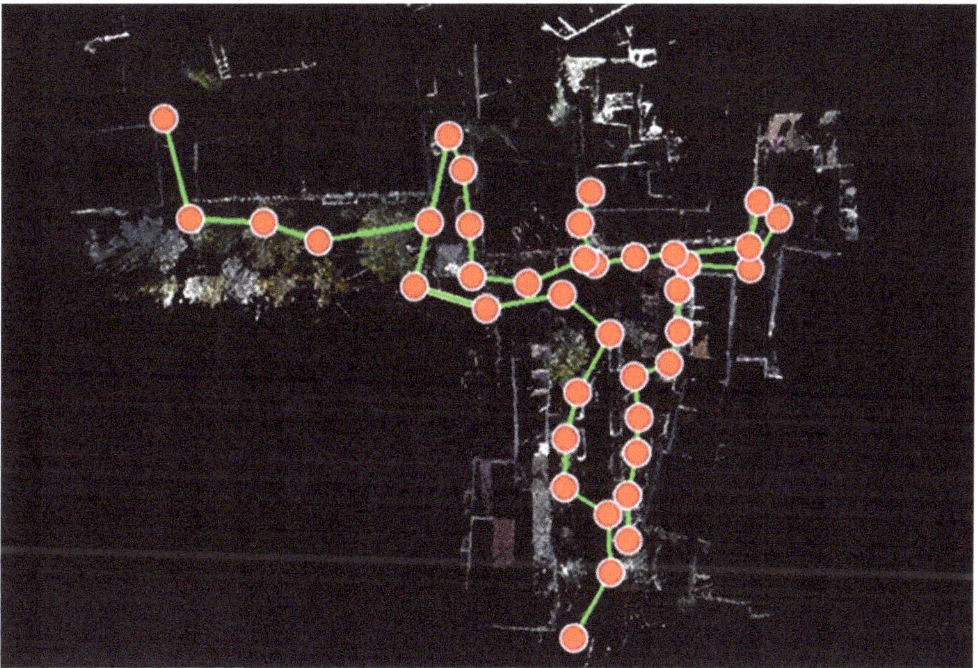

Figure 9. Plan view of the point cloud showing camera stations.

The first building (a) is public which dates back to the 15th century (Figure 8). Configuration is irregular both in height and plan, with one side being much longer than the other. The building has stone masonry walls and wooden floors and roof.

The second building (b) has a residential function. The building has brick masonry walls and wooden floors and roof. The construction period is between the 1920s and 1940s.

The third building (c) is very similar to the second, as they share the same construction period and construction features. There are no noticeable irregularities in plan and elevation; it has a wooden load-bearing structure with brick tiles to characterise the floors.

The fourth (d) and fifth buildings (e) are part of the same aggregate and have residential use. The walls were constructed using stone masonry, while the floors and roof are mainly wooden structures. In terms of configuration, no particular irregularities in height were found. The structures probably date back to the early 1900s and may not have been built simultaneously. The aggregate effect was not considered in this study.

Table 9 shows the results obtained by applying the method to these five buildings.

Table 9. Results of evaluation of building around the Podvorje square.

Building ID	Index of Structural Response	Masonry Categories	Most Probable Collapse Mechanism	Seismic Risk	Seismic Risk Category
Building a	III (0.43)	A, B, A	Horizontal deflection	0.59	Low
Building b	IV (0.68)	A, A, A	Horizontal deflection	0.37	Low
Building c	IV (0.56)	A, B, A	Vertical deflection	0.45	Low
Building d	III (0.45)	A, B, B	Horizontal deflection	0.56	Low
Building e	IV (0.65)	A, A, A	Horizontal deflection	0.39	Low

3.3. The Belgrade Case Study

The last case study site is located in Belgrade the capital of Serbia. Belgrade is one of the oldest settlements in Europe, and its considerable size (population 1,400,000) and rich history have resulted in a heterogeneous building heritage in terms of construction techniques. Seismic hazard of the area can be characterised as moderate, the PGA is 0.1 g, the lowest for the three selected case studies. Due to its geographical location [53], the city has not been affected by earthquakes since 1992.

The chosen area is close to the city centre, in immediate proximity to the Cyril and Methodius Park, as shown in Figure 10. It is a very busy area (resulting in a high exposure), surrounded by ancient buildings which were not designed to resist horizontal actions. This fact, together with the historical stratifications that characterise the site, make it the perfect location for applying the method. In this context, eight different load-bearing masonry buildings characterised by varied construction techniques and functions were assessed (Figure 11).

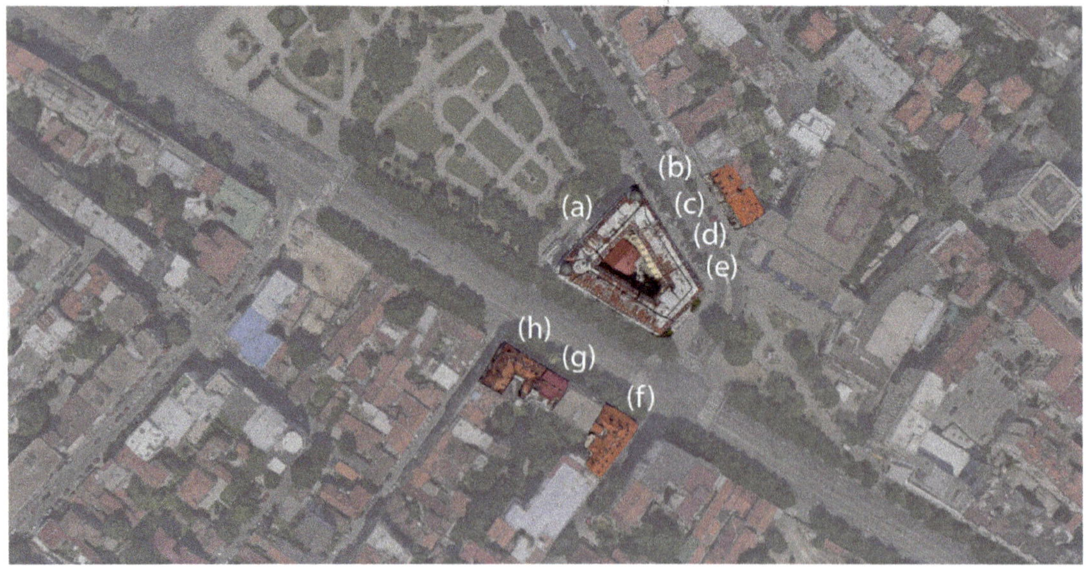

Figure 10. Belgrade case study and the eight buildings analysed.

Figure 11. The buildings assessed for the Belgrade study; © Google.

Similar to other case studies, expeditious assessment was initially conducted on the observation of structural elements, and archival material was used to reconstruct the construction history. Structural and architectural plans were also found for all the buildings. No surveys were carried out using advanced techniques.

Building (a) is currently used as the University canteen and dormitory (public building). It was built in 1926 and has a load-bearing structure composed of brickwork. The floors consist of RC cast-in-situ ribbed slabs. The building has a large inner courtyard and is developed as a trapezoid on the four perimeter sides. Building (b) is used for shops on the ground floor and as a residential building on the other levels. It has a highly irregular plan and is characterised by load-bearing brick masonry walls. The investigations carried out did not reveal any recent structural interventions. The other buildings have many common features: they were made of load-bearing masonry, have at least four floors above ground, and were built between 1930 and 1960. The main differences are related to the configuration and the technology used to construct the floors (precast concrete or cast-in-situ reinforced slab). Table 10 shows the results of the Belgrade case study.

Table 10. Result of the evaluation of buildings in Belgrade, Serbia.

Building ID	Index of Structural Response	Masonry Categories	Most Probable Collapse Mechanism	Seismic Risk	Seismic Risk Category
Building a	III (0.43)	B, A, A	Vertical Deflection	0.42	Low
Building b	II (0.33)	B, A, A	Cantonal Overturning	0.33	Medium
Building c	IV (0.67)	A, A, A	Vertical Deflection	0.22	Low
Building d	IV (0.65)	A, A, A	Vertical Deflection	0.23	Low
Building e	IV (0.67)	A, A, A	Vertical Deflection	0.22	Low
Building f	II (0.33)	B, A, A	Cantonal Overturning	0.45	Low
Building g	V (0.72)	A, A, A	Vertical Deflection	0.21	Low
Building h	IV (0.59)	A, A, A	Vertical Deflection	0.25	Low

3.4. Results and Discussion

In most cases, cadastral plans, storey heights, and an on-site inspection (or photographs) were sufficient to carry out the expeditious assessment. Required documentation is usually readily available and it is not required to use advanced equipment or perform detailed surveys, although, in some cases, obtaining all the necessary information could still be difficult and time-consuming. Some challenges were encountered while surveying buildings with plaster because the correct implementation of the M.Q.I. was difficult. In those situations, or even where little or no material is available, it is possible to assume the building construction period, the most common construction techniques, and the surveyor's experience. Furthermore, the approach is not accurate in the application on special structures (such as towers, campaniles, and churches), as it is designed to assess ordinary buildings such as residences, shops, etc. This method is particularly suitable for applications on a large scale when it is often not possible to find all the information. The loss of accuracy of the final results obtained in this way is partly compensated by the use of the categories (I to VI) associated with the index of structural response. The six categories are intended to simplify the outcome of verifications and to consider wide ranges of values, making inaccurate information input less significant.

The analysis of the study areas confirmed that the application of the method takes a few minutes per building (provided that the information was previously acquired) and gives sufficient time for the surveyor to derive the outputs and suggest structural intervention. The use of categories (I, II, III) simplifies the procedure, making it more user-friendly for operators with different cultural and educational backgrounds. The analysis of the options contained in the database, both in terms of proposed improvement measures and selected construction techniques, proved extensive enough to cover all the cases that arose in the application phases. In addition, the "notes" box allows the user to specify any details not present in the archive.

Figure 12 summarises the structural response categories recorded in the three case study areas. Analysis of the surveys form showed a good distribution of results, with an exception for the two extreme categories (I–VI). Category IV is the most populated (accounting for 43.8% of the results), followed by categories II and III (25%). Only a single case falls in category V. In general, the results showed that the Kaštela area appears to be the least susceptible to possible seismic actions, having recorded the highest structural behaviours (IV, III). In contrast, the Mirandola area showed generally poor structural responses (II and III).

The analysis of the out-of-plane kinematic mechanisms showed a good distribution of results: "vertical deflection" proved to be the most frequent (43.8%), followed by "horizontal deflection" (31.2%) and "cantonal overturning" (25%); while the kinematic mechanism of simple masonry overturning was not encountered.

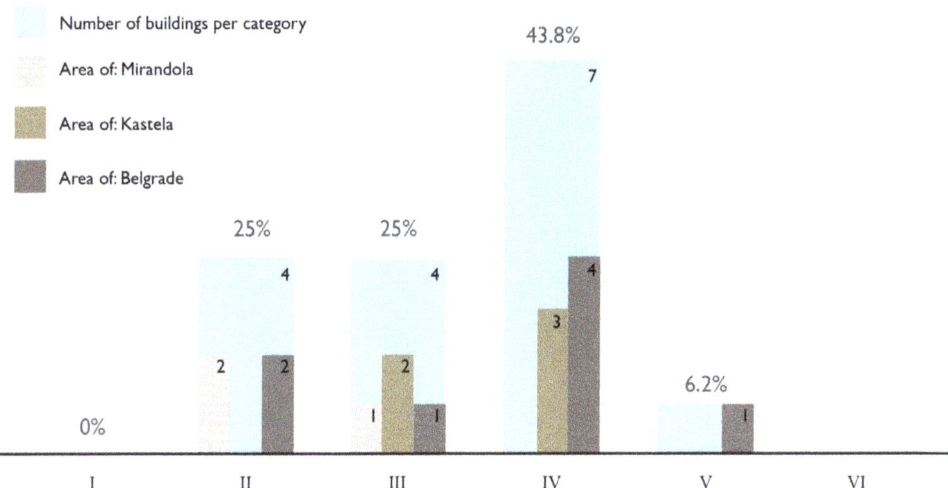

Figure 12. The index of structural response for the three study areas.

The results of the M.Q.I. method highlighted the importance of the masonry quality concerning the outcome offered by the structural response index. In general, high masonry values (majority of categories A and B) were associated with high response categories (as observed in the Kaštela area).

A significant variety of construction techniques, seismic hazards, and seismic exposure of the buildings in the three study areas were encountered in the results of the expeditious method.

4. Validation of the Adriseismic Method

4.1. A Comparison of Traditional Seismic Analysis and Expeditious Method

The procedure was verified using seismic analysis results for a sample of 25 buildings from Italy to understand the method's accuracy. The analysis results were compared with those obtained for the same constructions using the expeditious method. These buildings have different characteristics in terms of the construction period, the number of floors, plan articulation, the function of use, etc. Although limited in number, the sample covers a broad spectrum of structural types, but Appendix C lists the main characteristics.

The sample of structures, used for the validation, was analysed using two different methods: some buildings were assessed using the LV1 system [38] and others using multi-modal analysis. Multi-modal analysis is prescribed by Eurocode 8 and is one of the widely used methods, while the LV1 method is based on an Italian Directive and allows an expeditious assessment using a limited number of parameters. The validation aims to verify the results of the proposed expeditious method by comparing them with those obtained from a traditional analysis method and those obtained through an established expeditious method.

A result of the LV1 method is a seismic safety index (Is, ls), which is defined as the ratio between the return fundamental period of the seismic action that leads to the generic limit state and the corresponding reference period, which is calculated based on the code requirements. The Is value has a range from 0 (total inadequacy) to 1 (adequate) as a verification condition. This indicator is comparable to the index of structural response which is defined in the proposed Adriseismic method. The results obtained from the LV1 and from a traditional analysis method are based on the structural characteristics, the exposure (usually indicated as class of use), and seismic hazard at the site. On the other hand, the proposed method determines the structural response index based on the structural feature, and then the seismic risk is determined as a product of vulnerability, hazard, and exposure. However, the comparison of these results of the other two methods with the structural response index appears to be mathematically more effective than the estimated seismic risk determined.

The modal analysis applied to a structure produced a capacity/demand ration for each load-bearing element. This value is expressed in the same range as the index Is, Is of the LV1 method (values greater than 1 indicate satisfactory verifications). However, in the case of non-linear static analysis, there is no single parameter that summarises the global behaviour of the structure, but there are parameters for each load-bearing element. Therefore, the weighted average for all load-bearing elements, I_{ma}, is considered comparable to the structural response index and can be determined as follows:

$$I_{ma} = \frac{\sum_{i=1}^{n}(x_i \cdot p_i)}{\sum_{i=1}^{n} p_i} \tag{13}$$

4.2. Results and Discussion

Table 11 shows the results from the traditional seismic analysis methods in terms of the two indices Ima and Is, which are compared with the structural response index for 25 buildings.

Table 11. Results of the validation—traditional vs. expeditious methods.

Building ID (1)	Type of Analysis (2)	Results from Traditional Analysis, Ima, and Is (3)	Index of Structural Response, Isr (4)	Vulnerability Category (5)	$\lvert 1 - I_{sr}/I \rvert$ (6)
M001	LV1	0.85	0.74	V	0.13
M002	LV1	0.02	0.24	II	11.00
M003	LV1	0.35	0.25	II	0.29
M004	LV1	0.37	0.23	II	0.38
M005	LV1	0.38	0.25	II	0.34
M006	LV1	0.38	0.25	II	0.34
M007	LV1	0.51	0.46	III	0.10
M008	LV1	0.34	0.34	II	0.00
M009	LV1	0.10	0.17	I	0.70
M010	LV1	0.34	0.38	II	0.12
M011	LV1	0.27	0.34	II	0.26
M012	LV1	0.36	0.38	II	0.06
M013	LV1	0.51	0.40	III	0.22
M014	LV1	0.48	0.42	III	0.13
M015	LV1	0.77	0.43	III	0.44
M016	Modal	0.26	0.23	II	0.12
M017	Modal	0.51	0.28	II	0.45
M018	Modal	0.33	0.32	II	0.03
M019	Modal	0.82	0.78	V	0.05
M020	Modal	0.42	0.28	II	0.33
M021	Modal	0.28	0.34	II	0.21
M022	Modal	0.63	0.34	II	0.46
M023	Modal	0.69	0.48	III	0.30
M024	Modal	0.97	0.75	V	0.23
M025	Modal	0.92	0.63	IV	0.32

The last column shows the comparison between the values obtained from traditional analysis and the structural response index using the formula:

$$\left\lvert 1 - \frac{I_{sr}}{I} \right\rvert \tag{14}$$

Figure 13 shows the same results in a graphical manner. For each building, the results for traditional analysis and the index of structural response obtained from the expeditious methods have been compared.

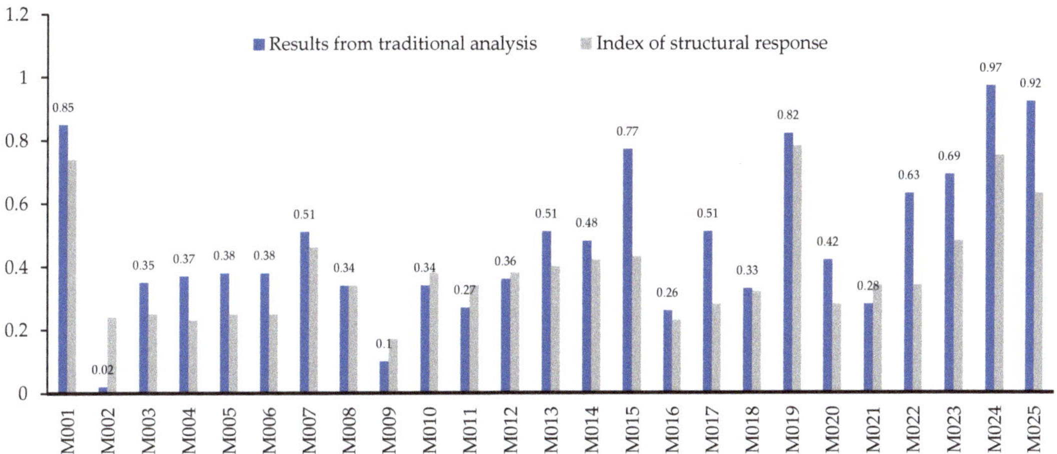

Figure 13. Results of the validation—traditional vs. expeditious methods.

A comparison of sample buildings analysed using the LV1 method shows a similarity between the structural response index Isr and Is. There is an average variation of about 25% between the results of the traditional and the proposed expeditious method, with a standard deviation of 0.18. Furthermore, in most cases, the expeditious method gives more conservative results compared to the LV1.

These considerations excluded building no. 2 (M002) at this stage because the difference between the two results (expeditious and LV1) is small in terms of absolute value (0.22). At the same time, it is important from a relative point of view, and with such a small sample, that the outlier would have affected the outcome of the collected data.

An average variation of 25% and a standard deviation of (0.15) were obtained for comparison with modal analysis. Out of ten analysed buildings, only once did the expeditious method overestimate the capacity of the building to a very limited extent (6% higher). The aggregation of the two methods (LV1 and modal) gives an average variation of 25% and a standard deviation of 0.167.

There are no substantial differences between the two methods in terms of the variation in results proposed by the expeditious method, which gives good accuracy for a qualitative system. This confirms that the method allows for the identification of the most critical buildings, effectively indicating the ones which require a more in-depth analysis, thereby ensuring a more efficient use of resources. However, a larger building sample will have to be analysed before definitive conclusions can be established.

5. Conclusions

The research study presented in this paper was focused on the development of the Adriseismic method for expeditious seismic assessment of URM buildings. The proposed method was developed to mitigate seismic risk associated with urban heritage buildings in six countries located close to the Adriatic and Ionian Seas and is one of the key deliverables of the Adriseismic project.

The first phase of the study involved the development of the expeditious seismic assessment system, with a choice of input parameters and assigned numerical values. The expected seismic performance of the building was estimated based on the proposed methodology through a Microsoft Excel spreadsheet, and the output included masonry category, index of structural response, most probable collapse mechanism, and seismic risk level.

Subsequently, the method was applied in three heterogeneous urban areas, namely Mirandola (Italy), Kaštela (Croatia), and Belgrade (Serbia). In parallel, the method was validated on 25 sample buildings, which were previously evaluated using the LV1 method (expeditious assessment) and modal analysis (traditional seismic analysis). In this manner, the method was validated both in terms of its applicability and the accuracy of results. The validation showed an average variation of 25% and a standard deviation of 0.167 between the index of structural response and the indices Ima and Is. These encouraging results will need to be confirmed by increasing the number of sample buildings and analysing the results of damage kinematics and pushover analyses in order to assess the accuracy of the method.

The application of the method on selected buildings in three urban areas in different countries proved to be very important for testing the practical applicability of the procedure. The results showed that information required at the preliminary stage was almost always easy to obtain. It was found that in large-scale applications (a large number of buildings and limited input data), the loss of accuracy associated with the output was relatively low, e.g., a category II building tends to remain in the same category for small variations in input data; therefore, the use of categories (e.g., I, II, III, etc.), rather than a range of numerical values, makes it possible to limit the incidence of errors or inaccuracies in data collection. The distribution of the results and their variety, in terms of response classes, collapse mechanism, and seismic risk, seem consistent with the case studies. However, it is necessary to perform similar tests on a larger number of buildings and other urban areas in order to collect data related to the distribution of results and the size of the database, especially regarding the construction characteristics and seismic intervention techniques.

The final aspect of the expeditious method was related to recommended structural improvement measures, which enables the user to identify a deficiency in the building and make a preliminary suggestion for intervention.

In spite of the cultural and technical differences, the development of common seismic assessment procedures is the first step towards increasing cooperation in the partner countries involved in the Adriseismic project, which can lead towards a general mitigation of seismic risk by combining typical earthquake engineering procedures with urban planning systems.

Author Contributions: Conceptualisation, G.P. and L.S.; methodology, L.S. and G.P.; software, L.S. and G.P.; validation, G.P., L.S., M.M. and M.S.; formal analysis, L.S., G.P. and S.B.; investigation, L.S. and G.P.; resources, L.S. and G.P.; data curation, L.S., G.P. and S.B., writing—original draft preparation, L.S. and G.P.; writing—review and editing, S.B. and L.S.; visualisation, G.P.; supervision, G.P. and S.B.; project administration, G.P.; funding acquisition, G.P. All authors have read and agreed to the published version of the manuscript.

Funding: This research was funded by the Adriseismic project, of the programme Interreg ADRION "https://www.adrioninterreg.eu/ (accessed on 20 December 2022)". INTERREG V-B ADRIATIC-IONIAN ADRION PROGRAMME 2014–2020: ADRISEISMIC 1019.

Data Availability Statement: Data regarding the expeditious method and validation are not available as they concern studies still in progress.

Acknowledgments: The authors would like to thank the Adriseismic project of the Interreg ADRION programme for the financial support.

Conflicts of Interest: The authors declare no conflict of interest.

Nomenclature

C_o	Corner overturning
$C_{o,n}$	Corner overturning, normalised
D_u	Designated use
E	Exposure
E_d	Expected ductility
F_n	Floors number
g	Thickness of mortar joints
H	Seismic hazard
H_b	Horizontal bending
$H_{b,n}$	Horizontal bending, normalised
H_j	Horizontality of mortar bed joints
I	Result obtained from traditional analysis (modal or LV1)
I_h	Irregularity in height
I_{ma}	Index derived from modal analysis
I_{nh}	Inter-floor height
I_p	Irregularity in plan
I_{sr}	Index of structural response
I_v	vulnerability index; $(1/I_s)$
m	Additional mortar quality coefficient
M_m	Quality of mortar
MQI_v	Quality masonry index for vertical actions
MQI_{op}	Quality masonry index for out-of-plane actions
MQI_{ip}	Quality masonry index for in-plane actions
P_a	Thrusts due to arches and vaults
P_{fw}	Permanent floor weight
P_i	Structural element verification
P_r	Thrusts due to roofs
P_{sd}	Parameters influencing seismic demand
$P_{sd,n}$	Parameters influencing seismic demand, normalised
r	Type of masonry units
S_d	Dimensions of the masonry units
S_h	Seismic risk
S_m	Mechanical characteristics and quality of masonry units
S_{mo}	Simple masonry overturning
$S_{mo,n}$	Simple masonry overturning, normalised
S_s	Shape of the masonry units
S_w	Stiffening wall distance
V_b	Vertical bending
V_j	Staggering of vertical mortar joints
$V_{b,n}$	Vertical bending, normalised
W_c	Level of connection between adjacent wall leaves/header
W_t	Wall thickness
x_i	Generic vertical element

Appendix A. Input Data

This section details all the inputs provided in the method divided by category, as seen in Table A1. In the first column is a brief description of the parameter; in the second is the values that can be entered. For the inputs that influence the final assessment results, the weights considered in the system are also shown.

Table A1. General data.

	Name and Description	Possible Values
1	Data The date of compilation of the evaluation sheet is requested	dd/mm/yy
2	Building Address The address of the building under evaluation is required	Alphanumeric entry
3	Presumed year of construction The user can enter a year of construction of the building, or a period	Alphanumeric entry
4	G.P.S. coordinates	Coordinates in WGS84 format

Table A2. Construction characteristics.

	Name and Description	Possible Values
5	**Foundation** The prevailing type of foundation detected or assumed must be entered	Stepped foundation, engraved in the rock Regular stone masonry foundation Irregular stone masonry foundation Stone rubble foundation with concrete binder Brick masonry foundations Continuous reinforced concrete foundations (strip footings) Wooden piles Reinforced concrete piles Inverted beams foundation Isolated footing Slab foundation
6	**Masonry** The prevailing type of masonry detected or assumed must be entered	Rubble stone masonry Rubble masonry with regular-sized stones Rubble masonry with bricks Cut stone with good bonding Masonry in rammed earth blocks Tuff masonry Dressed rectangular (ashlar) stone masonry Solid brick masonry with lime mortar Solid brick masonry with cement mortar Masonry in brick or cement blocks with cement mortar Reinforced masonry with distribution reinforcement Confined masonry with concentrated reinforcement Timber-reinforced masonry
7	**Floors** The prevailing type of floor detected or assumed must be entered	Single or double timber floors (beams and joists) with a simple wooden plank Single or double timber floors (beams and joists) with brick tiles Floors with metal beams and vaults made with brick tiles Floors with metal beams and hollow bricks Brick vaults Stone vaults Cast-in-situ reinforced concrete slab Hollow clay block floor without a reinforced concrete slab Hollow clay block floor with reinforced concrete slab Prefabricated reinforced concrete floor Hollow brick floor with prefabricated joists

Table A2. *Cont.*

	Name and Description	Possible Values
8	**Roof** The prevailing type of roof detected or assumed must be entered	Single or double timber floors (beams and joists) with a simple wooden plank Single or double timber floors (beams and joists) with brick tiles Floors with metal beams and vaults made with brick tiles Floors with metal beams and hollow bricks Brick vaults Stone vaults Cast-in-situ reinforced concrete slab Hollow clay block floor without a reinforced concrete slab Hollow clay block floor with reinforced concrete slab Prefabricated reinforced concrete floor Hollow brick floor with prefabricated joists
9	**Notes on input** Space is left to allow for specifications on the building under investigation	Alphanumeric entry

Table A3. Masonry characteristics.

	Name and Description	Possible Values	Weight Attributed by the System		
			Vertical Loads	Out-of-Plane Actions	In-Plane Actions
10	Type of masonry units The values are assigned according to the choice of quality of mortar (N.F., P.F., F.)	Stone Brick 	1 N.F. 0.2 P.F. 0.6 F 1	1 1 1 1	1 0.1 0.85 1
11	Horizontality of mortar bed joints	N.F. P.F. F.	0 1 2	0 1 2	0 0.5 1
12	Level of connection between adjacent wall leaves/header	N.F. P.F. F.	0 1 1	0 1.5 3	0 1 2
13	Shape of the masonry units	N.F. P.F. F.	0 1.5 3	0 1 2	0 1 2
14	Staggering of vertical mortar joints	N.F. P.F. F.	0 0.5 1	0 0.5 1	0 1 2
15	Dimensions of the masonry units	N.F. P.F. F.	0 0.5 1	0 0.5 1	0 1 2
16	Quality of mortar	N.F. P.F. F.	0 0.5 2	0 0.5 1	0 1 2
17	Mechanical characteristics and quality of masonry units	N.F. P.F. F.	0.3 0.7 1	0.5 0.7 1	0.3 0.7 1
18	Thickness of mortar joints	Large Standard	0.7 1	0.7 1	0.7 1
19	Additional mortar quality coefficient	Poor Standard	0.7 1	0.7 1	0.7 1

Table A4. Building characteristics.

	Name and Description	Possible Values	Weight Attributed by the System
20	Designated use	Residential Commercial Public	0.15 0.12 0.00
21	Floors above ground	1 2 3 4 5	0.11 0.00 0.00 0.01 0.20
22	Irregularity in plan	I II III	0 1.5 3
23	Irregularity in height	I II III	0.25 0.12 0.00
24	Expected ductility	I II III	0.00 0.33 0.67
25	Connections between structural elements	On Off	The parameter is not associated with a specific value but directly influences the formula used to define the index of structural response, as will be illustrated in the dedicated section

Table A5. Building characteristics (kinematics).

	Name and Description	Possible Values	Simple Masonry Overturning	Weight Attributed by the System		
				Vertical Bending	Horizontal Bending	Corner Overturning
26	Transversal wall distance	I II III	0.00 0.00 0.00	0.00 0.00 0.00	1.00 0.50 0.00	0.00 0.00 0.00
27	Wall thickness	I II III	0.00 0.50 1.00	0.00 0.50 1.00	0.00 0.50 1.00	0.00 0.50 1.00
28	Floor height	I II III	1.00 0.50 0.00	1.00 0.50 0.00	1.00 0.50 0.00	1.00 0.50 0.00
29	Permanent floor weights	I II III	0.00 0.00 0.00	1.00 0.50 0.00	0.00 0.00 0.00	0.00 0.50 1.00
30	Thrusts due to arches and vaults	I II III	1.00 0.50 0.00	1.00 0.50 0.00	1.00 0.50 0.00	0.00 0.00 0.00
31	Thrusts due to the roof	I II III	1.00 0.50 0.00	0.00 0.00 0.00	0.00 0.00 0.00	1.00 0.50 0.00

Table A6. Site data.

Name and Description	Possible Values	Weight Attributed by the System
Ag/g This is the ratio between the design ground acceleration, on type A soil, and the gravity acceleration	Numerical value	The number entered is replicated
Ground type The soil type is required according to the guidelines of Eurocode 8 Part 1, chapter 3.2.1. it is possible to enter a category from A to E (type 2 elastic spectrum)	A, B, C, D, E	A = 1.0; B = 1.35; C = 1.50; D = 1.80; E = 1.60
Building exposure The exposure is assessed using the importance classes given in Table 4.3 of Eurocode 8, Part 1	I, II, III, IV	I = 0.8; II = 1.0; III = 1.20; IV = 1.40

Appendix B. Building Characteristics

The values are derived from simplified formulations or general schematisation, defined with the specific purpose of evaluating the variation in output parameters due to the variation in input parameters. Sometimes it was analysed how a single parameter could influence the seismic demand and, other times, the structural response. The complexity of studying these variations in virtual situations (e.g., it is unlikely that in reality, one would see an increase in loads, hence greater mass, without an increase in the masonry section, thus an increase in seismic capacity) required a certain flexibility in the interpretation of the results, at the expense of rigorous formulations. General considerations for defining the weights of certain input parameters are summarised next:

1. Designated use: The parameter considers the typical loads assigned by Eurocode 1 [54]. Specifically, residential (200 Kg/sm), commercial (300 Kg/sm), and public (400 Kg/sm) were evaluated. Starting from invariant permanent loads, the mass variation was analysed using the coefficients provided for the seismic combination, Eurocode 8. The values were then correlated to obtain the percentage of variation relative to each other. The worst parameter for structural response (Public) was given a value of 0; the other two were higher than the first (Commercial 0.12 and Residential 0.15).

2. Floors above ground: The analysis of the variation in the structural response as the number of storeys varied imposed several parallel considerations. Firstly, the seismic design action was determined using the formula: $S_q = S_e \cdot W / q$. Where S_e is the spectral acceleration, assumed by imagining the building located in Bologna on ground A (the period is that resulting from the simplified formula in Section 4.3.2.2 of Eurocode 8), the mass depends on the number of floors, with fixed dead and live loads. The behaviour factor varies according to the structure's greater or lesser dissipative capacity.

Secondly, the response capacity of the building was determined using the simplified Mohr–Coulomb criterion [55], using the quality of the masonry "solid brick and lime mortar" as an unchanging parameter [56] and the thickness of the resisting panels (24 cm, 24 cm, 36 cm, 36 cm, and 48 cm, respectively) as a factor dependent on the number of floors. The values used in the method are the result of the capacity–demand ratio performed for each storey, then correlated with each other to provide, as for the designated use, the percentage of variation in one concerning the other (the worst always have value 0).

It is evident that the weights obtained result from choices made in advance and that different options could have led to slightly different results. Although considerations may be made in the future to make the procedure less dependent on specific decisions, it is nevertheless believed that the values provided can provide a guideline for expeditious analyses.

3. Irregularity in plan: The values derived from the indications given in paragraph 7.3.1 of Italian NTC [57], regarding determining the behaviour factor as the greater or lesser regularity in the plan of a building varies. Again, the three results obtained were correlated by formulating them as percentage variations (the less regular building has a value of 0; the regular one, 0.26). The Italian standard was used because no specific references were found in the European one.

4. Irregularity in height: The computation was carried out using Section 9.3 of Eurocode 8, part 1, in which guidance is given on reducing the q-factor by 20% for buildings that are not regular in height. This resulted in: regular buildings = 1.0, partially regular = 0.9, and irregular = 0.8. Using the formula, already used for the other parameters, to correlate the three values: $v_i = 1 - \frac{val_i}{val_{min}}$ where val_i denotes the generic value and val_{min} the lowest of the three; the three weights of 0 (worst case), 0.12, and 0.25 were obtained.

5. Expected ductility: For masonry buildings, it was assumed that the behaviour factor could be numerically equal to: (1.5; 2.0; 2.5). The three weights were obtained using the same procedure as illustrated above (and the same formula for correlating the values), which were then implemented in the expeditious method (0.0; 0.33; 0.67). The value 0 was attributed to the worst category (I) and 0.67 to the best (III).

6. Connections between structural elements: As already specified, the parameter is not associated with any specific numerical value, and its presence influences the formulae used in calculating the structural response index.

Appendix C. Summary of Building Characteristics

Appendix C shows the main characteristics of the 25 buildings analysed.

Table A7. Main characteristics of the sample buildings.

Building Code	Period of Construction	Designed Use	Height (Number of Storeys)	Type of Masonry	Type of Floor System
M001	1930–40	Public	3	Solid brick masonry with lime mortar	Hollow clay block floor without reinforced concrete slab
M002	1900–1910	Residential	5	Solid brick masonry with lime mortar	Floors with metal beams and vaults made with brick tiles
M003	Before 1900	Residential	3	Rubble stone masonry	Floors with metal beams and vaults made with brick tiles
M004	Before 1900	Residential	2	Rubble stone masonry	Floors with metal beams and vaults made with brick tiles
M005	1940	Public	4	Solid brick masonry with lime mortar	Hollow clay block floor without reinforced concrete slab
M006	1940	Public	4	Solid brick masonry with lime mortar	Hollow brick floor with prefabricated joists
M007	Before 1900	Public	4	Solid brick masonry with lime mortar	Floors with metal beams and vaults made with brick tiles
M008	Before 1900	Public	4	Solid brick masonry with lime mortar	Cast-in-situ reinforced concrete slab
M009	Before 1900	Public	1	Solid brick masonry with lime mortar	Brick vaults
M010	1920–1930	Residential	4	Solid brick masonry with lime mortar	Hollow clay block floor without reinforced concrete slab
M011	Before 1900	Public	4	Solid brick masonry with lime mortar	Hollow brick floor with prefabricated joists
M012	Before 1900	Public	3	Solid brick masonry with lime mortar	Single or double timber floors (beams and joists) with brick tiles

Table A7. Cont.

Building Code	Period of Construction	Designed Use	Height (Number of Storeys)	Type of Masonry	Type of Floor System
M013	Before 1900	Public	3	Solid brick masonry with lime mortar	Cast-in-situ reinforced concrete slab
M014	1930	Public	2	Solid brick masonry with lime mortar	Cast-in-situ reinforced concrete slab
M015	Before 1900	Public	3	Solid brick masonry with lime mortar	Floors with metal beams and vaults made with brick tiles
M016	Before 1900	Public	3	Rubble masonry with bricks	Brick vaults
M017	Before 1900	Public	3	Rubble masonry with bricks	Brick vaults
M018	Before 1900	Public	3	Rubble masonry with bricks	Brick vaults
M019	Before 1900	Public	4	Solid brick masonry with lime mortar	Hollow clay block floor without reinforced concrete slab
M020	Before 1900	Public	3	Solid brick masonry with lime mortar	Brick vaults
M021	Before 1900	Public	4	Solid brick masonry with lime mortar	Hollow clay block floor without reinforced concrete slab
M022	Before 1900	Public	4	Solid brick masonry with lime mortar	Single or double timber floors (beams and joists) with brick tiles
M023	Before 1900	Public	3	Solid brick masonry with lime mortar	Single or double timber floors (beams and joists) with brick tiles
M024	Before 1900	Public	2	Rubble masonry with bricks	Hollow clay block floor without reinforced concrete slab
M025	Before 1900	Public	4	Solid brick masonry with lime mortar	Hollow clay block floor without reinforced concrete slab

References

1. Albini, P. A survey of the past earthquakes in the Eastern Adriatic (14th to early 19th century). *Ann. Geophys.* **2004**, *47*, 675–703.
2. D'Agostino, N.; Avallone, A.; Cheloni, D.; D'Anastasio, E.; Mantenuto, S.; Selvaggi, G. Active tectonics of the Adriatic region from GPS and earthquake slip vectors. *J. Geophys. Res. Solid Earth* **2008**, *13*, B12413. [CrossRef]
3. Cao, X.Y.; Shen, D.; Feng, D.C.; Wang, C.L.; Qu, Z.; Wu, G. Seismic retrofitting of existing frame buildings through externally attached sub-structures: State of the art review and future perspectives. *J. Build. Eng.* **2022**, *57*, 104904. [CrossRef]
4. Bilgin, H.; Shkodrani, N.; Hysenlliu, M.; Ozmen, H.B.; Isik, E.; Harirchian, E. Damage and performance evaluation of masonry buildings constructed in 1970s during the 2019 Albania earthquakes. *Eng. Fail. Anal.* **2022**, *131*, 105824. [CrossRef]
5. Papadimitriou, P.; Kapetanidis, V.; Karakonstantis, A.; Spingos, I.; Kassaras, I.; Sakkas, V.; Kouskouna, V.; Karatzetzou, A.; Pavlou, K.; Kaviris, G.; et al. First Results on the Mw=6.9 Samos Earthquake of 30 October 2020. *Bull. Geol. Soc. Greece* **2020**, *56*, 251–279. [CrossRef]
6. Vlachakis, G.; Vlachaki, E.; Lourenço, P. Learning from failure: Damage and Failure of Masonry Structures, after the 2017 Lesvos Earthquake (Greece). *Eng. Fail. Anal.* **2020**, *117*, 104803. [CrossRef]
7. Stepinac, M.; Lourenço, P.; Atalić, J.; Kišiček, T.; Uroš, M.; Baniček, M.; Novak, M.Š. Damage classification of residential buildings in historical downtown after the ML5.5 earthquake in Zagreb, Croatia in 2020. *Int. J. Disaster Risk Reduct.* **2021**, *56*, 102140. [CrossRef]
8. Moretić, A.; Stepinac, M.; Lourenço, P.B. Seismic upgrading of cultural heritage—A case study using an educational building in Croatia from the historicism style. *Case Stud. Constr. Mater.* **2022**, *17*, e01183. [CrossRef]
9. D. o. A. University of Bologna. *Report on the State of the Art in Adriseismic Partner Countries Regarding Techniques of Interventions for Reducing Seismic Vulnerability-Deliverable D.T.2.1.2*; Adriseismic Project; University of Bologna: Bologna, Italy, 2021.
10. Moustafa, M.K.; Fadzli, M.N.; Ehsan, N.F. The seismic vulnerability assessment methodologies: A state-of-the-art review. *Ain Shams Eng. J.* **2020**, *11*, 849–864.
11. Calvi, G.; Pinho, R.; Bommer, J.; Restrepo-Vélez, L.; Crowley, H. Development of seismic vulnerability assessment methodologies over the past 30 years. *ISET J. Earthq. Technol.* **2006**, *43*, 75–104.
12. De Luca, F.; Verderama, G.M.; Manfredi, G. Analytical versus observational fragilities: The case of Pettino (L'Aquila) damage data database. *Bull. Earthq. Eng.* **2015**, *13*, 1161–1181. [CrossRef]

13. D'Altri, A.; Sarhosis, V.; Milani, G.; Rots, J.; Cattari, S.; Lagomarsino, S.; Sacco, E.; Tralli, A.; Castellazzi, G.; De Miranda, S. Modeling Strategies for the Computational Analysis of Unreinforced Masonry Structures: Review and Classification. *Arch. Comput. Methods Eng.* **2020**, *27*, 1153–1185. [CrossRef]

14. Rota, M.; Penna, A.; Magenes, G. A methodology for deriving analytical fragility curves for masonry buildings based on stochastic nonlinear analyses. *Eng. Struct.* **2010**, *32*, 1312–1323. [CrossRef]

15. Whitman, R.; Reed, J.; Hong, S. Earthquake Damage Probability Matrices. In Proceedings of the Fifth World Conference on Earthquake Engineering, Italy, Rome, 25–29 June 1973.

16. Braga, F.; Dolce, M.; Liberatore, D. A Statistical Study on Damaged Buildings and an Ensuing Review of the MSK-76 Scale. In Proceedings of the Seventh European Conference on Earthquake Engineering, Athens, Greece, 20–25 September 1982.

17. Medvedev, S.V. *Seismic intensity scale M.S.K.—76*; Publications of the Institute of Geophysics, Polish Academy of Sciences: Warsaw, Poland, 1977; Volume A-6, p. 117.

18. Giovinazzi, S.; Lagomarsino, S. A macroseismic method for the vulnerability assessment of buildings. In Proceedings of the 13th World Conference on Earthquake Engineering, Vancouver, BC, Canada, 1–6 August 2004.

19. Oliveira, C.; Ferreira, M.; de Sá, F.M. Seismic Vulnerability and Impact Analysis: Elements for Mitigation Policies. In Proceedings of the XI Congresso Nazionale on L'ingegneria Sismica in Italia, Genova, Italy, 25–29 January 2004.

20. Lantada, N.; Pujades, L.; Barbat, A. Risk Scenarios for Barcelona, Spain. In Proceedings of the 13th World Conference on Earthquake Engineering, Vancouver, BC, Canada, 1–6 August 2004.

21. Oliveira, C.; de Sá, F.M.; Ferreira, M. Application of Two Different Vulnerability Methodologies to Assess Seismic Scenarios in Lisbon. In Proceedings of the International Conference: 250th Anniversary of the 1755 Lisbon Earthquake, Lisbon, Portugal, 1–4 November 2005.

22. Federal Emergency Management Agency. *FEMA, 154: Rapid Visual Screening of Buildings for Potential Seismic Hazards: A Handbook*; Applied Technology Council: Redwood City, CA, USA, 2015.

23. Otani, S. Seismic Vulnerability Assessment Methods for Buildings in Japan. *Earthq. Eng. Eng. Seism.* **2000**, *2*, 47–56.

24. Hassan, A.; Sozen, M.A. Seismic vulnerability assessment of low-rise buildings in regions with infrequent earthquakes. *ACI Struct. J.* **1997**, *94*, 31–39.

25. Allen, D.E.; Rainer, J.H. Guidelines for the seismic evaluation of existing buildings. *Can. J. Civ. Eng.* **1995**, *22*, 500–505. [CrossRef]

26. Bernardini, A.; Gori, R.; Modena, C. Application of Coupled Analytical Models and Experimental Knowledge to Seismic Vulnerability Analyses of Masonry Buildings. In *Engineering Damage Evaluation and Vulnerability Analysis of Building Structures*; Omega Scientific: Tarzana, CA, USA, 1990.

27. D'Ayala, D.; Speranza, E. An Integrated Procedure for the Assessment of Seismic Vulnerability of Historic Buildings. In Proceedings of the 12th European Conference on Earthquake Engineering, London, UK, 9–13 September 2002.

28. Calvi, G.M. A displacement-based approach for vulnerability evaluation of classes of buildings. *J. Earthq. Eng.* **1999**, *3*, 411–438. [CrossRef]

29. Restrepo-Vélez, L.; Magenes, G. Simplified Procedure for the Seismic Risk Assessment of Unreinforced Masonry Buildings. In Proceedings of the 13th World Conference on Earthquake Engineering, Vancuver, BC, Canada, 1–6 August 2004.

30. Restrepo-Vélez, L. A Simplified Mechanics-Based Procedure for the Seismic Risk Assessment of Unreinforced Masonry Buildings. Ph.D. Thesis, European School for Advanced Studies in Reduction of Seismic Risk, Pavia, Italy, 2005.

31. Modena, C.; Lourenço, P.; Roca, P. *Structural Analysis of Historical Constructions—Possibilities of Numerical and Experimental Techniques*; Taylor and Francis: London, UK, 2005.

32. Grünthal, G. *European Macroseismic Scale 1998*; Centre Européen de Géodynamique et de Séismologi: Luxemburg, 1998.

33. Cao, X.Y.; Feng, D.C.; Li, Y. Assessment of various seismic fragility analysis approaches for structures excited by non-stationary stochastic ground motions. *Mech. Syst. Signal Process.* **2023**, *186*, 109838. [CrossRef]

34. Borri, A.; De Maria, A. Il metodo IQM per la stima delle caratteristiche meccaniche delle murature: Allineamento alla circolare n. 7/2019. In Proceedings of the XVIII Convegno ANIDIS L'ingegneria sismica in Italia, Ascoli Piceno, Italy, 15–19 September 2019.

35. Borri, A.; Corradi, M.; De Maria, A.; Sisti, R. Calibration of a visual method for the analysis of the mechanical properties of historic masonry. *Procedia Struct. Integr.* **2018**, *11*, 418–427. [CrossRef]

36. Adriseismic Project—Interreg ADRION, Interreg ADRION. 2022. Available online: https://adriseismic.adrioninterreg.eu/ (accessed on 20 December 2022).

37. Predari, G.; Stefanini, L.; Santangelo, A.; Marzani, G. A strategic-multidisciplinary approach to reduce the seismic risk. Ongoing activities within the ADRISEISMIC project. *Bologna*, 2022; in press.

38. S. G. Consiglio Superiore dei Lavori Pubblici. *Linee Guida per la Valutazione e Riduzione del Rischio Sismico del Patrimonio Culturale Allineate Alle Nuove Norme Tecniche per le Costruzioni (d.m. 14 Gennaio 2008)*; S. G. Consiglio Superiore dei Lavori Pubblici: Roma, Italy, 2011.

39. Predari, G.; Stefanini, L. The ADRISEISMIC Project: A survey on the building techniques. In *Colloqui.At.e 2022—Memoria ed Innovazione*; Enrico Dassori and Renata Morbiducci: Genova, Italy, 2022.

40. *EN 1998-1*; Design of Structures for Earthquake Resistance—Part 1: General Rules, Seismic Actions and Rules for Buildings. The European Committee for Standardization (CEN): Bruxelles, Belgium, 2004.

41. *EN 1998-3*; Design of Structures for Earthquake Resistance—Part 3: Assessment and Retrofitting of Buildings. The European Committee for Standardization (CEN): Bruxelles, Belgium, 2005.

42. Ministero Delle Infrastrutture e della Mobilità. *Decreto del Ministero delle Infrastrutture e dei Trasporti n. 58 del 28/02/2017*; Consiglio Superiore dei Lavori Pubblici: Roma, Italy, 2017.
43. *Decreto del Commisario Delegato per gli Interventi di Protezione Civile n.28 del 10 Aprile 2002, Repertorio dei Meccanismi di Danno, Delle Tecniche di Intervento e dei Relativi Costi Negli Edifici in Muratura, Regione Marche*; ITC-CNR: L'Aquila, Italy, 1997.
44. D'Ayala, D.F.; Speranza, E. Identificazione dei meccanismi di collasso per la stima della vulnerabilita'sismica di edifici in centri storici. In Proceedings of the Conference L'ingegneria Sisimica in Italia, Torino, Italy, 20–23 September 1999.
45. ReLuis. *Applicativo per le Verificazioni Sismiche dei Meccanismi di Colasso Locali Fuori Piano Negli Edifici Esistenti in Muratura Mediante Analisi Cinematica Lineare*; ReLuis: Napoli, Italy, 2009.
46. Office of the United Nations Disaster Relief Coordinator (Undro). *Natural Disasters and Vulnerability Analysis—Report on Expert Group Meeting*; UNDRO: Genova, Italy, 1979.
47. Government of Croatia. *World Bank Report: Croatia Earthquake–Rapid Damage and Needs Assessment*; Government of Croatia: Zagreb, Croatia, 2020.
48. Galić, J.; Vukić, H.; Andrić, D.; Stepinac, L. *Priručnik za Protupotresnu Obnovu Postojećih Zidanih Zgrada*; Arhitektonski Fakultet: Zagreb, Croatia, 2020.
49. Kišiček, T.; Stepinac, M.; Renić, T.; Hafner, I.; Lulić, L. Strengthening of masonry walls with FRP or TRM. *Građevinar* **2020**, *72*, 937–953.
50. Tehnički Propis o Izmjeni i Dopunama Tehničkog Propisa za Građevinske Konstrukcijetehnički Propis o Izmjeni i Dopunama Tehničkog Propisa za Građevinske Konstrukcije. 29 June 2021. Available online: https://narodne-novine.nn.hr/clanci/sluzbeni/2020_07_75_1448.html (accessed on 20 December 2022).
51. Stepinac, M.; Skokandić, D.; Ožić, K.; Zidar, M.; Vajdić, M. Condition Assessment and Seismic Upgrading Strategy of RC Structures—A Case Study of a Public Institution in Croatia. *Buildings* **2022**, *12*, 1489. [CrossRef]
52. Kuula. 30 November 2022. Available online: https://kuula.co/share/NPrhh/collection/7ktnT?logo=1&info=1&fs=1&vr=0&zoom=1&autorotate=0.45&autop=30&thumbs=1 (accessed on 20 December 2022).
53. Marović, M.; Djoković, I.; Pesić, L.; Radovanović, S.; Toljić, M.; Gerzina, N. Neotectonics and seismicity of the southern margin of the Pannonian basin in Serbia. *Stephan Mueller Spec. Publ. Ser.* **2002**, *3*, 277–295. [CrossRef]
54. *EN 1991-1-1*; Actions on Structures—Part 1-1: General Actions—Densities, Self-Weight, Imposed Loads for Buildings. The European Committee for Standardization (CEN): Bruxelles, Belgium, 2002.
55. *EN 1996-1-1*; Design of Masonry Structures–Part 1-1: General Rules for Reinforced and Unreinforced Masonry Structures. The European Committee for Standardization (CEN): Bruxelles, Belgium, 2005.
56. Il Ministero Delle Infratrutture e Dei Trasporti. *CIRCOLARE 21 Gennaio 2019, n. 7 C.S.LL.PP—Istruzioni per L'applicazione Dell'«Aggiornamento Delle "Norme Tecniche per le Costruzioni"» di cui al Decreto Ministeriale 17 Gennaio 2018*; Il Ministero Delle Infrastrutture e Dei Trasporti: Roma, Italy, 2019.
57. Il Ministero Delle Infrastrutture e Dei Trasporti. *DECRETO 17 Gennaio 2018—Aggiornamento Delle «Norme Tecniche per le Costruzioni»*; Il Ministero Delle Infrastrutture e Dei Trasporti: Roma, Italy, 2018.

Article

Seismic Retrofitting of Dual Structural Systems—A Case Study of an Educational Building in Croatia

Mario Uroš [1], Marija Demšić [1,*], Maja Baniček [2] and Ante Pilipović [1]

[1] Department of Engineering Mechanics, Faculty of Civil Engineering, University of Zagreb, 10000 Zagreb, Croatia
[2] Croatian Centre for Earthquake Engineering, Faculty of Civil Engineering, University of Zagreb, 10000 Zagreb, Croatia
* Correspondence: mdemsic@grad.hr

Abstract: On 29 December 2020, a devastating Mw6.4 earthquake struck near the town of Petrinja, Croatia. The main earthquake was preceded by a Mw4.9 foreshock the day before. The earthquakes caused extensive damage to buildings, especially historic buildings made of unreinforced masonry but also to buildings of other typologies and to critical infrastructure. Today, recovery efforts in Croatia focus primarily on reconstruction and seismic retrofitting. Family homes and public, cultural, educational, and other facilities are top priorities. In this paper, a comprehensive study of existing building in the educational sector is presented as a case study. The seismic performance of the building is evaluated using numerical methods, first for the as-built condition and then for the retrofitted building. For each condition, the collapse mechanisms of the building were determined and critical structural elements were identified. The presented retrofit strategy of the dual structural system consisting of RC frame system and masonry walls aims to reduce the displacements of the RC frame system to a level sufficient to prevent the early brittle failure of the concrete. Additionally, the discrepancies when using different modelling approaches are discussed.

Keywords: earthquake; Petrinja; reinforced concrete; masonry; pushover analysis; case study; seismic retrofitting

check for
updates

Citation: Uroš, M.; Demšić, M.;
Baniček, M.; Pilipović, A. Seismic
Retrofitting of Dual Structural
Systems—A Case Study of an
Educational Building in Croatia.
Buildings **2023**, *13*, 292. https://
doi.org/10.3390/buildings13020292

Academic Editor: Marco
Di Ludovico

Received: 17 December 2022
Revised: 6 January 2023
Accepted: 12 January 2023
Published: 18 January 2023

1. Introduction

In 2020, Croatia was hit by two strong earthquakes that caused significant property damage to tens of thousands of buildings and to the essential infrastructure. The first strong earthquake, measuring 5.5 on the Richter scale, occurred on Sunday, 22 March 2020, predominantly affecting the capital city of Zagreb [1,2]. Just nine months after this earthquake, a devastating earthquake measuring 6.2 on the Richter scale occurred near the town of Petrinja on 29 December 2020. The highest intensity was given as VIII on the EMS scale. The greatest building damage, including the complete collapse of several buildings, occurred in Petrinja, and significant property damage was recorded in the neighboring towns of Sisak and Glina, as well as in the wider area of Sisak-Moslavina County. Due to the force of the impact, the earthquake also caused damage to buildings in neighboring counties, including progressive damage to buildings damaged in the Zagreb earthquake [3].

These two major earthquakes occurred in the region of northwestern Croatia, the seismic activity of which is described as moderate, with rare occurrences of strong events, but highly vulnerable due to the economic importance and concentration of population centers, including the capital Zagreb [4]. The analysis of the earthquake catalogue presented in [5] indicates that the continental part of Croatia can generate on average one Mw = 5.0 earthquake per year or one Mw = 6.4 event per century.

The December 2020 Petrinja earthquake series occurred within the Petrinja–Zrinska gora seismic zone, which includes the Petrinja fault system [5–8]. The maximum expected moment magnitude there is estimated to be M6.5. The probabilistic seismic hazard analyses performed for the local type A soil conditions predict a peak horizontal ground acceleration (PGA) for Petrinja and Glina of 0.11 g for a return period of 225 years and 0.15 g for 475 years [9,10]. However, the Petrinja and Glina communities are mainly located on recent alluvial sediments, so the actual PGA value is likely to be higher [3].

The 2020 Petrinja earthquake series began with moderately strong foreshocks on December 28, 2020, with epicenters near Strašnik, about 5 km southeast of Petrinja city center with the strongest shock of ML = 5.1 at 5:28 UTC. The main earthquake occurred the next day, 29 December 2020 at 11:19 UTC, with an epicenter near the foreshocks—its magnitude was estimated at ML = 6.2 [11] and MW = 6.4 [12]. The main earthquake had a focal depth of about 6–7 km. The earthquake series was recorded by the National Strong Motion Network, which consists of seven stations in the Zagreb metro area. All stations are located north-northwest of Petrinja within a narrow backazimuthal range at epicentral distances between 45 and 60 km. At one of the stations with an epicentral distance of Repi = 48 km, the measured peak ground acceleration is PGA = 0.13 g. The approximate shakemaps [13] of the main shock perceived PGA at the ground surface in the epicentral area around 0.5 g, while in Zagreb, PGA was estimated to be 0.1–0.15 g, which agrees well with the recorded data [3].

According to official data, the December 2020 earthquake affected about 50,000 buildings, which were inspected by March 2022. Most of the significant damage affected older buildings of unreinforced masonry built before the adoption of the first official seismic regulations [3], introduced in 1964 after the devastating 1963 Skopje earthquake. According to the Rapid Damage and Needs Assessment document [14] prepared by the Croatian government with technical assistance from the World Bank, the total economic impact of the Petinja earthquake based on the adopted international DaLA methodology was approximately EUR 4.8 billion (Figure 1), with Sisak-Moslavnina County being the hardest hit, with a share of about 80%. This includes damage (buildings and infrastructure) in the form of the replacement value of damaged or destroyed physical assets. Estimated losses are based on changes in economic flows resulting from the temporary absence of damaged assets or the disruption of access to goods and services in the form of lost sales, higher operating costs, and risk reduction measures [14].

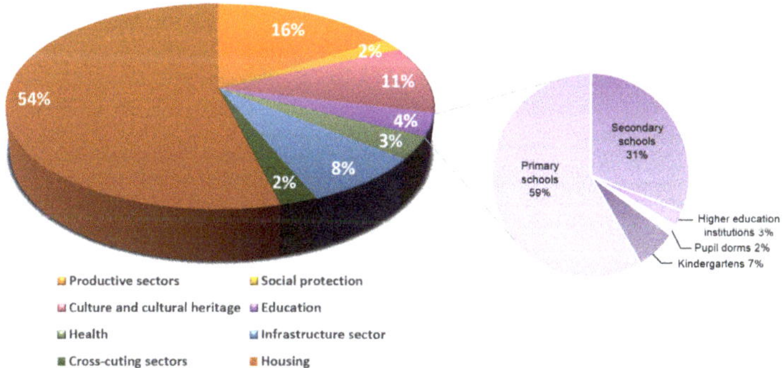

Figure 1. Share of economic estimates of damage and loses after Petrinja earthquake (adopted from [14]).

It is worth mentioning that the seismic risk in Croatia is considered high [15], but there are no systematic strategies for its mitigation [16]. Risk mitigation activities in the country are insufficient and are mostly based on individual initiatives of research institutes, local authorities, and civil protection teams. There is no program for the systematic seismic assessment and retrofitting of buildings of strategic importance, and there are few studies involving hospitals and schools, bridges, and other important facilities [15–17]. Another major problem is that the building stock in Croatia is very old and more than 40% of the buildings were built more than 50 years ago.

With the development of technical regulations in Croatia, seismic requirements for buildings have gradually increased (Figure 2). However, it is important to emphasize that the regulations relate primarily to the construction of new buildings and those undergoing major reconstruction. Inadequacies related to earthquake-damaged buildings in the technical regulation itself, as well as deficiencies in specific knowledge in engineering practice, proved to be key problems after the 2020 earthquakes. As a result of these, the Technical Regulations for Building Structures [18] was amended in order to include the reconstruction of buildings damaged in an earthquake. Additionally, the Act on the Reconstruction of Earthquake-Damaged Buildings [19] was issued and adopted after the Petrinja earthquake in order to regulate the procedures of reconstruction and seismic retrofitting, as well as the removal of damaged buildings in the areas of counties affected by earthquakes in March and December 2020.

The engineering and scientific community became heavily involved in activities to mitigate the effects of the earthquakes. Many experts from the academic community were directly involved in the assessments and preparation of studies on damaged buildings and the retrofitting techniques. This includes detailed guidelines for conducting post-earthquake damage inspections of buildings [20] and bridges [21,22], the development of methodologies for the assessment and retrofitting strategies of masonry buildings [23], RC buildings [24], heritage buildings [25–27], and vulnerability assessments of historical building aggregates [28].

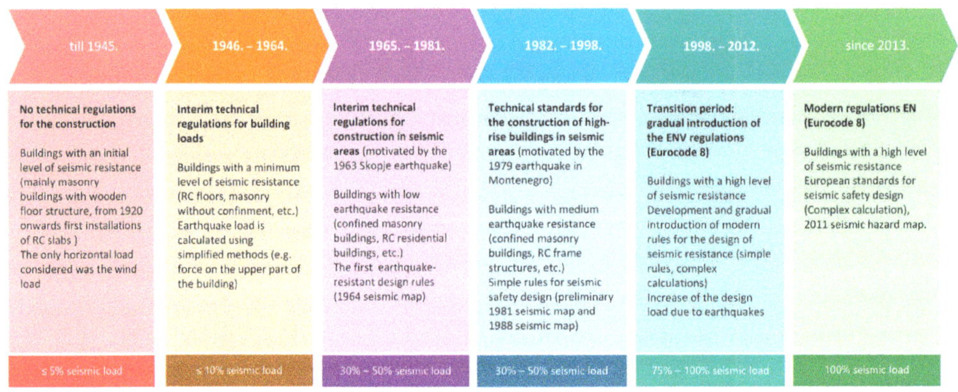

Figure 2. Development of technical regulations for seismic design in Croatia (adopted from [23]).

Detailed seismic performance assessments of individual buildings are essential tools for obtaining more reliable data to determine the seismic vulnerability of a specific building typology. At minimum, it should be available for all buildings of strategic importance, such as hospitals [17,29,30], that must remain fully functional after an earthquake. However, given the diversity of structural systems and irregular structural geometry, it can be considered scarcely reported. Moreover, the use of field monitoring data is strongly recommended when applying numerical analyses. The methodology involving the derivation of "time-building specific" fragility curves for the existing eight-story RC hospital building is reported in [29]. The importance of on-site investigation to evaluate actual structural conditions is emphasized in the seismic vulnerability assessment of an existing RC school building in Italy using a variety of alternative formulations to calculate the vulnerability index [31]. In another case study, where an irregular RC hospital building was investigated using a nonlinear static method based on a multimodal distribution of lateral loads, it was shown that conventional load patterns may not be conservative with respect to the inelastic behavior of the irregular RC building [30]. Other important issues in evaluating the seismic performance of the buildings concern the numerical modelling approach itself. A detailed overview of the challenging issues related to the use of different modelling strategies with equivalent frame models and models with a more "refined" discretization with 2D or 3D elements that do not strictly require any a priori identification of piers and spandrels are discussed in [32]. The paper emphasizes that perfect calibration between the simplified and refined models is generally not possible in all regions of the panel failure domain and that a solid expertise in the seismic response of the building is still required given the variety of options available in modelling approaches. Numerical epistemic uncertainties that arise when using the equivalent frame model (EFM) approach to model unreinforced masonry buildings are discussed in [33]. Based on the analysis of two existing masonry buildings, it was concluded that the choice of conservative deterministic parameters aims at a conservative approximation of the PGA at failure; however, in that case, there is a certain risk of overlooking certain damage mechanisms and their locations.

This paper presents a case study on the building of a higher education institution located in Petrinja. Among educational facilities, a total of 271 buildings were affected by the earthquake (Figure 1), of which 109 are located in Sisak-Moslavina County, of which 18 were marked as temporarily unusable and 14 as unusable. It should be emphasized that in the affected region, much of the important infrastructure, such as hospitals and schools, was built after World War II and is usually a mixture of reinforced concrete frames and brick walls [2]. The floor structures of these buildings often consists of a reinforced concrete fine-rib floor that is load-bearing in only one direction and cannot be fully considered a rigid diaphragm. A case study building consists of a system of RC columns and beams in one direction, while the main load-bearing system in the other direction consists of brick walls. For this period of construction (until 1965), it is typical that RC columns do not have a minimum transverse reinforcement, so brittle shear failure in the concrete is to be expected. This type of dual system is peculiar and rarely found in the literature. The building under consideration has an irregular plan only on the ground floor, which further complicates the structural response during an earthquake.

This study presents a comprehensive methodology for seismic performance assessment and retrofit strategy for this type of structural system, which are commonly present in the territory of the countries of the former Yugoslavia. In the following sections, the structural system of the case study building damaged by the Petrinja earthquake is described in detail. A brief overview of the earthquake damage is provided, followed by a numerical evaluation of the seismic performance of the as-built state of the building based on a pushover analysis. Two modeling approaches are investigated and a numerical model using finite elements and macroelements is used. The second part of the paper gives an overview of the relevant strengthening techniques and the evaluation of the seismic performance of the retrofitted building. When considering the seismic retrofit strategy, the frames of the RC façade are affected by minimal structural strengthening measures since the building was energy retrofitted in 2018. Therefore, the retrofit strategy aims to reduce the displacements of the original frame system to a level that prevents early brittle shear failure of the RC columns. In the last section, relevant conclusions and discussions are given concerning the numerical methods used and the retrofit strategy.

2. Case Study Building

2.1. General Information about Case Study Building

The building of the higher education institution in Petrinja, Croatia, was built in 1963 and underwent energy renovation in 2018. The building consists of the ground, 1st, and 2nd floors; the area on the ground floor is 1294.3 m^2 and on the upper floors 603.4 m^2. The total gross floor area of the building is 2501.1 m^2, while the useful heated area of the building is approximately 2130 m^2. The building has an irregular T-shaped layout and consists of three volumes connected into one whole (Figure 3). In the northwest–southeast direction (hereafter X-dir), there are two volumes that have only the ground floor. The volume marked as V1 is 3.84 m high, while the volume V2 in the southeastern part of the building is 5.52 m high and consists of a one large hall. The central volume V3 of the building is arranged in a northeast–southwest direction (hereafter Y-dir) and has three stories with a total height of 11.52 m.

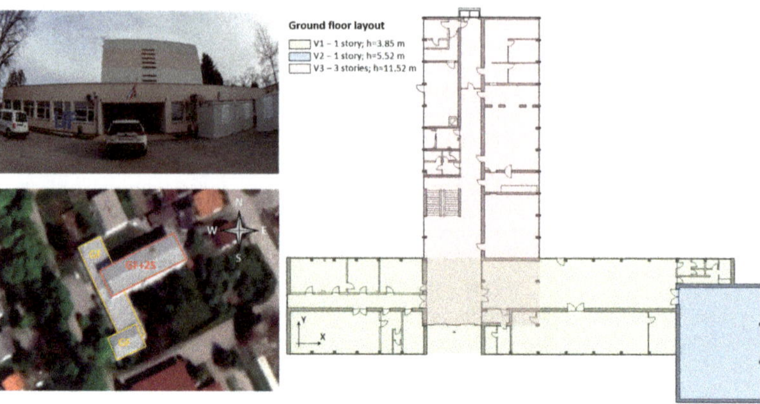

Figure 3. Case study building.

The load-bearing structure consists of interconnected reinforced concrete (RC) columns and beams that form façade frame systems at the building edges. Within the floor plan, the load-bearing system in the longitudinal direction consists of masonry brick walls with a thickness of 38 and 30 cm. In the transverse direction, the system consists of brick walls 30 cm thick, while the partition walls are 20 cm thick and are arranged irregularly within the floor plan. Columns and walls have RC strip foundations, most of which are interconnected. The floor structure is monolithic reinforced concrete, the so-called thin ribbed RC floor.

2.2. Detail Description of the Structure

The dimensions of the structural elements were obtained from the drawings accompanying the original static design (Figure 4), an architectural drawing of the as-built condition from 2015, the energy retrofit project from 2018, and an on-site survey. A detailed inspection of the load-bearing structure revealed that the building was not built entirely in accordance with the project documentation. The arrangement of columns was altered so that they were not constructed in some locations when a brick wall was present (Figure 5). Other subsequent interventions, alterations, or modernizations carried out over time did not have a significant impact on the load-bearing capacity of the structure. These were mainly offsets of the partition walls.

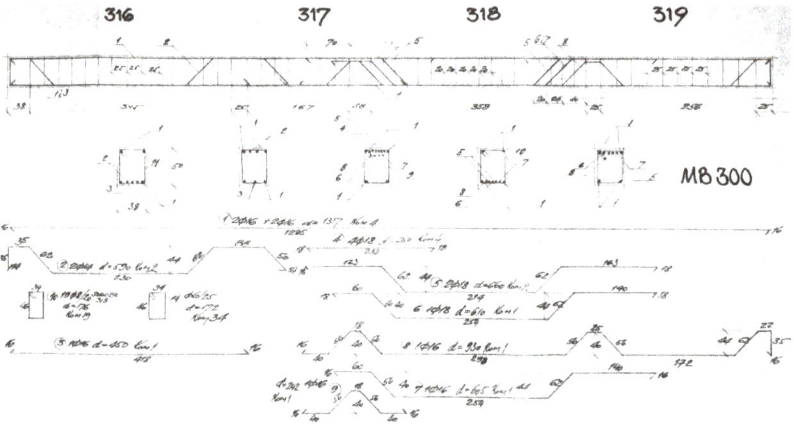

Figure 4. Reinforcement layout form original static design.

Figure 5. Layout of the building stories and markings of gravity and lateral load resisting system dimensions.

The detailed building inspection determined the type, layers, and dimensions of the floor structures, verified the location of the reinforced concrete columns, determined the dimensions of the brick walls, and identified other information important for understanding the structural system of the building. Based on the results of experimental testing of concrete quality on the taken samples from columns and beams and with a sclerometer, it is estimated that the concrete embedded in RC elements is of compressive strength, corresponding to concrete class C20/25 (MB25). All reinforcement bars embedded in RC elements are of smooth steel with a yield strength of 220 MPa and ultimate tensile strength of 360 Mpa (approximately corresponding to BSt 22/34 GU). The reinforcement of some columns was also confirmed on-site. The load-bearing walls of the building are masoned with solid bricks of size $29 \times 15 \times 6.5$ cm. The results of the testing of the shear strength of the mortar in the solid brick walls showed that the average value of the shear strength of the mortar at all locations is 0.685 MPa, i.e., the shear strength with the contribution of vertical stress.

Reinforced concrete frames consist of columns with dimensions 25/38 cm and beams with dimensions 38/43 cm. The columns are provided with four reinforcing bars of 12, 14, and 16 mm diameter, depending on their position. Transverse reinforcement consists of 6 mm diameter bars on the spacing of 25–30 cm. The frame beams are located along certain façades of the building (volumes V1 and V3). The reinforcement of the beams is smooth, as mentioned before, and it is bent diagonally from the upper zone to the lower zone of the section (Figure 4), i.e., at the zones with the highest shear stresses. The frame parapets are made of 19 cm thick hollow concrete blocks.

In the central part of V1, there are load-bearing brick walls in the X-dir with a thickness of 38 cm, which serve as a mid-support for the ribbed concrete slab. In the Y-dir, there are slab ribs 40 cm high and 12 cm thick, spaced 50 cm apart. The concrete slab is 6 cm thick. There are also ribs for stiffening in one third of the span of the slab. In volume V3, the load-bearing walls in Y-dir are 38 cm thick that are manly continuous in height and partly interrupted by openings. There is also a brick walls 20 cm thick in Y-dir, these are partition walls. Since they also have stiffness and load-bearing capacity, they are included in the numerical models. The 30 and 20 cm thick brick walls in X-dir are only partially continuous in height. In this part of the building there is also a two-flight staircase.

Along the edge of the floor plan in the southeastern part of the building there is a hall (volume V2), which is a single volume with a height of 5.52 m. It is bounded on three sides by 38 cm thick brick walls, which are not adequately connected to the RC elements. On the fourth side, there is a RC frame consisting of columns with dimensions 25/38 cm and beams with dimensions 38/43 cm. The edge supports of the beams rest directly on the transverse brick walls. The floor structure also consists of an RC ribbed slab, the ribs are 64 cm high and 12 cm thick at 50 cm intervals, while slab is 4 cm thick. The span of the floor structure is 10.0 m, and there are ribs for stiffening in Y-dir.

2.3. Damage in Petrinja Earthquake

The building was damaged during the 29 December 2020 Petrinja earthquake. After a preliminary inspection, it was labelled as temporarily unusable with the need for emergency intervention measures due to the suffered damage. There was a risk of plaster and installations falling from the callings and the out-of-plane failure of severely damaged partition walls, so access was restricted in some parts of the building. The inspection revealed slight damage to the floor structures, moderate damage to the vertical elements and severe damage to the staircase and partition walls. The overall damage to the structure was classified as Level 3, according to the EMS scale. Characteristic damage can be seen in the Figure 6 with indicated location information.

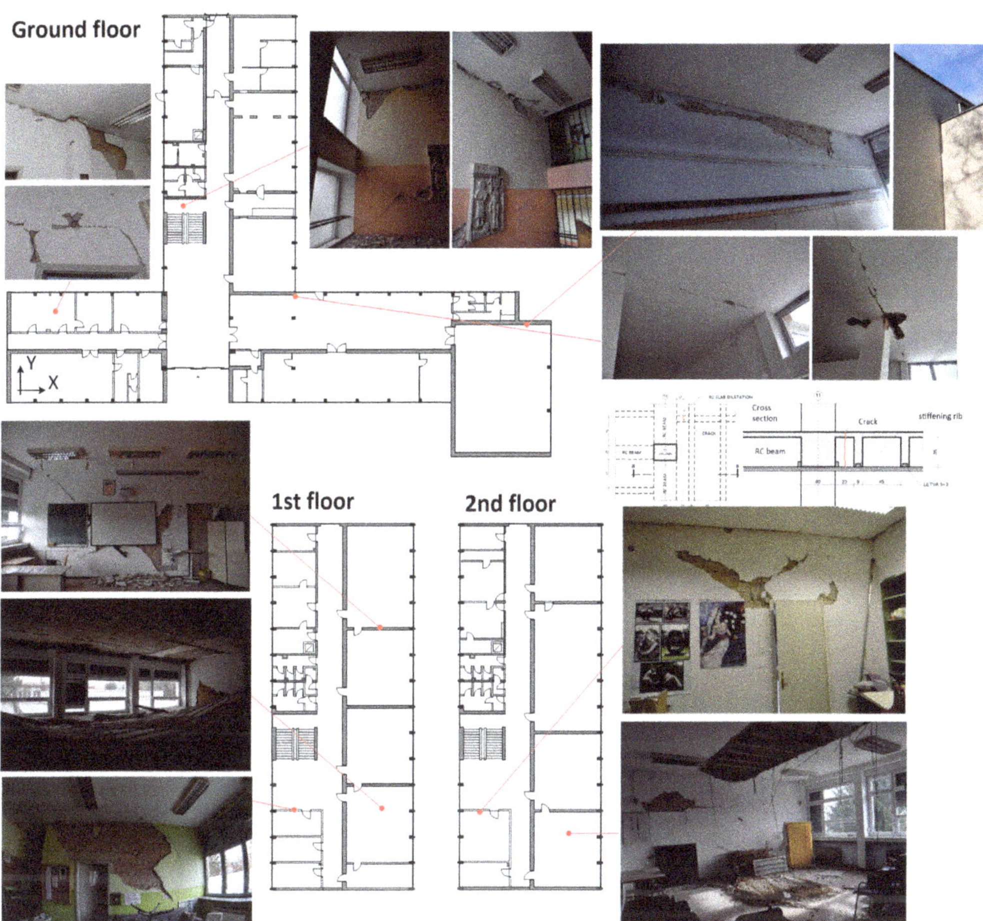

Figure 6. Photographs of damage to the building (photos: courtesy of Joško Krolo and Karlo Jandrić, 2021; and CCEE, 2021).

Further detailed inspection in particular revealed that the damage to the brick wall to which the staircase landing is connected is significant and possess a considerable risk for usage. It is the only staircase in the building. Due to the structural irregularities, special attention was given to the locations where different volumes connect. Although there is no dilatation, the detailed inspection showed that the slabs are partially separated and that the stiffening ribs connect the different units of the building. In this location, between the lower and upper parts of the building, damage to the floor structure was observed. To determine the extent of the damage found, the lower layers of the ceiling were removed, and cracks were also found on the stiffening ribs of the slab in this area. Furthermore, in the section connecting volumes V1 and V3, cracks were recorded on the part of the wall located above the ground floor. Other observed damage included minor damage to the load-bearing masonry walls, extensive cracking of partition walls and major damage to the lower layers of ceiling structures, with cracking and the falling of plaster on almost all walls.

3. Assessment of the Seismic Performance of the Existing Building

3.1. Numerical Models

In order to verify the methodology and to evaluate the applicability of different modelling approaches, two types of numerical model were employed for the assessment of the building's seismic performance (Figure 7). The first model is based on the finite element method (FEM) and is designed using the CSI ETABS [34] software package that is intended for the numerical analysis of buildings in seismically active areas. The modeling of the load-bearing structure consisting of columns and beams was carried out using frame elements, while the masonry walls were modeled with shell elements. The horizontal structures were modeled with shell elements, and they were additionally assigned the property of rigid diaphragms. The cracking of cross-sections during an earthquake was taken into account in the model according to the guidelines from Croatian standards (adopted Eurocode regulation [35,36]), so that the bending stiffness of reinforced concrete beams, columns, and walls was reduced by 50%. The shear stiffness of all elements is taken also with 50% reduction of the initial stiffness.

Finite element model Macroelement model

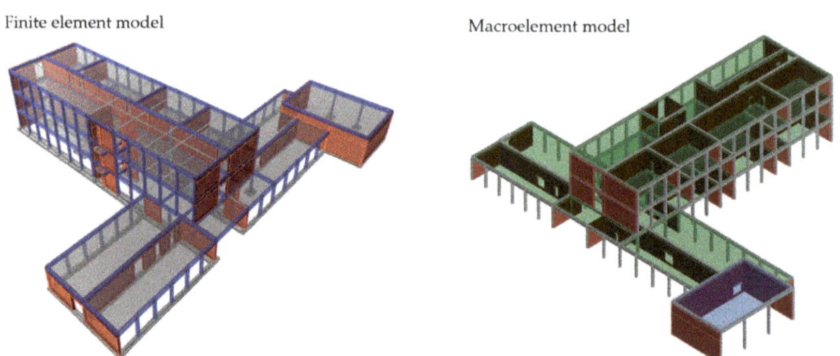

Figure 7. Numerical models of the building.

The nonlinear behavior of the elements in FEM model was accounted for by the assumption of local opening of the plastic hinges in the relevant load-bearing elements of the structure. To determine the capacity curve of a load-bearing masonry walls (Figure 8), different failure mechanisms were accounted for and the predominant one was selected. The shear failure by the development of diagonal cracks in the wall or shear failure by sliding of the wall is proved to be the prevailing mechanism of damage. There was also a failure mechanism due to crushing the material at the element edges, which is initiated by wall rocking.

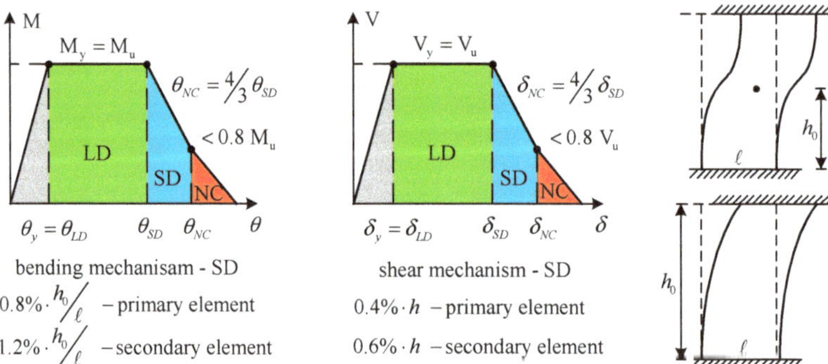

Figure 8. Typical force–deformation relation for masonry elements.

The deformation capacity of elements that include columns, walls, lintels, and beams is taken according to [34]. Force-deformation relations (Figure 9) for reinforced concrete columns and beams were calculated on the basis of the strength of the concrete and the installed reinforcement, by considering several code guidelines [34–38].

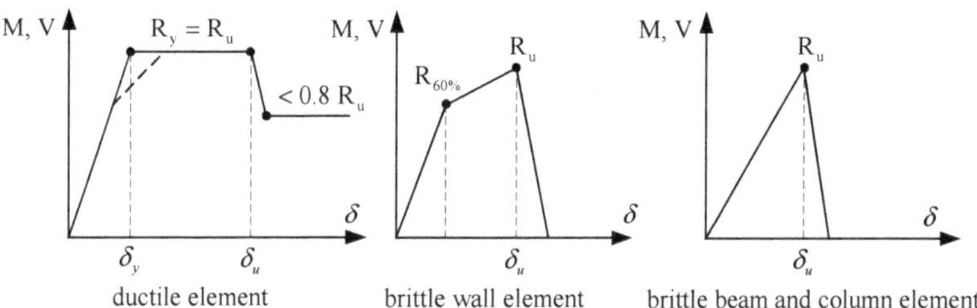

Figure 9. Typical force–deformation relation for RC elements.

The second model is based on the macroelements and formation of equivalent frames model (EFM) for which the 3Muri [39] software package was employed. In this model, the previously stated assumptions about stiffness reduction and rigid diaphragms were also used. In this numerical model the nonlinear masonry wall elements are taken into account by bilinear behavior with maximum values of shear force or bending moment, depending on the prevalent failure mechanism automatically determined by the program. The failure of an element by bending under pressure is determined by the relationship that connects the normal stress and the peak value of the moment, under the assumption of a material with no tensile strength. The shear failure mechanism is defined in the model according to the Mohr–Coulomb criterion, i.e., shear cracking along the mortar joints. The resistance mechanisms of the RC elements considered are ductile bending (with or without normal forces) for each of end with the consequent formation of a plastic hinge and fragile to shears, in conformity with the criteria found in the code [39]. With this model, the element damage mechanism is automatically calculated by the program, depending on internal forces.

The characteristics of the materials used in the calculation were adopted based on experimental tests and estimated conservative values. Based on the material testing program and review of all significant details in the structure, knowledge level 3 was selected, and therefore the confidence factor in the calculation of the structure is equal to 1.0. In the calculation, the mean values of the mechanical characteristics of the materials were used, which were taken as slightly smaller than the measured values due to the unreliability of the measurement method and the dispersion of the results (Table 1).

Table 1. Mechanical properties of the materials.

Material Characteristics	Value		
	Masonry	Concrete	Rebar Steel
Modulus of elasticity	1500 N/mm^2	29,000 N/mm^2	200,000 N/mm^2
Shear modulus	500 N/mm^2	12,083 N/mm^2	76,923 N/mm^2
Specific weight	18 kN/m^3	25 kN/m^3	77 kN/m^3
Mean compressive strenght	3.4 N/mm^2	24 N/mm^2	
Initial shear strength	0.16 N/mm^2	-	
Yielding strength	-	-	140 N/mm^2

The software packages automatically take into account the self-weight (W) of all assigned elements of the structure. The additional permanent load (G) was partially taken from the original design and evaluated during the detailed inspection of the building. Additional permanent loads of 2.6 kN/m^2 are applied to all floor structures and 3.3 kN/m^2 to the roof level. Additional loads from parapet and façade weight are applied to the edge frames at approximately 5 kN/m. The imposed load (Q) is considered according to the current regulations, which is 3 kN/m^2 for areas with tables, such as in schools, and for the roof level 1 kN/m^2. The load combinations considered are summarized in Table 2.

Table 2. Load combinations.

	Load Case	Scale Factor
ULS	W	1.35
	G	1.35
	Q	1.5
MASS	W	1.0
	G	1.0
	Q	0.3

As for the seismic actions on the site, Figure 10 shows the peak ground acceleration values for soil class A (a_{gR}). Since this is an educational building, the importance factor γ_I is 1.2. For the calculation of the seismic action, the MASS combination of vertical forces is used.

Figure 10. Horizontal peak ground accelerations for soil class A at the location [Google maps].

3.2. Limit States

The survey of damage to the buildings after the earthquakes in 2020 [1–3,20] has shown that the extensive retrofitting of buildings or replacement with new buildings will be difficult to achieve and will require significant investment to ensure modern earthquake safety standards. Therefore, in order to ensure that reconstruction measures will lead to a certain level of seismic safety, the Croatian Technical Regulation for Building Structures (CTRBS, [18]) established various seismic safety requirements depending on the purpose of the building and the level of damage, which applies only to buildings being rehabilitated due to seismic damage [18]. In contrast to the three limit states defined in code [36], i.e., Near Collapse (NC), Significant Damage (SD), and Damage Limitation (DL), the CTRBS refers only to the SD limit state. The CTRBS defines the significant structural damage index (SDI) as a ratio of the design seismic resistance of the structure and the structural requirement for the significant damage limit state. Design seismic resistance is defined by the value of the seismic action for the design peak ground acceleration on type A ground for which the structure reaches the limit state of significant damage. The design ground acceleration on type A ground (a_g) is equal to the reference peak ground acceleration (a_{gR}) on type A ground for the return period of 475 years (probability exceeding 10% in 50 years) times the importance factor. The seismic resistance levels are:

- Level 2: The level of structural retrofitting should reach a significant structural damage index (SDI) of at least 0.5.
- Level 3: The level of structural retrofitting should reach a significant structural damage index (SDI) of at least 0.75.

- Level 4: The level of structural retrofitting should reach a significant structural damage index (SDI) of at least 1.0.

Level 3 is a mandatory minimum level for educational buildings, such as the one in this study. Therefore, according to the CTRBS, for the building in question, the design acceleration value of $a_{gR} = SDI \times PGA \times \gamma_I = 0.75 \times 0.15$ g $\times 1.2 = 0.135$ g on type A ground was used for the definition of demand requirement.

3.3. Finite Element Model Results

First, the calculation of the structure for permanent vertical load was carried out. Relevant requirements for structural elements are carried out by means of the combinations of static loads, based on the provisions of the current codes. Furthermore, these results represent the basis for further non-linear analysis, and in addition, it is important to know the internal stress state of the elements before applying the seismic load in order to gain insight into other structural faults and week parts of the structure itself. Due to the large amount of data, only the most significant results, which are important for the proposal to strengthen the structure, are shown.

Figure 11 shows normal stresses of masonry walls. As can be seen, the stress in the walls of the ground floor is up to 0.75 MPa on the central longitudinal brick wall, which is about 22% of the compressive strength of the walls.

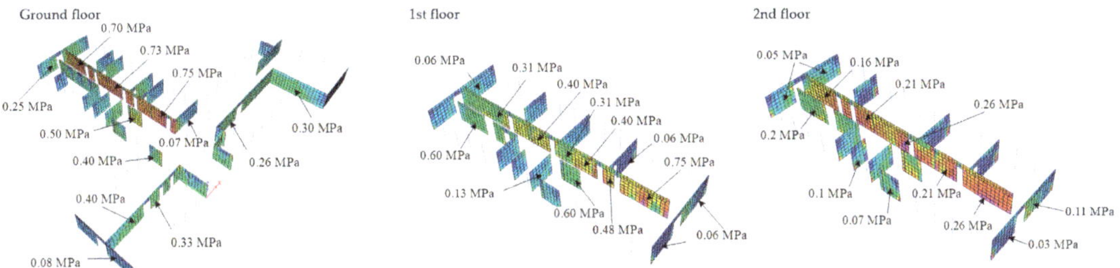

Figure 11. Normal stresses in masonry walls due to permanent loads.

In addition, it was found that the existing reinforcement mostly meets the requirements for permanent vertical action. However, there were three critical positions in structural system that need to be strengthened. These are the beams on the ground floor in the axis marked in Figure 12. They are all located in the entrance area of the building. Their exceeding load capacity is the result of the displaced position of the columns as well as the existence of masonry walls in the upper floors, which has interrupted the flow of forces in the structure. It is mainly a matter of lack of transverse reinforcement, but in some places the longitudinal reinforcement is also lacking.

Numerical results of the dynamic properties, i.e., eigenperiods and mode shapes, are presented below. Figure 13 shows the results of a model that has a limited crack state since it is loaded only by the weight of the structural elements and other permanent and service loads. Further results for eigenmodes are obtained for the state of the structure near the SD limit state (Figure 14). In this case, the stiffness of the structure is significantly reduced, which is reflected in the prolonged eigenperiods. It should be noted that the irregularity of the building on the ground floor can cause torsional effects that increase the requirements for the relative displacements of the perimeter walls, which could lead to their greater vulnerability during earthquake.

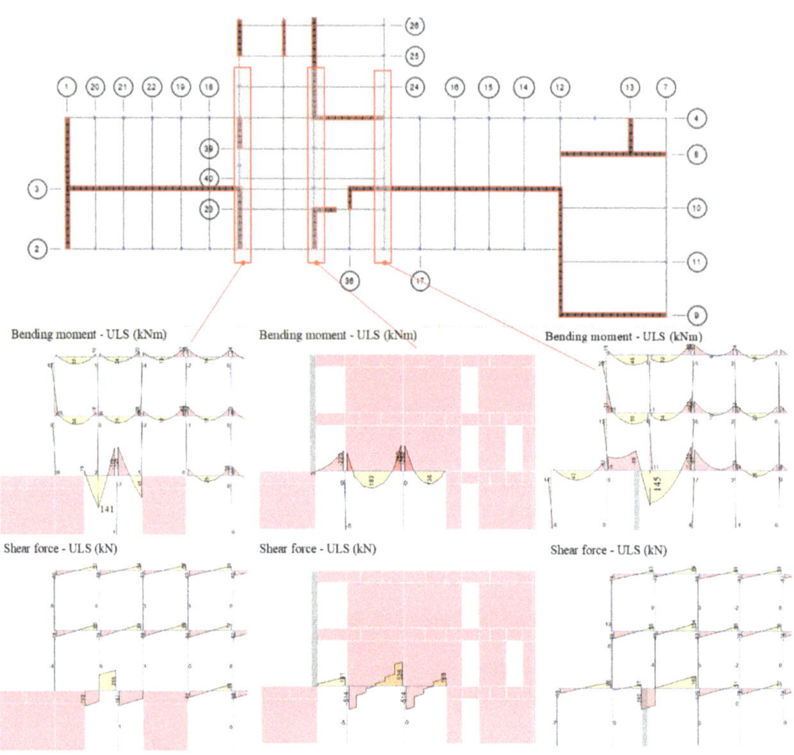

Figure 12. Weak positions in the structure due to vertical loads.

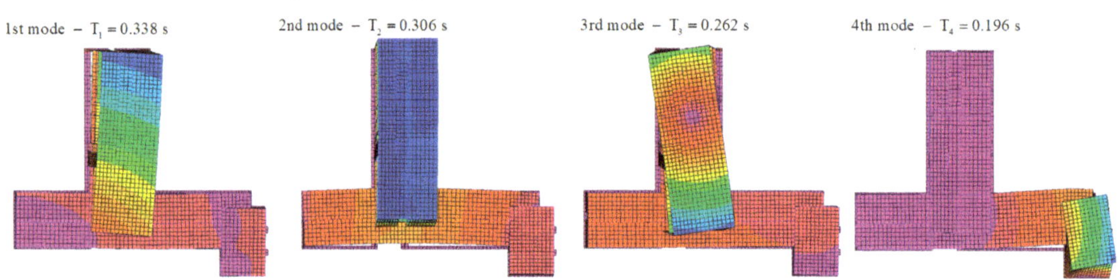

Figure 13. Eigenmodes and corresponding periods values for the as-built state of the building.

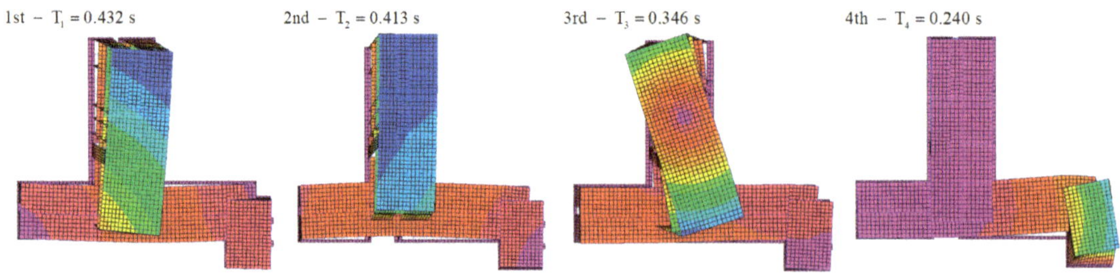

Figure 11. Eigenmodes and corresponding periods values for cracked sections of the as-built state of the building.

A pushover analysis for various horizontal loading patterns was performed and critically analyzed, but only the most important results for determining the lateral capacity of the structure are presented here. The uniform distribution (UNIF) and the distribution according to the lateral force method (LF), i.e., the triangular distribution, are considered as load patterns. Pushover curves were determined separately for each horizontal direction, considering both positive and negative loading directions, and assuming an eccentricity of ±5% with respect to the center of mass (CM) of the floor. The CM on the 2nd floor was chosen as the control point. The evolution of the damage mechanisms along the bearing capacity curve of the building was analyzed by following the distribution of the internal forces and the bearing capacity of the structural elements.

It was found that the same patterns of lateral force distribution were relevant for both load directions. The pattern of load distribution that corresponds to the vertical distribution of horizontal forces according to the lateral force method (LF) proved to be the most critical.

The most significant pushover curves for the action of the lateral force in the X-dir are shown in Figure 15. It can be seen that some elements of the structure come to a state of limited damage, which is reflected on the curve as a decrease in stiffness, i.e., the slope at a force of 1800 kN (base shear coefficient, B.S. = 5.6%) and a displacement of 7.4 mm. Then there is the initial opening of cracks and a reduction in the stiffness of individual elements. The diagram shows elements on the ground floor where cracks appear, even though they are present on all floors. The beginning of the failure of the structural elements occurs at a force of 2900 kN (B.S. = 9.1%) and at a displacement of 17.5 mm. The columns on the 1st floor reach the SD limit state due to exceeding the shear capacity. Since such damage of the columns is non-ductile, exceeding their ultimate capacity means that they soon reach the NC limit state.

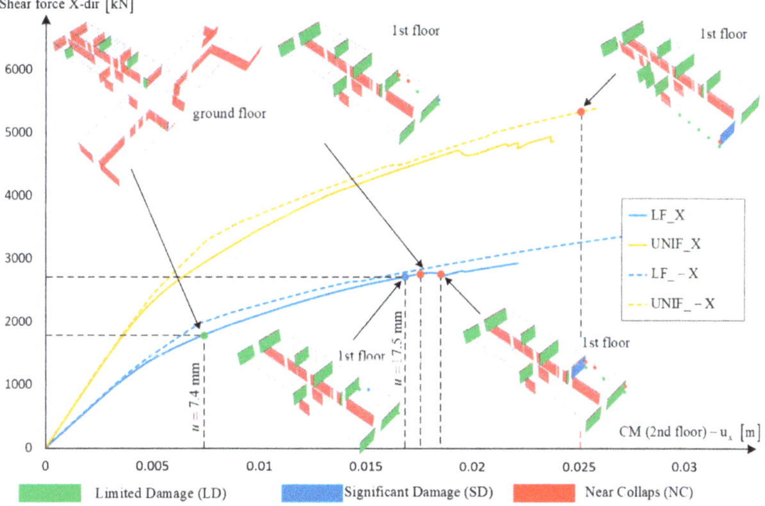

Figure 15. Pushover curves in X-dir for as-built structure.

The next important point is the failure of these columns at a lateral force of 2900 kN and for a displacement of 18 mm. The masonry wall on the 1st floor next to the critical columns is also in a SD limit state, but it still has not utilized total load-bearing capacity. At this point, it can be said that the bearing capacity of part of the building is reached and that a local failure of the building is possible. As a result, the local failure of the floors may occur. It can already be said that the building is in a NC limit state, and it can be classified as having the highest degree of damage. However, even in the case of local failure, it is unlikely that the rest of the structure will collapse at this level of loading. The walls and columns of the ground floor and the 2nd floor are not critical for action in the X-dir, since none have reached the SD limit state.

Figure 16 shows the most significant pushover curves for the action of the lateral force in the positive Y-dir with markings of the points, indicating the state of structural damage. At a force of 3900 kN (B.S. = 12.1%) and a displacement of 12 mm, limited damage occurs on several elements. Significant damage in the structural system occurs at a force value of 5500 kN (B.S. = 17%) and a displacement of 26 mm. The critical elements that first reach the SD limit state are the masonry walls in the central axis on the 1st floor. Subsequently, other walls in the same axis reach the SD limit state. Eventually, the failure of several walls occurs, and the bearing capacity of the structure decreases significantly with a maximum displacement of about 33 mm. The walls on the ground floor and the 2nd floor have not been shown to be critical to the action in the Y-dir.

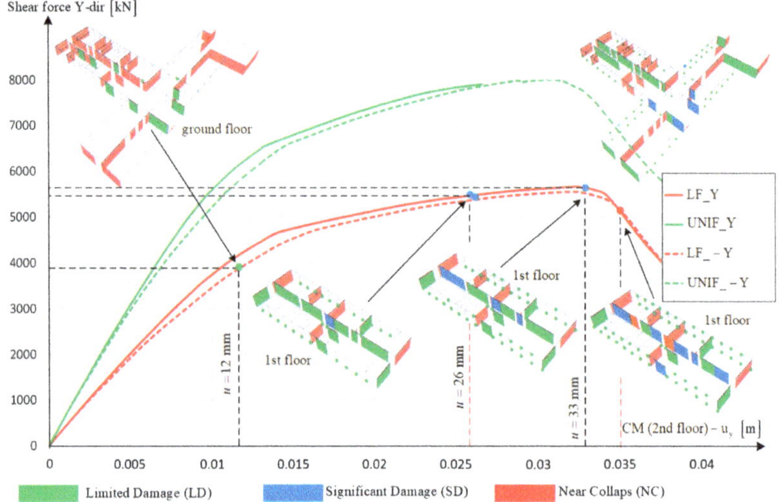

Figure 16. Pushover curves in Y-dir for as-built structure.

Figure 17 shows a diagram that cumulatively represents the number of elements that have reached the limit state of near collapse. For the action in the X-dir, the elements begin to enter this region at a displacement of about 2 cm, while in the Y-dir, this occurs at a value of about 2.5 cm.

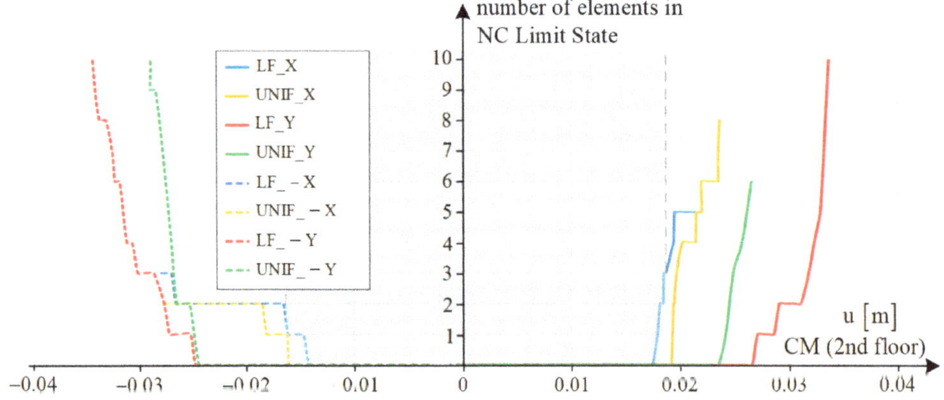

Figure 17. Cumulative number of elements in NC limit state.

Bearing capacity and building deformation requirement by reducing the system to an equivalent system with one-degree-of-freedom (1DOF) was also determined. The procedure is carried out according to the N2 method [40,41]. The Figure 18 shows the relevant idealized capacity curves correspond to ground acceleration on the bedrock of 0.08g for X-dir and LF load pattern, while for the UNIF load pattern, the value equals 0.11g. On the right, the idealized curves show the idealized capacity curve for Y-dir, which results in 0.14g and 0.15g of peak ground acceleration on bedrock for LF and UNIF patterns of lateral load, respectively. Corresponding critical value of SDI is 0.44 and 0.78, for X-dir and Y-dir, respectively.

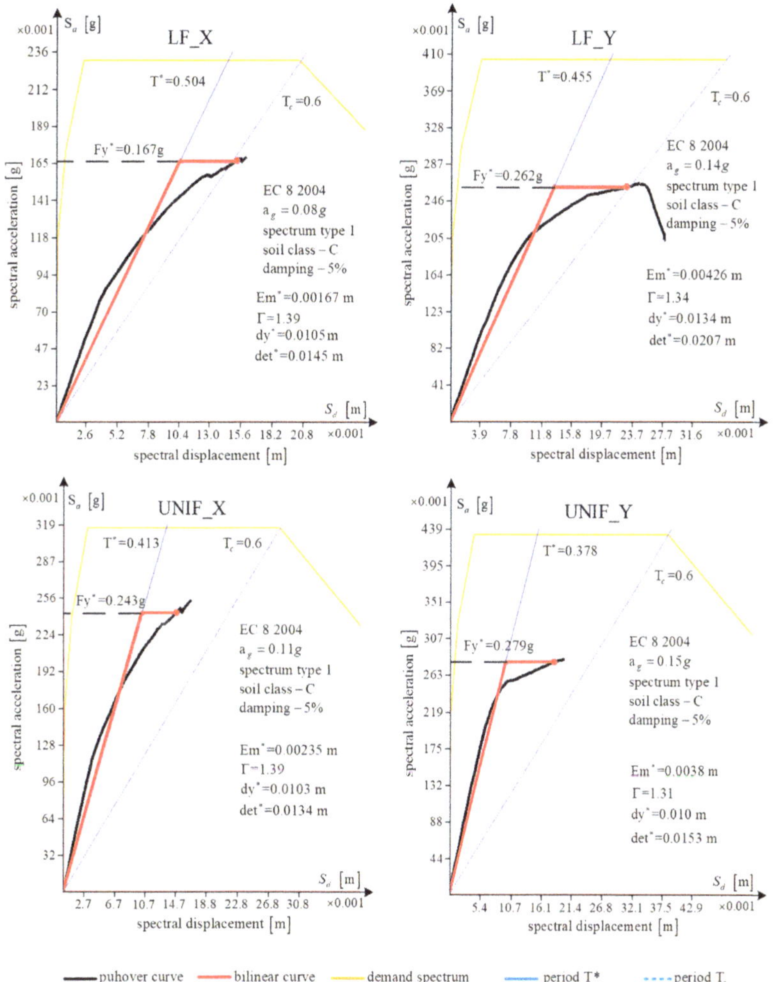

Figure 18. Capacity curves and displacement requirement of an equivalent 1DOF system.

3.4. Macroelement Model Results

The second model was created with the program 3Muri [39], as mentioned before. First, the eigenmodes and the corresponding periods are determined for the structure, taking into account the crack state of the elements (Table 3 and Figure 19). The first period of the structure is 0.45 s and corresponds to the translation in the X direction, activating 62.5% of the total mass, while the translation mode in the Y direction with a period value of 0.356 s occurs in the third mode of vibration with the participation of 63.1% of the mass, noting that in this mode the torsional deformation is also coupled with the translation Y. Another relevant vibration mode with a period of 0.388 s is the torsion resulting from the deformation of the structure by the floors. However, the program does not calculate the participation of the moment of inertia, so the contribution of the moment of inertia to the corresponding vibration mode is not known.

Table 3. Eigenmode periods obtained by EFM.

Mode	T [s]	mx [t]	Mx [%]	my [t]	My [%]
1	0.45100	1914.61	62.49	17.00	0.55
2	0.38775	65.22	2.13	488.40	15.94
3	0.35639	0.22	0.01	1934.63	63.14
4	0.23324	9.31	0.30	110.62	3.61
5	0.17896	791.46	25.83	18.67	0.61
6	0.17141	177.95	5.81	94.66	3.09
7	0.15139	4.40	0.14	303.61	9.91
8	0.14414	16.57	0.54	61.11	1.99
9	0.12400	4.74	0.15	0.01	0.00
10	0.12198	47.38	1.55	4.34	0.14

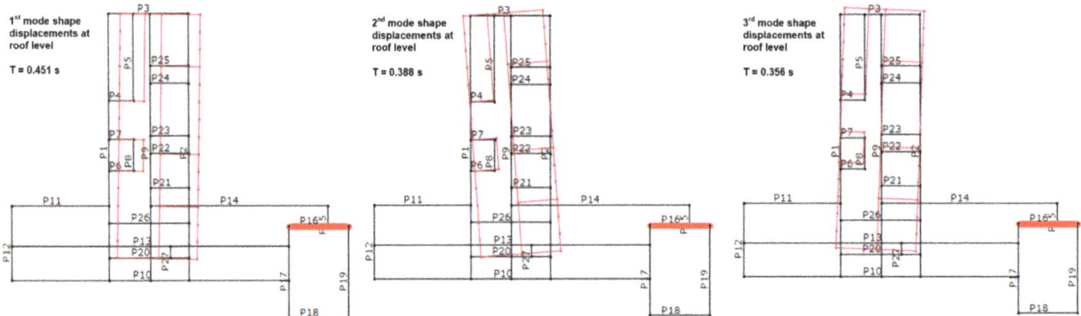

Figure 19. Relevant mode shapes obtained by EFM.

Pushover curves were obtained separately for each horizontal direction, considering both positive and negative directions of loading and the eccentricity option. The final failure mechanism of the building is presented for the relevant analysis.

In the case of uniform distribution (Uniform, Figure 20) in the X-dir, the maximum value of the lateral force ranges from 3700 kN to 4500 kN, and the relevant analysis indicates a failure mechanism caused by the shear failure of the perimeter walls on the P20 axis, and of the walls that are not continuous in height on the axes P14, P21, P23, and P24, as well as the wall of the staircase P7 and the walls on the 2nd floor on the axis P3. Damage also occurs on the elements of the edge frames, most of which is damage to beams and columns due to bending, while the shear failure of beams is visible on axes P1, P2, and P19. The critical value of SDI for the X-dir is 0.52. The pushover curves obtained in the Y direction indicate the structure has a significantly higher load capacity, which is approximately 7400 kN, and the relevant damaging mechanisms indicate that the walls of the ground floor on axis P9 and the walls of the large hall on axis P17 are particularly critical. Furthermore, damage to the walls on axis P1 and P12 due to the shear mechanism is also noticeable, and in addition to this, there is shear damage to the walls in the X-direction on axes P3 and P20. On the edge frames, the failure of the beams at the ground floor level due to shear is critical. The smallest value SDI is 0.71 and corresponds to the +Y-dir and a positive eccentricity value.

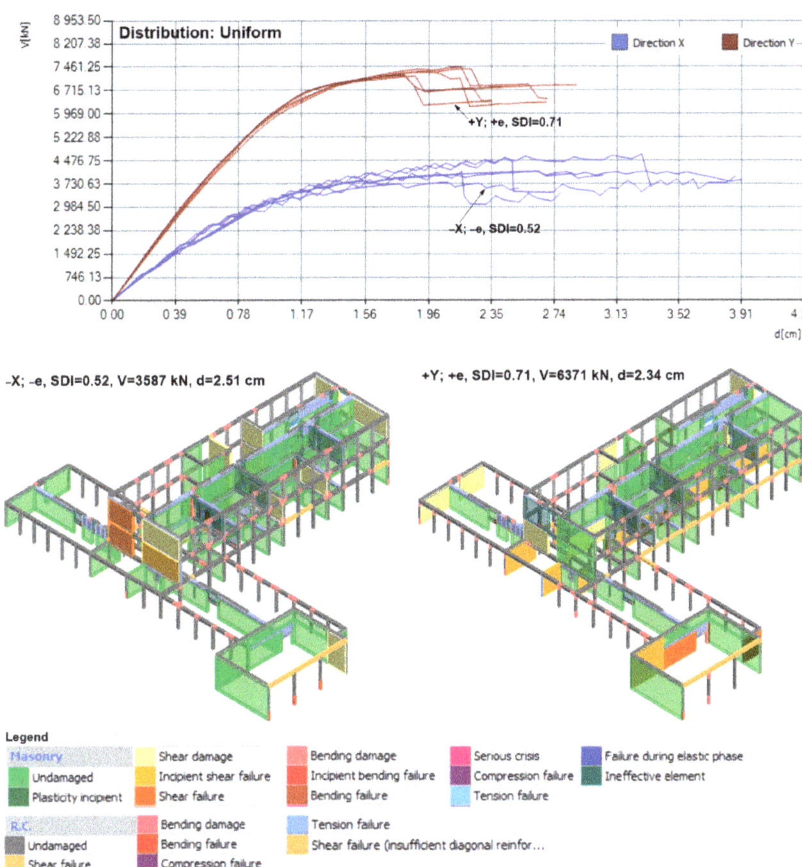

Figure 20. EFM results: pushover curves and corresponding damage mechanism of the building obtained using the uniform load pattern.

The analysis performed for the load distribution pattern according to the lateral force method (Static forces, Figure 21) proves to be less favorable for the structure than the previous case. Although the relevant structural failure mechanisms remain almost unchanged for the X-dir, the peak lateral force is around 2900 kN and the corresponding SDI is 0.38. In the Y-dir, significant damage is observed on a large number of walls on all floors, and in addition to the shear failure mechanism, there is also damage due to bending. However, the critical elements are the walls on the P9 axis, whose damage is noticeable on all floors. The main analysis performed for this direction has an SDI of 0.71.

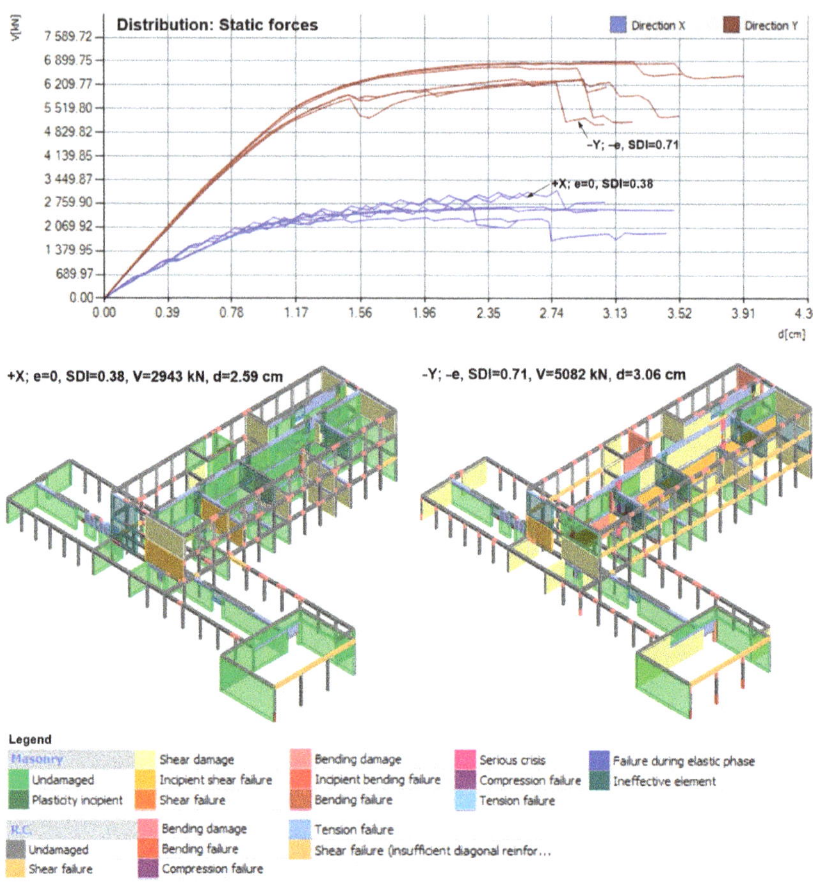

Figure 21. EFM results: pushover curves and corresponding damage mechanism of the building obtained using the static forces load patterns.

4. Structural Strengthening

4.1. An Overview of the Relevant Strengthening Techniques

Recent structural rehabilitation efforts are aimed at using modern techniques and new materials, as well as techniques that have proven to be very effective in increasing the load-bearing capacity, stiffness, and ductility of existing buildings [42,43]. The structural strengthening strategy must address the existing structural deficiencies while improving the capacities of the critical elements. Increasing the stiffness of the weakest structural element usually only leads to an increased vulnerability of the adjacent elements or structural connections, which may affect the box-like behavior of the building [44].

The dual structural system of the subject building requires consideration of strengthening techniques that include RC and masonry elements. One of the most commonly used techniques for both types of elements is concrete jacketing. This technique can be used to improve the insufficient load-bearing capacity, confinement, and reinforcement of the existing RC beams and columns, as well as the load-bearing capacity of in-plane and out-of-plane masonry walls [42].

The RC-jacketing of beams and columns has proven to be an effective strengthening technique. However, many experimental studies show that the detailing of the connections and the preparation of the surface are critical for improving the strength and ductility properties. In an experimental study [45], different methods for RC beam jacketing were investigated, considering the influence of dowel connectors, microconcrete, and bonding agents on the RC beams with smooth and chipped surfaces. The experimental results clearly showed that jacketing can improve the structural properties of RC beams, and the application of different jacketing methods was found to be more beneficial for RC beams with a chipped surface compared to beams with a smooth surface. In the experimental study [46], the comparison of performance of RC-jacketed and CFRP-strengthened RC beams showed that both techniques noticeably improved the strength and energy dissipation capacity, but the post-yield strength of RC-jacketed beams was noticeably higher than that of CFRP-strengthened beams. Furthermore, the effectiveness of details in strengthening RC columns with smooth surface with RC-jacketing is studied in [47]. In this study, it was demonstrated that the behavior of the elements can be significantly improved by strengthening, even when the jacket is constructed with no treatment at the interface. In a recent experimental study [48], the cracked RC columns were strengthened with RC jackets. Fifteen specimens of RC columns with different cross-sections were analyzed. It was found that strengthening after cracking affects the column bearing capacity, with 15.7%, 14.1%, and 13.5% lower bearing capacity for square, rectangular, and circular columns, respectively, compared to specimens that were not cracked before strengthening. The effectiveness of seismic strengthening methods for soft-story RC frames using buckling-restrained braces and concrete jacketing is investigated and compared by [49]. Both techniques were shown to be effective in mitigating an excessive soft-story response but on the basis of the conducted parametric analysis the authors noted that relative strength between the RC frame and buckling-restrained braces can significantly affect the response particularly in the short period range.

RC-jacketing is also often used to reinforce existing masonry walls. When this technique is carried out with a thin layer of cement mortar, it is called reinforced plaster [42]. The retrofitting of existing masonry wall foundations is usually required in combination with the RC-jacketing of walls. The retrofitting of masonry by RC-jacketing with shotcrete has been well studied, and we mention here some relevant research papers on this strengthening technique. For the experimental tests in [50], three half-scale walls were built using half-scale brick masonry units and weak mortar, one as a reference specimen, one with 40 mm shotcrete on one side, and two with 20 mm shotcrete on both sides. The tests showed that retrofitting with shotcrete increase the lateral strength of the specimens by factor of approximately three and that the specimens with shotcrete on both sides exhibited more ductile failure and better energy dissipation. The effectiveness of a one-sided shotcrete layer for reinforcing URM walls considering the height-to-length ratio of the walls and the efficiency of the connection of the shotcrete reinforcement layer to the foundation RC was studied by [51]. The study showed that the shotcrete layer increased the strength of the reinforced short wall, which exhibited rocking failure mode, while for the reinforced long walls, the shear sliding capacity was increased and the rocking capacity was not increased or increased only slightly, changing the failure mode from shear sliding to rocking. Therefore, shotcrete strengthening can improve the energy dissipation of a reinforced wall of short length due to yielding and fracture of the steel bars anchored in the foundation. The energy dissipation of reinforced long walls was lower than that of the reference wall because its failure mode changed. However, the anchorage system of strengthened long length walls delayed the occurrence of rocking and the energy dissipation was less than the specimen without anchorage system. In the study [52], a series of shake table tests carried out on a half-scale single-story unreinforced masonry building with asymmetric openings, first on an unretrofitted building and then on a building rehabilitated by using steel mesh and shotcrete. Three cases of interior-to-interior, interior-to-exterior, and exterior-to-exterior shotcrete connections are considered at the intersections of perpendicular walls. The rehabilitation method enhanced the overall strength and integrity of the specimen. The shotcrete layers covered the previously damaged areas and postponed the collapse of the specimen to higher excitation levels. The results showed that the fixity of shotcrete vertical rebars to the foundation played a crucial part in the deformation of the specimen.

Another very effective seismic retrofit technique is the installation of new RC shear walls. It is expected that the addition of new structural elements will change the dynamic properties and seismic response due to the increased stiffness and seismic forces [42]. However, this technique can also be used to address some other structural deficiencies, such as reduce the effects of plan and vertical irregularities and improve torsional response. It is important to emphasize that new RC foundations are required to support the added RC shear wall and, where possible, integrate them with the foundations of the existing structure. Ensuring adequate connections between existing RC beams or floor slabs is crucial for the transfer of seismic forces. This is usually accomplished by vertical reinforcement passing through drilled holes in the existing beams and slabs or by steel anchors embedded in the existing elements. In addition, new shear walls can be integrated with the existing RC columns by also providing adequate connections. The seismic behavior of mixed RC–URM wall structures has been studied experimentally and additionally modelled with shell elements and macroelements to investigate appropriate numerical models [53]. The study showed that the advantages of this retrofit technique are an increase in strength capacity and a change in deformed shape. The latter provides the combined contribution of existing URM walls and new RC walls failure mechanisms with larger top displacements for the same level of inter-story drift at the ground floor. As a consequence, for such mixed structures the damage in the URM walls is not concentrated on the first story—as for URM buildings—but it spreads to the stories above.

All of the above mentioned techniques are quite invasive and require considerable time for installation. In contrast, seismic retrofitting with fiber-reinforced polymers (FRP) and textile-reinforced mortars (TRM) offers a relatively rapid strengthening technique that provides satisfactory levels of increase in ductility and flexural and/or shear capacity for both RC and masonry elements. Today, there are various composites that differ in the fiber material (carbon, glass, aramid, basalt) and bonding agent (organic or inorganic). In this case study, the focus is on the TRM system, also known as fabric-reinforced cementitious matrix (FRCM), which consists of a fiber grid embedded in the inorganic matrix. The full composite action of the TRM material is achieved by the mechanical interlocking of the grid structure and the mortar protruding from the openings of the grid [54]. The advantages of using the TRM system are especially recommended for heritage masonry buildings, since it is a reversible method of strengthening. In the case of brick masonry wallets subjected to out-of-plane cyclic bending, as reported in [55], TRM overlays outperform their FRP counterparts on the basis of maximum load and displacement at failure when failure is controlled by damage to the masonry, whereas the effectiveness of TRM over FRP decreases slightly when the failure mechanism involves the tensile failure of the textile reinforcement. In another experimental study [54], the specimens were subjected to cyclic loading, causing in-plane bending combined with axial force, out-of-plane bending, and in-plane shear with an axial force. For out-of-plane loading, similar results were obtained, while for in-plane loading, strengthening with TRM resulted in lower effectiveness in strength (but not more than 30%) compared to the FRP strengthening technique. However, in terms of deformation capacity, overlays with TRM was found to be more effective than FRP, with an increased effectiveness of about 15–30% for shear walls. It is also reported that strength generally increases with the number of layers and axial load at the expense of deformation capacity. Furthermore, the strengthening of concrete elements with TRM has also proven to be an efficient technique to increase the ultimate flexural or shear capacity of RC elements with typical geometries. TRM increases their stiffness and thus their performance under service loads. Moreover, cracking is better controlled [56]. It is worth mentioning that experimental results have shown that TRM have a much better effectiveness than FRP in increasing the flexural capacity of RC beams subjected to high temperatures [57]. Further experimental studies on TRM composites are ongoing in order to perceive all benefits and drawbacks as well as in the purpose of design guidelines development

4.2. Proposed Retrofitting of Existing Building

The concept of strengthening the case study building is designed to meet modern seismic standards. In addition to increasing the ductility and load-bearing capacity of the overall structure, the deficiencies in load-bearing capacity due to permanent vertical loading identified during the structural analysis are also addressed. In addition, it should be emphasized that the building underwent energy renovation in 2018, so the strengthening measures should interfere as little as possible with the exterior façades, which are mainly integrated into the edge frame system. Therefore, it is important to prevent, as much as possible, the non-ductile failure of frame columns because they do not have a sufficient transverse rebars. Another critical aspect of the structure are unreinforced masonry walls, which have low in-plane bearing capacity. Due to the presence of RC slabs, the out-of-plane failure of these walls is not considered to be critical. Furthermore, the partial discontinuity of the floor structure at the junction of volumes V1 and V3 is certainly a critical location where major damage could occur in case of a future stronger earthquake. Finally, some interventions are planned in order to improve the functionality of the building.

The proposed structural strengthening measures are shown in Figure 22. The most significant retrofitting intervention concerns the addition of new RC walls. Along the staircase masonry walls, new RC walls with a thickness of 15 cm are added. On the ground floor, RC walls with a thickness of 15 cm are added next to each brick wall face. In fact, this axis is significantly strengthened because it is a position of eccentricity of the columns on the floor above. Next, new 20 cm thick RC walls are added over the entire height in the building volume V3. These include perimeter walls next to the existing brick walls and walls inside the structure at the positions of the partition walls. In addition to providing additional functionality to the building, it is planned to install a new elevator RC core inside the building. To further connect these walls to the RC slab, new RC beams must be added as the new opening will weaken it. Other retrofit measures include the RC-jacketing of the columns and beams in the ground floor entrance hall and the RC-jacketing of the central masonry wall along the entire height of volume V3, and to the masonry walls of the large hall (volume V2). In order to avoid the formation of other weak elements, FRCM overlay is provided for other masonry walls that were not found to be critical in the analysis of the seismic performance of existing structure.

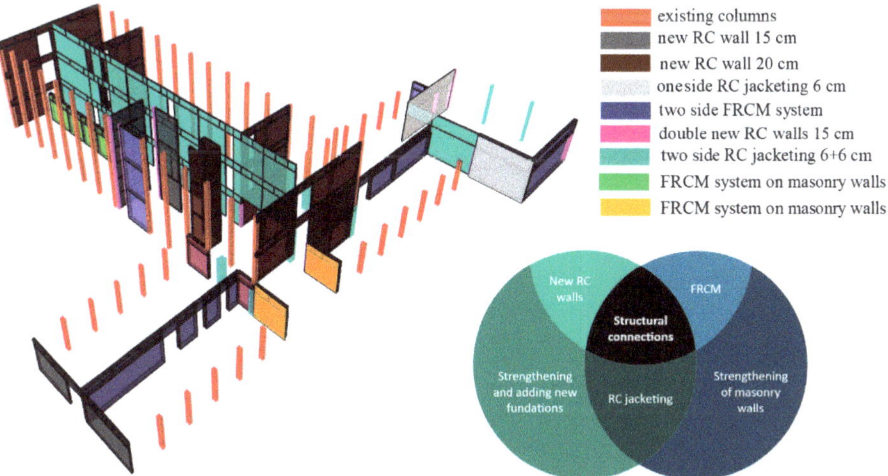

Figure 22. Scheme of strengthening for vertical elements.

It is important to emphasize that the connections with the existing structural elements must be ensured and the continuation of longitudinal rebars through the existing RC slab for all new shear walls and an RC-jacketing system. In addition, an adequate foundation must be ensured for new shear walls as well as for the RC-jacketing of vertical elements (columns and masonry walls).

In the numerical models of the case study building, the specified reinforcement techniques were considered in such a way that the new RC walls are assumed to fully support the shear force, and only the weight contribution is considered for the adjacent brick walls.

In addition, for the elements reinforced with the RC-jacketing system and the FRCM system, the joint response of the existing elements with reinforcements is considered. For new materials, the concrete quality C30/37 is assigned and B500B steel class is used for the rebars and Q503 reinforcement mesh. RC-jacketing is considered, assuming composite section and the substitute stiffness and mass as if a layer of concrete had been placed on the brick wall. Plasticization and ductility are assumed according to the assumed reinforcement in the concrete layer. It is also assumed that the cross-section is compact, which is ensured by the fact that the two layers of are adequately connected to each other ensuring equal displacements. For the FRCM composite with AR glass fibers on one or both sides of the masonry wall in one layer, this reinforcement is taken according to the Italian code [39].

4.2.1. Finite Element Model Results

Due to the presence of new and strengthened elements the dynamic properties of the building have changed, i.e., the period values have reduced notably which is the result of significant increase of structural stiffness manly due to addition of new RC walls. The Figure 23 shows relevant modes, and it is noticeable that coupling of mode shapes is evident what can be even more emphasized since relevant mode shapes all have near values of periods.

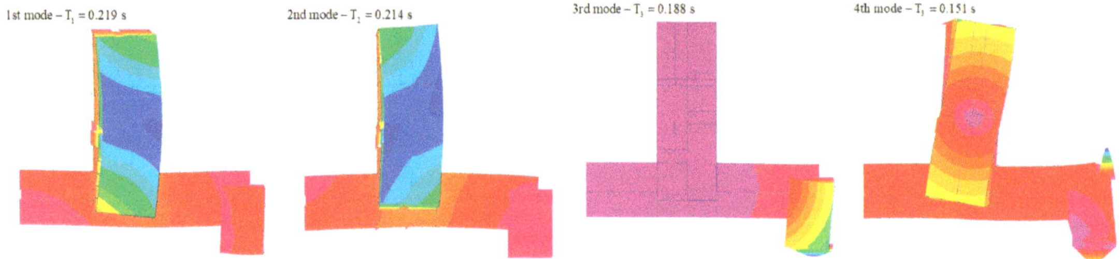

Figure 23. Eigenmodes and corresponding periods values for cracked sections of the retrofitted building.

The main results of the pushover analysis for strengthened structure obtained by modelling with ETABS are given below. The limit states of the critical elements and the development of the failure mechanisms are marked on the curves. In Figure 24, it can be seen that some elements of the structure reach a state of limited damage, which is reflected on the curve as a decrease in stiffness, i.e., the slope at a force of 11,000 kN (B.S. = 32%) and a displacement of 9 mm. In addition, there is the initial opening of cracks and a reduction in the stiffness of the individual elements in all floors. These elements still did not utilize bearing capacity and continue to deform. The beginning of the failure of the structural elements occurs at a force of 15,000 kN (B.S. = 44%) and a displacement of 40 mm. The columns on the ground floor on the north side reach the SD limit state due to the exceeding of the shear force capacity. Since such a fracture of the columns is non-ductile, exceeding their ultimate capacity means failure very soon. The walls of the large hall on the ground floor are also in an SD limit state. The next important stage is the failure of these columns (NC limit state) at a force of 16,000 kN and a displacement of 50 mm. At this stage, it can be said that the load-bearing capacity of this part of the building is utilized and a partial failure of the building takes place. As a result, the local failure of the floor slab may occur. It can already be said that the building is in a NC limit state and can be classified as having the highest degree of damage. However, even in the case of local failure, it is unlikely that the rest of the structure will collapse at this load level. The walls and columns of the 1st and 2nd floors are not critical for action in the X-dir and have not reached the SD limit state.

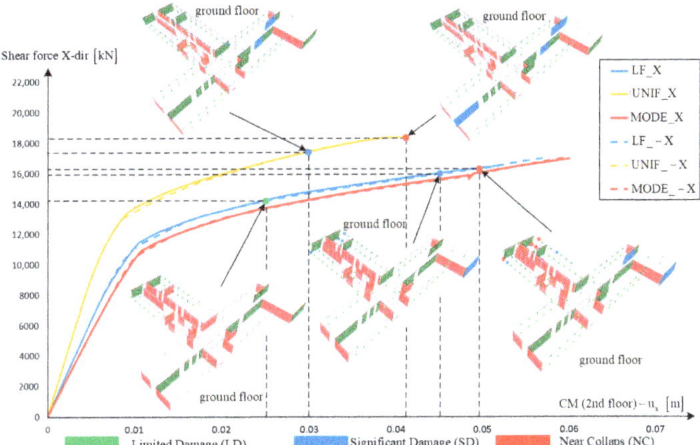

Figure 24. Pushover curves in X-dir for retrofitted structure.

Figure 25 shows the relevant pushover curves for the Y-dir. Individual elements reach a state of limited damage at a force value of 3900 kN (B.S. = 12%) and a displacement of 5 mm. The diagram shows the elements on the ground floor where the cracks appear first, but the limited damage of elements is present on all floors.

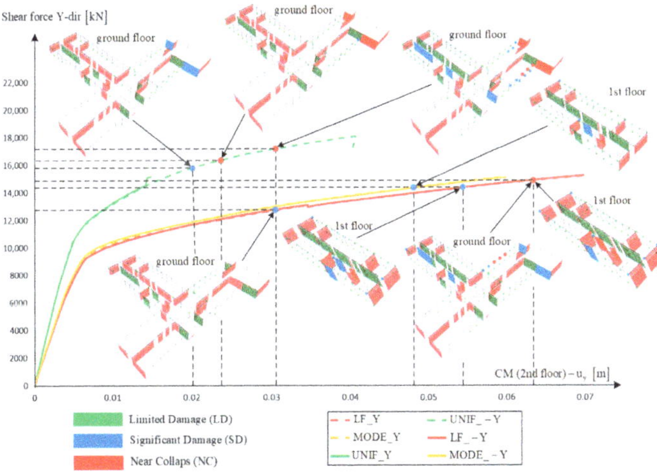

Figure 25. Pushover curves in Y-dir for retrofitted structure.

The beginning of the failure of the structural elements occurs at a force of 14,000 kN (B.S. = 41%) and a displacement of 50 mm. The critical elements that first reach the SD limit state are the walls next to the staircase on the 1st floor, followed by the exceeding of the SD limit state for the adjacent walls and for the walls of the large hall on the ground floor. The final result is the failure of these walls and the drop of the bearing capacity in the pushover curve. However, it should be mentioned that the walls on the ground and on the 2nd floor were not found to be critical in the Y-dir for the lateral force.

The diagrams that cumulatively show the number of elements that have reached the SD and NC limit state for different lateral load patterns are displayed in the Figure 26.

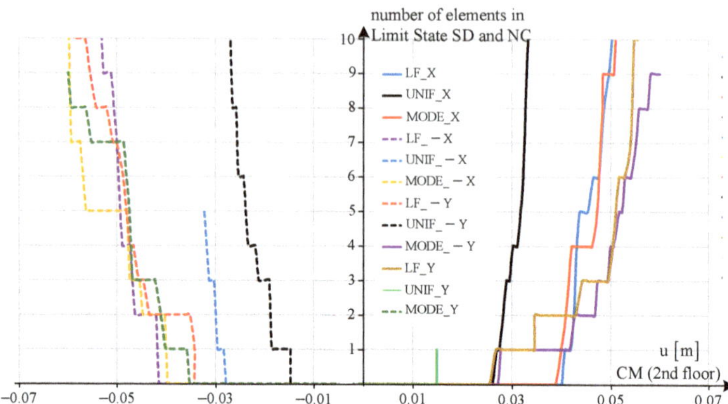

Figure 26. Cumulative number of elements in SD and NC limit state.

Figure 27 shows the relevant idealized capacity curves of the equivalent system with 1DOF. The criterion is evaluated here using the spectra of type 1 and type 2 [35]. The results show that the building meets the criterion of maximum ground acceleration in both directions, according to the current regulation with an SDI of 1.06, which is higher than the minimum level 3 requirements according to CTRBS.

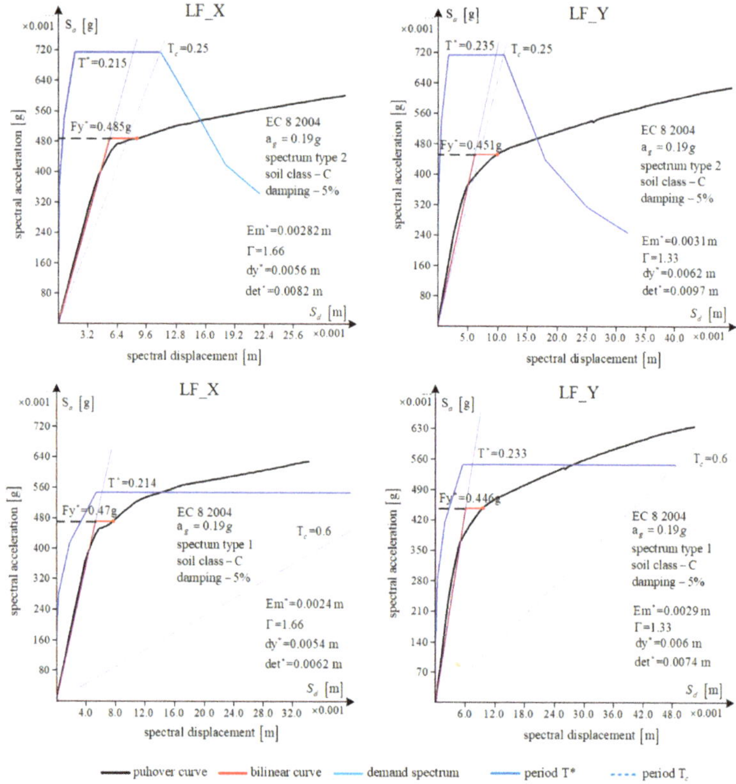

Figure 27. Capacity curves and displacement requirement of an equivalent 1DOF system.

4.2.2. Macroelement Model Results

The results of the EFM with the program 3Muri also demonstrate that the reinforcements caused a significant change in the dynamic properties of the structure, which is a consequence of the stiffening of the structural system (Table 4). In this model for the state of cracked sections (Figure 28), the first mode of vibration with a period of 0.204 s is a local one and affects the deformation of the elements of the large hall with a participation of 2.1% of the mass in the Y-dir and a visible torsional influence. The translation mode in the X-dir appears in the second period with a value of 0.181 s and a mass participation of 70.2%, while the translation in the Y-dir direction maintains the influence of torsion and has a period value of 0.16 s with a mass participation of 58.5%. In the fourth oscillation mode, the torsion is noticeable, and the value of the period is 0.145 s.

Table 4. Eigenmode periods obtained by EFM for the retrofitted structure.

Mode	T [s]	mx [t]	Mx [%]	my [t]	My [%]
1	0.20220	21.99	0.65	69.70	2.07
2	0.18130	2362.83	70.17	7.99	0.24
3	0.16048	29.90	0.89	1969.90	58.50
4	0.14534	30.84	0.92	515.58	15.31
5	0.11175	6.13	0.18	16.11	0.48
6	0.09405	38.51	1.14	260.12	7.73
7	0.07867	676.09	20.08	13.04	0.39
8	0.07422	8.80	0.26	0.02	0.00
9	0.07357	0.19	0.01	0.01	0.00
10	0.07278	0.08	0.00	0.01	0.00

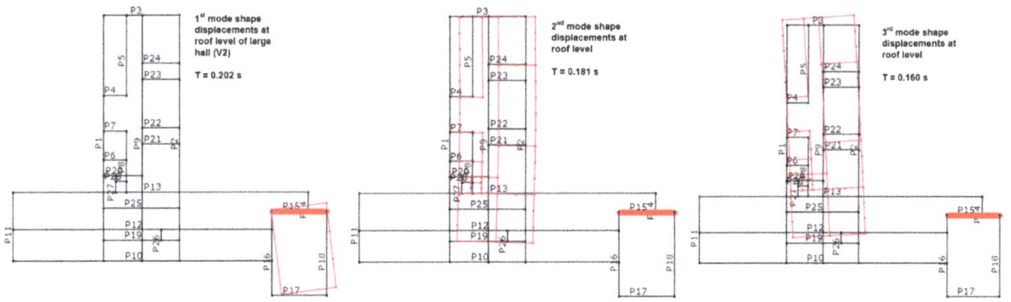

Figure 28. Relevant mode shapes obtained by EFM for the retrofitted structure.

The results of the numerical model of the retrofitted structure show a significant improvement in the seismic performance. In the case of uniform distribution of the lateral force (Uniform, Figure 29), a significant increase in the load capacity in both directions is evident. The least favorable analysis for the X-dir of the lateral force gives an SDI of 1.72, which has a maximum load capacity of approximately 18,700 kN. This is almost four times the load capacity compared to the existing structure and the same lateral load pattern. Significant damage first occurs in the RC elements, walls on axes P7, P13, and P19 on the ground floor and then P24. On the edge frames, significant damage is mainly related to the beams on axes P2 and P13 on the ground floor. Furthermore, there is damage to the walls on the P12 axis, which are reinforced with the FRCM system, and the wall on the P15 axis, which is reinforced with RC-jacketing. Again, it is important to note that the program for RC elements uses ductile bending mechanism and fragile for the shear mechanism. This is the reason that the damaging mechanism differs in comparison to the finite elements model created in ETABS. In the analysis carried out for the Y-dir, it can be noted that the damage is mainly related to the elements on the ground floor of the building. The corresponding calculations show that the first significant damage occurs to the new RC walls on the P1 axis, followed by the damage to the walls on the P8 axis, i.e., the walls of the staircase and the elevator core. These elements are the first to be damaged because they have high stiffness and a large part of the lateral force is supported by these elements. However, ductility is ensured by the walls on the P9 axis, which are strengthened with RC-jacketing, and to which a lateral force is redistributed after the failure of the RC elements. A significant drop in the bearing capacity is observed at a value of the lateral force of about 20,900 kN and a displacement of 1.27 cm, when the edge walls on axis P11 are significantly damaged, and damage also occurs in the walls of the large hall on axis P16. The most unfavorable SDI in the Y-dir is 2.32.

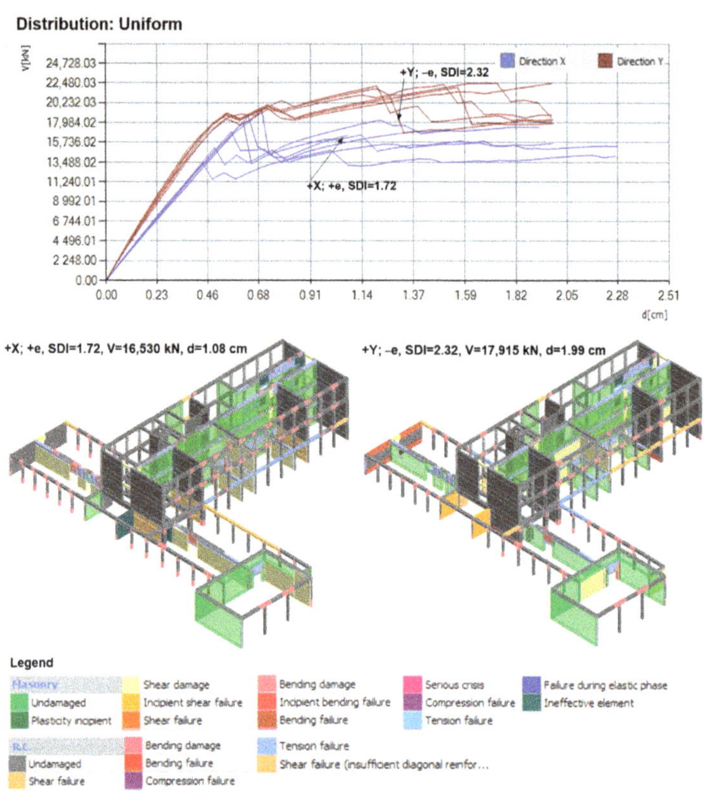

Figure 29. EFM results: pushover curves and corresponding damage mechanism of the retrofitted structure obtained using the uniform load pattern.

The pattern of load distribution by the lateral force method (Static forces, Figure 30) results in somewhat less favorable safety index values compared to a uniform distribution pattern. The most significant analysis for the X-dir results in an SDI of 1.37. The first significant damage occurs to the 1st floor RC wall on axis P13 and continues with the failure of the ground floor walls on axes P13 and P19. Additionally, in this case, the shear force is redistributed on the walls reinforced with the FRCM on axes P12 and P15, reaching a maximum lateral force value of about 17,400 kN with a displacement of 1.01 cm. The edge frames are mostly damaged in the end nodes due to bending and there are also beams with shear failure. These are mainly beams connected with new and reinforced rigid elements. The corresponding analysis for the Y-dir shows almost the same failure mechanism as in the previous case, starting with the failure of the RC walls on the P1 axis, followed by the walls on the P8 and P11 axes. Ductility is maintained by the walls on the P9 axis, which are reinforced with the RC-jacketing. The minimum value of the SDI is 2.06, which is the lowest value of all the analyzes performed with macroelements of the reinforced structural model in the Y-dir. The final utilization of the load-bearing capacity of the structure occurs at a force of approximately 17,500 kN and a displacement of 2.23 cm.

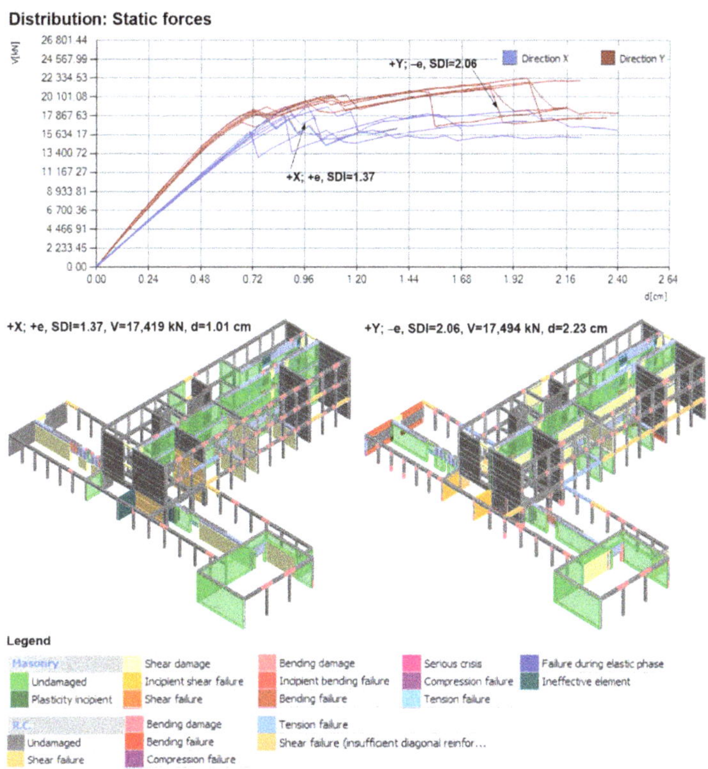

Figure 30. EFM results: pushover curves and corresponding damage mechanism of the retrofitted structure obtained using the static forces load pattern.

The load pattern corresponding to the dominant mode of vibration (modal distribution, Figure 31) for the strengthened structure was also found to be critical for the X-dir of lateral force. However, the pushover curves show that the structure responds with a low degree of ductility and a high level of lateral force capacity. The lowest value of the SDI is 1.24. The damage mechanism of the structure was caused by a relatively early phase failure of the walls on the P3 and P24 axes. This failure mechanism is partly the result of a few load-bearing elements compared to the front part of the structure. However, the seismic safety condition is fulfilled and the lateral force carrying capacity exceeds 16,000 kN in all analyses. Moreover, the structure maintains greater ductility in the Y-dir for the distribution according to the dominant mode, which is visible in all pushover curves. The most informative analysis has an SDI of 2.12. The failure mechanism develops due to the shear failure of the RC walls on the P1 axis at the ground floor, followed by the walls on the P8 axis. In addition, the force is distributed to other walls on the P9 axis. Furthermore, torsional effects are visible in case, since the walls on the P3, P12, P13, and P19 axes are also damaged at the ground floor level. The collapse of the structure was caused by the failure of several walls reinforced with RC-jacketing on axis P9 on the ground floor.

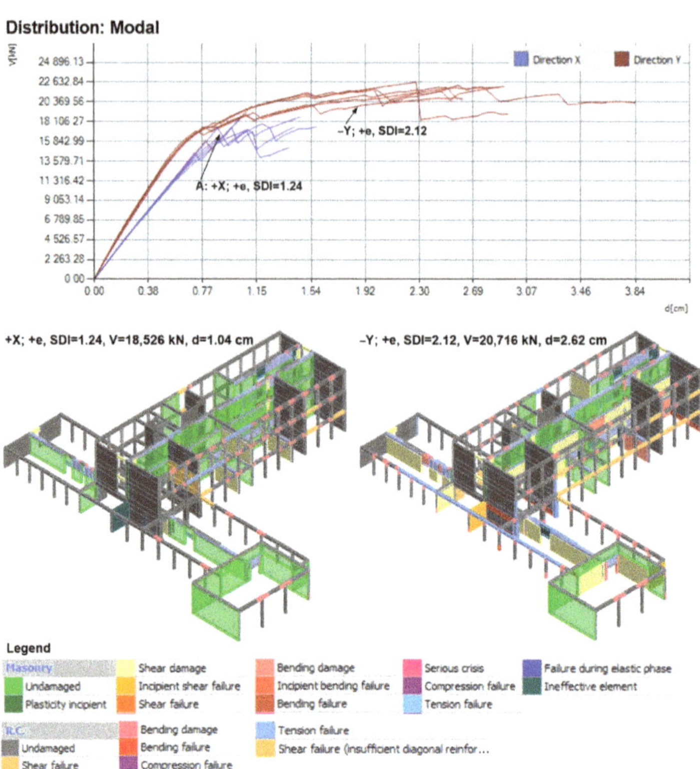

Figure 31. EFM results: pushover curves and corresponding damage mechanism of the retrofitted structure obtained using the modal load pattern.

5. Conclusions and Discussion

The recent earthquakes in Croatia have led to a series of activities and adaptations of technical regulations aimed at creating a legal framework for the assessment of seismic safety and the level of the rehabilitation of damaged buildings. Reconstruction priorities are certainly housing and critical infrastructure, but other sectors are also crucial for the functioning of society, including health, education, the productive sector, and others. It is extremely important to prevent the displacement of people from the affected area, and for this it is necessary to ensure the functioning of the social community.

The case study building is one of the few higher educational institutions in Sisak-Moslavina County. This paper describes the current condition of the building and its seismic performance of its as-built state. A concept for strengthening the structure was proposed and the seismic performance additionally analyzed. In the numerical analysis, two modeling approaches are employed.

In general, the numerical models showed that the stiffness of the retrofitted structure increased significantly, primarily due to the introduction of new elements but also due to the strengthening of the existing elements. This is evident when comparing the periods of the models of the existing and the retrofitted structure but also when analyzing the slope of the initial part of the pushover curves. The load-bearing capacity of the building has also increased significantly, since the strengthened elements have a higher peak resistance force. The good positioning of the new reinforced concrete elements prevents the local failure of the brittle elements. The ductility of the strengthened system has obviously increased according to the pushover curves, but it should be emphasized that the brittle failure of the columns occurs at the same displacement value as before strengthening. Crucially, the strengthening has significantly reduced the demand displacement. The level of demand displacement is such that there is no failure of the non-ductile elements in the structure.

By using finite element modelling for the as-built state of the building, it can be said that the maximum ground acceleration on the bedrock at which the displacement requirement is satisfied is approximately 0.08g for the X-dir and 0.14 g for the Y-dir. The X-dir proved to be critical (SDI = 0.44), since the dominant bearing system in that direction is frame consisting of columns with insufficient transverse reinforcement that cannot achieve the required ductility. The Y-dir has a higher seismic resistance level (SDI = 0.78), and it meets the requirement for Level 3 according to CTRBS. The model of the retrofitted building showed that the requirement of the modern building code can be achieved. In addition, the building has significant reserve capacity at the prescribed load, and some of the elements are in a state of limited damage, which is to be expected at such an intensity of seismic action. The strengthening made the building relatively stiff, and the displacements of the floors were significantly reduced compared to the existing condition. As a result, most of the columns did not utilize their ductility capacity and remained in a state of limited damage even though they were not directly strengthened.

In another model, by using macroelements damaging mechanism for the as-built state of the building, the results proved to be rather similar. For the same load pattern that corresponds to the distribution according to lateral force method, the maximum ground acceleration is 0.07g (SDI = 0.38) and 0.13g (SDI = 0.71) for X-dir and Y-dir, respectively. For the retrofitted structure, the relevant damaging mechanisms show that in the X-dir the system has less ductility, which is a consequence of the more RC walls. Given that the strengthening in the Y-dir is mainly provided by reinforcing the walls with the RC-jacketing of the masonry walls on the P9 axis, a more ductile behavior of the critical walls is expected with a significant increase in their bearing capacity. It is important to note that in the model of the retrofitted structure, the difference in the bearing capacity of the structure with regard to the X and Y-dir is also significantly reduced. The damage is mainly concentrated on the elements on the ground floor, where the value of the lateral force is the highest, and therefore it is important that all load-bearing elements have certain strengthening measures to ensure the stability of the structure. Additionally, it is important to emphasize that the program has a limitation because it does not assume the ductile behavior of the new RC walls, i.e., the bilinear load-bearing diagram for shear mechanism, which leads these elements to brittle failure.

The eigenperiods and mode shapes do not completely match in the two program packages. There are a number of reasons for these discrepancies, but the most important reason is that they are different methods of numerical calculation. Each pier or spandrel modeled as a finite element has its own discretization and six degrees of freedom per node, while the tree-dimensional rigid nodes of equivalent fame has a total of five degrees of freedom, which are a collection of two-dimensional rigid nodes identified in each of the incident walls, and the nodes belonging to only one wall remain two-dimensional and retain only three degrees of freedom, instead of five. The influence of the out-of-plane stiffness of the elements is present in the finite element case but has been kept to a minimum in order to make the calculation conservative. The deformation of the elements is different in both software packages, which is a consequence of the discretization and formulation of the element. A similar situation is observed when a wall is modeled as an equivalent frame element in a finite element model. Equivalent frame has rigid zones at the junctions of piers and spandrels, which is also an assumption. At the slightest irregular geometry and irregular arrangement of elements in the joint, these influences multiply and discrepancies occur. Nevertheless, it can be said that these differences in the given results are minimal, since in both models it has been shown that the 2nd and 3rd modes are quite close. During an earthquake, the cross sections crack unevenly and the stiffness decreases globally and locally within the element. Therefore, the coupling of the 2nd and 3rd modes is to be expected, even if they do not completely coincide in the different numerical programs.

In the analysis of the as-built state of the structure, both models had very similar results in terms of the significant damage index. However, the damage mechanisms did not completely agree, although they were similar in certain parts of the structure. The difference in damage mechanisms was even more evident in the models for the retrofitted structure. In FEM model, the utilization of the load-bearing capacity of the structure is caused by the failure of the strengthened walls at the ground floor and the capacity utilization of the existing columns, while only a few new RC walls reached the state of limited damage. In the second EFM model, the brittle failure of the new RC walls was critical, but the additional ductility of the structure was provided by redistributing the lateral force to other reinforced elements. However, it is worth mentioning that the 3Muri program is primarily intended for the analysis of masonry structures. Various numerical models and methods are a necessary tool to evaluate the seismic performance of buildings. Therefore, engineers should critically consider the assumptions used in certain models and the results of the analysis to evaluate the applicability of certain modeling approaches.

In addition to the aforementioned discrepancies in the use of different numerical models, this paper provides a methodology for evaluating the seismic performance and retrofit strategy of a dual structural system consisting of RC frames and masonry walls. Taking into account the previously mentioned peculiarities, mainly related to the geometric irregularity, the retrofit strategy aims to reduce the interventions in the original frame system, which is the critical structural part of the building due to the expected brittle failure of the RC elements. One of the solutions presented here is to systematically increase the stiffness of the structure so that the ductility of the existing elements is sufficient. This would allow the façade to be mostly preserved. This solution reduces the previously calculated displacements at key points of the structure. The arrangement of the new reinforcement elements is also important and results from the analysis of the failure mechanism. Another important condition is the ability of the diaphragm to efficiently transfer forces to the new elements, so their arrangement should correspond to the stiffness of the diaphragm. This approach leads to the use of different reinforcement systems for certain parts of the building. Thus, in the ground-floor area, the strengthening was carried out with the addition of new walls, RC-jacketing, and the FRCM system, while on the upper floors, new RC walls and RC-jacketing techniques were mainly used in key positions. It is also possible to carry out the solution in steel (bracing, the steel jacketing of RC elements), but it is important to arrange them in such a way that there is no critical displacement of the frame system. The result of this approach is an optimal retrofit solution that provides satisfactory safety and requires a minimum of construction work.

Finally, it can be said that the problem of retrofitting key infrastructure buildings, such as hospitals and schools, is particularly pronounced in earthquake-prone areas. Many of these structures were built after World War 2, when the use of concrete was significant, but the knowledge of seismic design and the ductility of reinforced concrete elements was not yet so well known. Therefore, when retrofitting such buildings, the problem of the low ductility of most load-bearing elements should be solved systematically. Increasing the ductility of columns and beams individually incurs high costs, and retrofitting may be unprofitable. Moreover, such buildings have often been renovated for energy efficiency reasons, which usually means that the façade has been renovated and its removal would cause additional high costs. The analysis of the seismic performance of such buildings should be nonlinear in order to determine the actual load-bearing capacity of the building and the failure mechanisms that can occur during an earthquake. Due to the importance of the building and the sensitivity of the failure mechanism to the initial conditions, it is recommended that the analysis be performed in two different software packages.

The analysis of irregular structures with different structural systems requires a comprehensive approach, since the numerical assessment of seismic performance is not straightforward and requires a high level of engineering judgment. The response of the structure should be monitored and critically considered in order to ensure that the load-bearing elements perform optimally during an earthquake. Further investigations concerning this type of structural system should focus on targeted retrofitting strategies and aim to minimize the cost of retrofitting or consider some constraints, such as the use of certain materials or the preservation of the originality of certain parts of the building—a factor that is commonly required for heritage buildings.

Author Contributions: Conceptualization, M.U. and M.D.; methodology, M.U.; software, M.U. and M.D.; validation, M.U., M.D. and M.B.; formal analysis, M.U., M.D. and M.B.; investigation, M.U., M.D., M.B. and A.P.; resources, M.U., M.D. and M.B.; data curation, M.U. and M.D.; writing—original draft preparation, M.D.; writing—review and editing, M.U. and M.B.; visualization, M.U., M.D., M.B. and A.P.; supervision, M.U.; project administration, M.U.; funding acquisition, M.U. All authors have read and agreed to the published version of the manuscript.

Funding: This research was funded by Croatian Science Foundation, grant number UIP-2020-02-1128 (2BeSafe project—New vulnerability models of typical buildings in urban areas: Applications to seismic risk assessment and target retrofitting methodology).

Institutional Review Board Statement: Not applicable.

Informed Consent Statement: Not applicable.

Data Availability Statement: The data presented in this study are available on request from the corresponding author. The data are not publicly available due to privacy reasons.

Acknowledgments: The authors acknowledge the assistance of the administration of the Faculty of Teacher Education Department in Petrinja for providing the archival documentation of the building. We also thank Joško Krolo for the photos of the building damage and the data of the experimental tests and Karlo Jandrić for the photos of the building damage.

Conflicts of Interest: The authors declare no conflict of interest.

References

1. Novak, M.Š.; Uroš, M.; Atalić, J.; Herak, M.; Demšić, M.; Baniček, M.; Lazarević, D.; Bijelić, N.; Crnogorac, M.; Todorić, M. Zagreb Earthquake of 22 March 2020—Preliminary Report on Seismologic Aspects and Damage to Buildings. *Gradjevinar* **2020**, *72*, 843–867. [CrossRef]
2. Atalić, J.; Uroš, M.; Šavor Novak, M.; Demšić, M.; Nastev, M. The Mw5.4 Zagreb (Croatia) Earthquake of March 22, 2020: Impacts and Response. *Bull. Earthq. Eng.* **2021**, *19*, 3461–3489. [CrossRef]
3. Atalić, J.; Demšić, M.; Baniček, M.; Uroš, M.; Dasović, I.; Prevolnik, S.; Kadić, A.; Šavor Novak, M.; Nastev, M. The December 2020 Magnitude (Mw) 6.4 Petrinja Earthquake, Croatia: Seismological Aspects, Emergency Response and Impacts. *Res. Sq.* 2022; *preprint*. [CrossRef]
4. Herak, D.; Herak, M.; Tomljenović, B. Seismicity and Earthquake Focal Mechanisms in North-Western Croatia. *Tectonophysics* **2009**, *465*, 212–220. [CrossRef]
5. Ivančić, I.; Herak, D.; Herak, M.; Allegretti, I.; Fiket, T.; Kuk, K.; Markušić, S.; Prevolnik, S.; Sović, I.; Dasović, I.; et al. Seismicity of Croatia in the Period 2006–2015. *Geofizika* **2018**, *35*, 69–98. [CrossRef]
6. Ivančić, I.; Herak, D.; Markušić, S.; Sović, I.; Herak, M. Seismicity of Croatia in the Period 1997–2001. *Geofizika* **2002**, *18–19*, 17–29.
7. Ivančić, I.; Herak, D.; Markušić, S.; Sović, I.; Herak, M. Seismicity of Croatia in the Period 2002–2005. *Geofizika* **2006**, *23*, 87–103.
8. Markušić, S.; Herak, M. Seismic Zoning of Croatia. *Nat. Hazards* **1999**, *18*, 269–285. [CrossRef]
9. Herak, M.; Allegretti, I.; Herak, D.; Ivančić, I.; Kuk, K.; Marić, K.; Markušić, S.; Sović, I. *Seismic Hazard Maps for the Republic of Croatia*; Department of Geophysics, Faculty of Science, University of Zagreb: Zagreb, Croatia, 2011; Available online: http://seizkarta.gfz.hr/hazmap/ (accessed on 28 December 2022).
10. Herak, M. *Seismic Hazard Maps for the Republic of Croatia for the Return Period of 225 Years*; Department of Geophysics, Faculty of Science, University of Zagreb: Zagreb, Croatia, 2020; Available online: http://seizkarta.gfz.hr/hazmap/ (accessed on 28 December 2022).
11. Croatian Seismological Survey; Department of Geophysics; Faculty of Science; University of Zagreb. Preliminary Results on the Earthquake Series near Petrinja for Period 28 December 2020–28 January 2021. (In Croatian). Available online: https://www.pmf.unizg.hr/geof/seizmoloska_sluzba/mjesec_dana_od_glavnog_petrinjskog_potresa (accessed on 28 December 2022).
12. European-Mediterranean Seismological Centre. M6.4 CROATIA on December 29th 2020 at 11:19 UTC. Available online: https://www.emsc-csem.org/Earthquake/264/M6-4-CROATIA-on-December-29th-2020-at-11-19-UTC (accessed on 28 December 2022).
13. USGS United States Geological Survey. Shakemap M 6.4–2 Km WSW of Petrinja, Croatia. Available online: https://earthquake.usgs.gov/earthquakes/eventpage/us6000d3zh/shakemap/intensity (accessed on 28 December 2022).

14. Government of the Republic of Croatia; The World Bank. Croatia December 2020 Earthquake—Rapid Damage and Needs Assessment. 2021. Available online: https://mpgi.gov.hr/UserDocsImages/dokumenti/Potres/RDNA_2021_07_01_web_ENG.pdf (accessed on 12 December 2022).
15. Atalić, J.; Šavor Novak, M.; Uroš, M. *Updated Risk Assessment of Natural Disasters in Republic of Croatia—Seismic Risk Assessment*; Faculty of Civil Engineering in collaboration with Ministry of Construction and Physical Planning and National Protection and Rescue Directorate: Zagreb, Croatia, 2018. (In Croatian)
16. Atalić, J.; Šavor Novak, M.; Uroš, M. Seismic Risk for Croatia: Overview of Research Activities and Present Assessments with Guidelines for the Future. *Gradevinar* **2019**, *71*, 923–947. [CrossRef]
17. Uros, M.; Prevolnik, S.; Novak, M.S.; Atalic, J. Seismic Performance Assessment of an Existing Rc Wall Building with Irregular Geometry: A Case-Study of a Hospital in Croatia. *Appl. Sci.* **2020**, *10*, 5578. [CrossRef]
18. Technical Regulation for Building Structures. Official Gazette 17/17, 75/20, 7/22 2022. (In Croatian). Available online: https://narodne-novine.nn.hr/clanci/sluzbeni/2022_01_7_72.html (accessed on 12 December 2022).
19. Act on Reconstruction of Earthquake-Damaged Buildings in the City of Zagreb, Krapina-Zagorje County, Zagreb County, Sisak-Moslavina County and Karlovac County. Official Gazette 102/20, 10/21, 117/21 2021. (In Croatian). Available online: https://narodne-novine.nn.hr/clanci/sluzbeni/2021_10_117_2004.html (accessed on 12 December 2022).
20. Uroš, M.; Novak, M.Š.; Atalić, J.; Sigmund, Z.; Baniček, M.; Demšić, M.; Hak, S. Post-Earthquake Damage Assessment of Buildings—Procedure for Conducting Building Inspections. *Gradevinar* **2021**, *72*, 1089–1115. [CrossRef]
21. Vlašić, A.; Srbić, M.; Skokandić, D.; Ivanković, A.M. Post-Earthquake Rapid Damage Assessment of Road Bridges in Glina County. *Buildings* **2022**, *12*, 42. [CrossRef]
22. Skokandić, D.; Vlašić, A.D.; Marić, M.K.; Srbić, M.; Ivanković, A.M. Seismic Assessment and Retrofitting of Existing Road Bridges: State of the Art Review. *Materials* **2022**, *15*, 2523. [CrossRef] [PubMed]
23. Uroš, M.; Todorić, M.; Crnogorac, M.; Atalić, J.; Šavor Novak, M.; Lakušić, S. (Eds.) *Earthquake Engineering—Retrofitting of Masonry Buildings*; University of Zagreb, Faculty of Civil Engineering: Zagreb, Croatia, 2021.
24. Stepinac, M.; Skokandić, D.; Ožić, K.; Zidar, M.; Vajdić, M. Condition Assessment and Seismic Upgrading Strategy of RC Structures—A Case Study of a Public Institution in Croatia. *Buildings* **2022**, *12*, 1489. [CrossRef]
25. Salaman, A.; Stepinac, M.; Matorić, I.; Klasić, M. Post-Earthquake Condition Assessment and Seismic Upgrading Strategies for a Heritage-Protected School in Petrinja, Croatia. *Buildings* **2022**, *12*, 2263. [CrossRef]
26. Moretić, A.; Stepinac, M.; Lourenço, P.B. Seismic Upgrading of Cultural Heritage—A Case Study Using an Educational Building in Croatia from the Historicism Style. *Case Stud. Constr. Mater.* **2022**, *17*, e01183. [CrossRef]
27. Balić, I.; Smoljanović, H.; Trogrlić, B.; Munjiza, A. Seismic Analysis of the Bell Tower of the Church of St. Francis of Assisi on Kaptol in Zagreb by Combined Finite-Discrete Element Method. *Buildings* **2021**, *11*, 373. [CrossRef]
28. Moretić, A.; Chieffo, N.; Stepinac, M.; Lourenço, P.B. Vulnerability Assessment of Historical Building Aggregates in Zagreb: Implementation of a Macroseismic Approach. *Bull. Earthq. Eng.* **2022**. [CrossRef]
29. Karapetrou, S.; Manakou, M.; Bindi, D.; Petrovic, B.; Pitilakis, K. "Time-Building Specific" Seismic Vulnerability Assessment of a Hospital RC Building Using Field Monitoring Data. *Eng. Struct.* **2016**, *112*, 114–132. [CrossRef]
30. Ferraioli, M. Case Study of Seismic Performance Assessment of Irregular RC Buildings: Hospital Structure of Avezzano (L'Aquila, Italy). *Earthq. Eng. Eng. Vib.* **2015**, *14*, 141–156. [CrossRef]
31. Domaneschi, M.; Zamani Noori, A.; Pietropinto, M.V.; Cimellaro, G.P. Seismic Vulnerability Assessment of Existing School Buildings. *Comput. Struct.* **2021**, *248*, 106522. [CrossRef]
32. Cattari, S.; Calderoni, B.; Caliò, I.; Camata, G.; de Miranda, S.; Magenes, G.; Milani, G.; Saetta, A. Nonlinear Modeling of the Seismic Response of Masonry Structures: Critical Review and Open Issues towards Engineering Practice. *Bull. Earthq. Eng.* **2022**, *20*, 1939–1997. [CrossRef]
33. Tomić, I.; Vanin, F.; Beyer, K. Uncertainties in the Seismic Assessment of Historical Masonry Buildings. *Appl. Sci.* **2021**, *11*, 2280. [CrossRef]
34. CSI. *CSI Analysis Reference Manual for SAP2000, ETABS, SAFE and CSiBridge*; CSI: Berkeley, CA, USA, 2011.
35. *EN 1998-1:2004*; Eurocode 8: Design of Structures for Earthquake Resistance—Part 1: General Rules, Seismic Actions and Rules for Buildings. European Committee for Standardization: Brussels, Belgium, 2004.
36. *EN 1998-3:2004*; Eurocode 8: Design of Structures for Earthquake Resistance—Part 3: Assessment and Retrofitting of Buildings. European Committee for Standardization: Brussels, Belgium, 2004.
37. *ASCE/SEI 41-13*; Seismic Rehabilitation of Existing Buildings. American Society of Civil Engineers: Reston, VI, USA, 2014.
38. *ACI 318-14*; Building Code Requirements for Structural Concrete and Commentary. American Concrete Institute: Farmington Hills, MI, USA, 2014.
39. S.T.A. DATA. *3Muri User Manual*; S.T.A. DATA: Torino, Italy, 2022; Available online: https://www.3muri.com/documenti/brochure/en/3Muri12.2.1_ENG.pdf (accessed on 10 December 2022).
40. Fajfar, P. Capacity Spectrum Method Based on Inelastic Demand Spectra. *Earthq. Eng. Struct. Dyn.* **1999**, *28*, 979–993. [CrossRef]
41. Fajfar, P.; Gašperšič, P. The N2 Method for The Seismic Damage Analysis of RC Buildings. *Earthq. Eng. Struct. Dyn.* **1996**, *25*, 31–46. [CrossRef]
42. Brzev, S.; Begaliev, U. *Practical Seismic Design and Construction Manual for Retrofitting Schools in the Kyrgyz Republic*; World Bank Group: Washington, DC, USA, 2018.

43. Raza, S.; Khan, M.K.I.; Menegon, S.J.; Tsang, H.H.; Wilson, J.L. Strengthening and Repair of Reinforced Concrete Columns by Jacketing: State-of-the-Art Review. *Sustainability* **2019**, *11*, 3208. [CrossRef]

44. Ferretti, E.; Pascale, G. Some of the Latest Active Strengthening Techniques for Masonry Buildings: A Critical Analysis. *Materials* **2019**, *12*, 1151. [CrossRef] [PubMed]

45. Raval, S.S.; Dave, U.V. Effectiveness of Various Methods of Jacketing for RC Beams. *Procedia Eng.* **2013**, *51*, 230–239. [CrossRef]

46. Tahsiri, H.; Sedehi, O.; Khaloo, A.; Raisi, E.M. Experimental Study of RC Jacketed and CFRP Strengthened RC Beams. *Constr. Build. Mater.* **2015**, *95*, 476–485. [CrossRef]

47. Vandoros, K.G.; Dritsos, S.E. Concrete Jacket Construction Detail Effectiveness When Strengthening RC Columns. *Constr. Build. Mater.* **2008**, *22*, 264–276. [CrossRef]

48. Sayed, A.M.; Rashwan, M.M.; Helmy, M.E. Experimental Behavior of Cracked Reinforced Concrete Columns Strengthened with Reinforced Concrete Jacketing. *Materials* **2020**, *13*, 2832. [CrossRef]

49. Leelataviwat, S.; Warnitchai, P.; Tariq, H. Comparison of Seismic Strengthening Methods for Soft-Story Rc Frames Using Buckling-Restrained Braces and Concrete Jacketing. In Proceedings of the International Conference in Commemoration of 20th Anniversary of the 1999 Chi-Chi Earthquake, Taipei, Taiwan, 15–19 September 2019.

50. Elgawady, M.A.; Lestuzzi, P.; Badoux, M. Retrofitting of Masonry Walls Using Shotcrete. In Proceedings of the 2006 NZSEE Conference, Napier, New Zeland, 10–12 March 2006; Available online: http://db.nzsee.org.nz/2006/Paper45.pdf (accessed on 14 December 2022).

51. Shabdin, M.; Attari, N.K.A.; Zargaran, M. Experimental Study on Seismic Behavior of Un-Reinforced Masonry (URM) Brick Walls Strengthened with Shotcrete. *Bull. Earthq. Eng.* **2018**, *16*, 3931–3956. [CrossRef]

52. Ghezelbash, A.; Beyer, K.; Dolatshahi, K.M.; Yekrangnia, M. Shake Table Test of a Masonry Building Retrofitted with Shotcrete. *Eng. Struct.* **2020**, *219*, 110912. [CrossRef]

53. Paparo, A.; Beyer, K. Modeling the Seismic Response of Modern URM Buildings Retrofitted by Adding RC Walls. *J. Earthq. Eng.* **2016**, *20*, 587–610. [CrossRef]

54. Papanicolaou, C.; Triantafillou, T.; Lekka, M. Externally Bonded Grids as Strengthening and Seismic Retrofitting Materials of Masonry Panels. *Constr. Build. Mater.* **2011**, *25*, 504–514. [CrossRef]

55. Papanicolaou, C.G.; Triantafillou, T.C.; Papathanasiou, M.; Karlos, K. Textile Reinforced Mortar (TRM) versus FRP as Strengthening Material of URM Walls: Out-of-Plane Cyclic Loading. *Mater. Struct. Mater. Constr.* **2008**, *41*, 143–157. [CrossRef]

56. Koutas, L.N.; Tetta, Z.; Bournas, D.A.; Triantafillou, T.C. Strengthening of Concrete Structures with Textile Reinforced Mortars: State-of-the-Art Review. *J. Compos. Constr.* **2019**, *23*, 03118001. [CrossRef]

57. Raoof, S.M.; Bournas, D.A. TRM versus FRP in Flexural Strengthening of RC Beams: Behaviour at High Temperatures. *Constr. Build. Mater.* **2017**, *154*, 424–437. [CrossRef]

Article

Post-Earthquake Condition Assessment and Seismic Upgrading Strategies for a Heritage-Protected School in Petrinja, Croatia

Aida Salaman [1], Mislav Stepinac [2,*], Ivan Matorić [3] and Mija Klasić [3]

[1] Elea iC d.o.o.—Zagreb Branch Office, 10000 Zagreb, Croatia
[2] Faculty of Civil Engineering, University of Zagreb, 10000 Zagreb, Croatia
[3] Peoples Associates Structural Engineers—Zagreb Office, 10000 Zagreb, Croatia
* Correspondence: mislav.stepinac@grad.unizg.hr

Abstract: Following the Zagreb earthquake in March of 2020, a destructive 6.2 magnitude earthquake struck Croatia again in December of 2020. The Sisak-Moslavina county suffered the most severe consequences; many historical and cultural buildings were badly damaged. In the education sector, 109 buildings were damaged. One such building is the case study of this research. The heritage-protected building of the First Primary School in Petrinja is an unreinforced masonry structure, constructed using traditional materials and building techniques. The historical background of the building and the results of the post-earthquake assessment are presented. A numerical calculation of three strengthening methods was performed in 3Muri software: FRCM, FRP, and shotcrete. Non-linear pushover analysis was performed for each model. Finally, the strengthening methods are compared based on the achieved earthquake capacity, cost, and environmental impact.

Keywords: Petrinja; school; earthquake; FRP; FRCM; shotcrete; renovation; assessment; 3Muri; pushover

Citation: Salaman, A.; Stepinac, M.; Matorić, I.; Klasić, M. Post-Earthquake Condition Assessment and Seismic Upgrading Strategies for a Heritage-Protected School in Petrinja, Croatia. *Buildings* **2022**, *12*, 2263. https://doi.org/10.3390/buildings12122263

Academic Editor: Tiago Miguel Ferreira

Received: 14 November 2022
Accepted: 15 December 2022
Published: 19 December 2022

Publisher's Note: MDPI stays neutral with regard to jurisdictional claims in published maps and institutional affiliations.

1. Introduction

Recent earthquakes in the Mediterranean once again confirmed the high seismic vulnerability of unreinforced masonry buildings. Thousands of buildings were damaged or demolished by earthquakes in the last 3 years in Albania [1], Greece [2,3], Turkey [4,5], and Croatia [6,7]. The year 2020 in Croatia was marked by two catastrophic earthquakes which caused enormous socioeconomic and material damage in the capital city of Zagreb and surrounding counties. The consequences of the earthquakes are severe: eight fatalities, hundreds of families displaced across the country, and many buildings of great historical and cultural importance damaged or collapsed.

The devastating consequences of the Zagreb earthquake are explained in greater detail in [6,8]; however, the focus of this paper will be on the impact of the 2020 Petrinja earthquake on the historic district of Petrinja, particularly on one typical heritage-protected masonry building. A case study of a heavily damaged primary school in Petrinja is used to showcase the most common types of building damage, as well as seismic strengthening methods for unreinforced masonry buildings.

On 29 December 2020, the Sisak-Moslavina county was struck by a 6.2 M_L earthquake. The maximum intensity of the earthquake in the epicenter was estimated to be VIII–IX on the European Macroseismic Scale [9]. The Petrinja earthquake intensity is shown in Figure 1. The earthquake caused enormous material damage. In Sisak-Moslavina county, where the most affected cities, Petrinja, Glina, and Sisak, are located, the damage is estimated at EUR 4.8 billion and the cost of reconstruction at nearly double—EUR 8.4 billion [10]. According to the data collected by the Croatian Center for Earthquake Engineering (HCPI—in Croatian), more than 57,000 buildings were damaged [11].

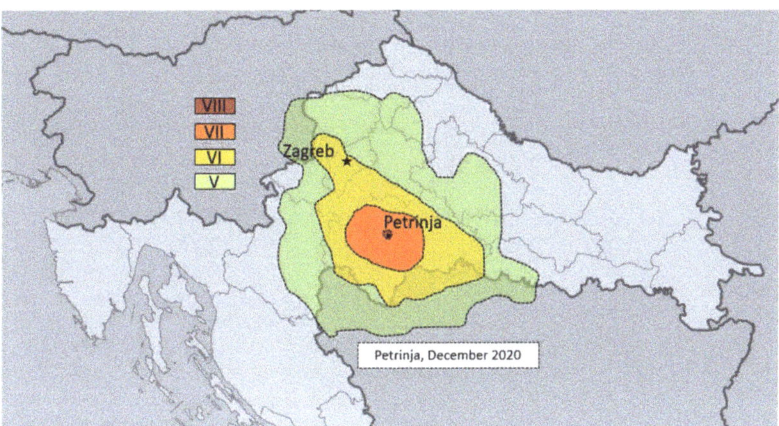

Figure 1. Petrinja earthquake intensity map according to EMS-98.

The Petrinja earthquake was devastating for buildings in the education sector, which suffered the greatest damage in Zagreb and Sisak-Moslavina counties. According to the data from the World Bank report [10] and HCPI [11], as many as 271 buildings were damaged in the earthquake, including 70 kindergartens, 160 primary schools, 32 secondary schools, 3 higher education buildings, and 6 student dormitories in the Sisak-Moslavina, Zagreb, Karlovac, Krapina-Zagorje Counties, and the City of Zagreb. All buildings are public assets, except for six kindergartens in Zagreb County. Most of the buildings were designated as usable or usable with a recommendation. In Sisak-Moslavina County, 109 buildings were damaged. In total, 18 were marked as temporarily unusable and 14 as unusable (1 kindergarten, 9 primary and 4 secondary schools). Some of these buildings require complete reconstruction or should be demolished and rebuilt again. Two photos of damaged buildings after the Petrinja earthquake in the historical center of the city of Petrinja are shown in Figure 2.

Figure 2. Damaged buildings in the historical center of Petrinja.

The total amount of losses and damage in the education sector from the Petrinja earthquake is estimated at EUR 174 million. Out of the total amount, the damage caused to buildings is estimated at EUR 154 million. The remaining EUR 20 million is equal to losses which include the costs of demolition and removal of unusable buildings, protection of cultural heritage buildings, transportation of students to other schools during reconstruction, etc. Over 3000 students were relocated to other schools after the earthquake to continue their education.

Although there were some initiatives for defining seismic risk and vulnerability of buildings in Croatia [12,13], the research was mainly carried out just from a scientific point of view, and the results were not respected by policy makers. Recently, several researchers were dealing with the vulnerabilities of existing educational buildings [14–20] and case studies were presented [21,22]. However, that kind of studies does not exist for Croatian building typologies.

As seismic preparedness was absent, the legislative procedures for the reconstruction were brought very late. The Law on Reconstruction of Earthquake-Damaged Buildings in the City of Zagreb, Krapina-Zagorje County, and Zagreb County [23] was issued six months after the Zagreb earthquake, but it wasn't well prepared and focused solely on the damaged areas in the first earthquake in 2020. As the Petrinja earthquake occurred 3 months after the "well-prepared" Law, modifications to the Law needed to be made. In the end, in February 2021, the new Law on Reconstruction of Earthquake-Damaged Buildings in the City of Zagreb, Krapina-Zagorje County, Zagreb County, Sisak-Moslavina County and Karlovac County was issued [24]. The Law and the Amendment to the Technical Regulation for Building Structures (Official Gazette 75/2020) [25] define four different levels of reconstruction of earthquake-damaged structures in relation to the achieved mechanical resistance and stability. In renovating, every building has to achieve the level of earthquake resistance that is required by HRN EN1998 [26]. The levels of reconstruction according to the Technical Regulation for Building Structures [25] depend on the degree of damage, the importance and purpose of the building, and the financial backing by the investor. The relationship between the selected reconstruction level and the duration/price of the reconstruction is shown in Figure 3. More details about the renovation levels can be found in [7]. It must be said that the rapid post-earthquake assessments and all the following procedures regarding the renovation, developments of legislation, etc. were based on Italian experiences [27–32].

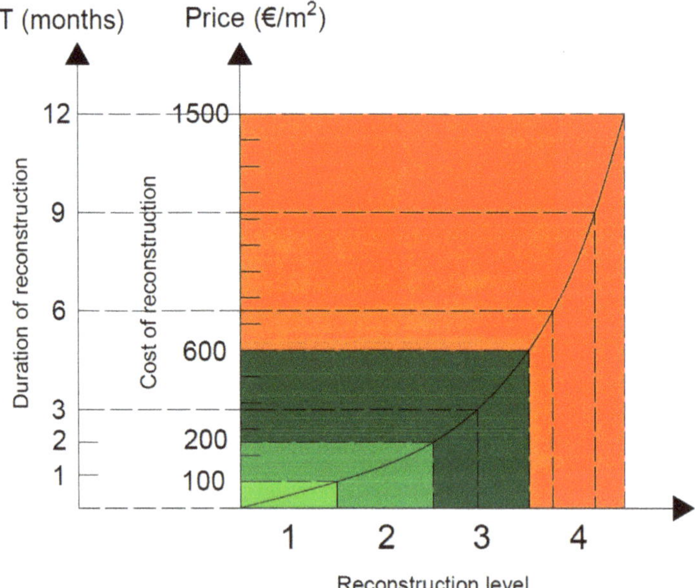

Figure 3. The relationship between the selected reconstruction level and the duration/cost of the reconstruction.

Figure 4 shows the difference between the actual seismic resistance level of the building and the seismic resistance ensured by different levels of reconstruction.

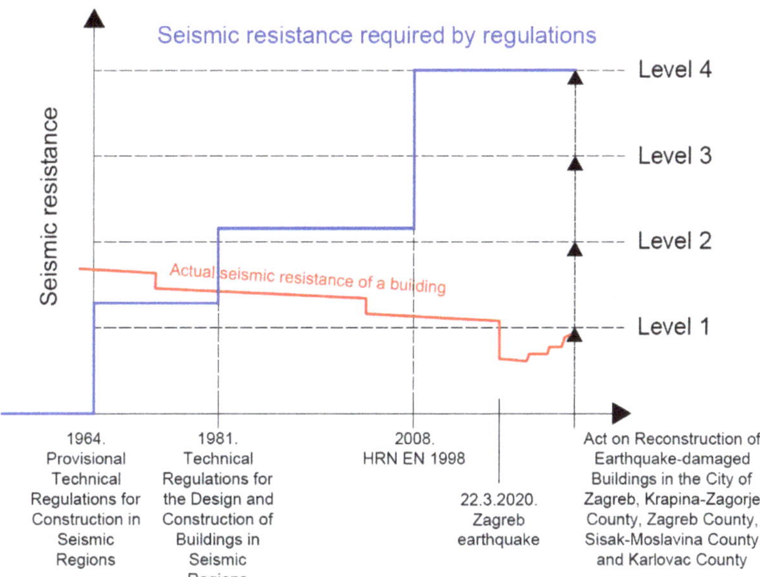

Figure 4. Schematic representation of the reconstruction levels in relation to the actual level of the building's seismic resistance capacity.

In Croatia, unreinforced or partially reinforced masonry buildings make up a large percentage of its building stock. Due to a lack of confining elements, the walls of these buildings almost exclusively transfer compressive loads, while tension causes the appearance of cracks [33,34]. Figure 5 shows some flaws of URM (unreinforced masonry) which cause its poor seismic performance.

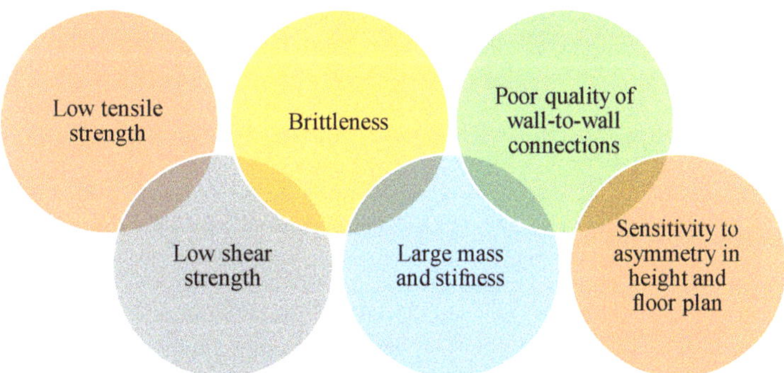

Figure 5. Typical characteristics of unreinforced masonry structures in Croatia.

In Sisak-Moslavina county the most common form of housing is low-rise, usually two-story residential buildings which were built during the post-war reconstruction period in the mid-to-late 1990s. A large percentage of these low-rise residential buildings are URM buildings whose walls were constructed using solid clay bricks and lime mortar of poor quality. The foundations are assumed to be either thin concrete slab or strip foundations. Based on the available, albeit limited, data, it can be assumed the soil is composed of layers of sand, gravel, or thick clay with low bearing capacity and large deformability. The ground floor of these buildings very often lacks vertical confining elements. The facades usually aren't protected by plaster. Horizontal confining elements are either missing or have insufficient height. The floor structure is usually a thin reinforced-concrete slab or a semi-precast masonry/concrete floor. The roof structures are mainly made of timber. The poor quality of construction and lack of confining elements, which are essential in seismic areas, can be linked to the lower financial status of the residents in the earthquake-affected areas. The most commonly observed types of damage in URM buildings include a partial collapse of gable walls, wall failure due to lack of confining elements, and in-plane shear failure.

In this paper, one heavily damaged school will be presented as a case study. A simple overview of construction practices in Croatia with the historical background of the educational sector in the region is very briefly shown. The assessment procedure and damage classification are explained. However, the focus of this research is to show different strengthening methods for a structural upgrading of a heritage-protected masonry building. A decent number of novel technologies for the seismic upgrading of existing buildings exist on the market. They can be divided as local or global measures as follows:

- Local: FRP (fiber-reinforced polymers)-based systems [35–37], FRCM (fabric-reinforced cementitious matrix)-based systems [38,39], concrete shotcreting [40], reinforcing with steel members and connection retrofitting [41,42];
- Global: capacity increase (new bracing systems, new reinforced shear walls, different additions of shear walls), integrity enhancement (especially for masonry buildings), demand reduction (base isolation, seismic dampers, energy dissipation systems) [43].

Three different, but most common, strengthening strategies are analyzed: strengthening with FRP, FRCM, and concrete shotcrete. The strategies are compared in points of seismic resistance, costs, and very basic environmental impacts.

2. Case Study

2.1. The Founding of the First Primary School in Petrinja—Historical Background

Near the end of the 18th and during the 19th century, the Military Frontier, a military province that the Habsburg Monarchy founded to defend itself against the Ottoman Empire, lost its defensive role and became a burden for the impoverished imperial treasury. The authorities in Vienna attempted to improve conditions by founding new, privileged Frontier cities where the development of crafts and trade was encouraged. These cities/towns soon became economic centers with a large influx of people and goods not only from the surrounding area but also from the more distant parts of the Habsburg Monarchy. It was these immigrants who participated in the work of schools and cultural institutions. The construction practices were similar to those in other major cities in the Monarchy [6,13,44,45]: URM buildings (brick masonry of format 29 × 14 × 7 cm) with timber floors and roofs.

The educational reform of Maria Theresa in 1764, which ordered that a German school must be founded in all important cities of the Military Frontier, was extremely important for this area. The main task of these schools was to train a certain number of literate people for the Frontier army. Up until then, schools were founded by nobles and church orders, and the church authorities oversaw the schools.

As a result of these reforms, the public school in Petrinja became a Normal School (*dt. Normalschule*) in 1777. The building of the Normal School was built in 1780. In 1861 the old, dilapidated timber building was demolished, and a new single-story masonry building was built in its place. The plans for the building were made by Croatian architect Bonifacio Cettola. The construction of the building was completed in 1862, and the new building (which today houses the First Primary School) was said to be the most beautiful school in the Military Frontier. In 1871, the construction of a neighboring two-story neo-Renaissance building was completed, which today houses the Petrinja High School (Figure 6).

Figure 6. The building of the First Primary School in Petrinja in the 1930s [46].

During the Croatian War of Independence (1991–1995), the school was used as a warehouse for clothes and food. After the liberation of Petrinja in 1995, the building was in poor condition due to aging and lack of maintenance. It had also been damaged during the war. Renovation works lasted until June 1996, when students began to attend classes again. More than a decade later, in June 2007, a much-needed reconstruction of the building structure was carried out since the previous works were mainly for aesthetic purposes. The building was restored to its former glory and classes continued regularly until the devastating Petrinja earthquake in December 2020, when the building suffered considerable damage.

Today the building of the First Primary School is located in the cultural district of the city of Petrinja and, due to its long history and tradition, is classified as a cultural heritage building. The building is listed in the Register of Cultural Properties of the Republic of Croatia and is protected by the Act on the Protection and Preservation of Cultural Goods.

2.2. Description of the Case Study

The building of the First Primary School consists of two units: the main building and the auxiliary building, which is detached from the main building. In the second half of the 20th century, a vestibule was built between the courtyard wings as a single-story, reinforced concrete structure, with a second floor added in 2007. The main building has a U-shaped floor plan; the external dimensions are 36.45 m × 31.10 m. The dimensions of the auxiliary building are 33.39 m × 6.30 m. The dimensions of the courtyard wings are 13.30 m × 16.60 m. The total height of the main building from the ground level is approx. 15.50 m. The height of the auxiliary building is approx. 4.00 m above ground level. The height of the flat roof of the vestibule between the courtyard wings is 8.42 m above ground level. The main building was built in 1862 as an unreinforced masonry structure, and the auxiliary building was built in 2017 as a reinforced concrete structure. The auxiliary building was built in accordance with current regulations; it is detached from the main building and was not damaged in the 2020 Petrinja earthquake. Therefore, it is not the subject of this paper.

The main building achieved its current floor plan (Figure 7) in the second half of the 20th century when the wings on the eastern side were upgraded, and their length increased by 7 m. Various adaptations, upgrades, and reconstructions of parts of the building have been carried out over the years.

On the ground floor the ceiling structure is a masonry barrel vault, reinforced with steel beams. The floor structure in the sanitary areas in both upgraded wings, as well as the vestibule, is a concrete slab. In the classrooms on the first floor, the floor structure is timber—the floor joists are 90 cm apart. Walls are constructed using solid brick elements 30 × 15 × 7 cm and lime mortar. The total thickness of the walls varies from 50 to 100 cm, including layers of plaster. The thickness of the outer load-bearing walls varies from 81 to 96 cm, and of the inner load-bearing walls from 51 to 67 cm. The staircases are U-shaped and made of timber. The load-bearing roof structure is timber. The original plans of the foundations, and the structural analysis and design of the building are not available. The structure of the vestibule is constructed using reinforced concrete columns and beams with a semi-precast masonry/concrete ceiling. Figure 8 shows the 3D models of the floors of the main building without the added vestibule, which are modeled and calculated in this paper. Grey parts represent secondary walls without a load-bearing function.

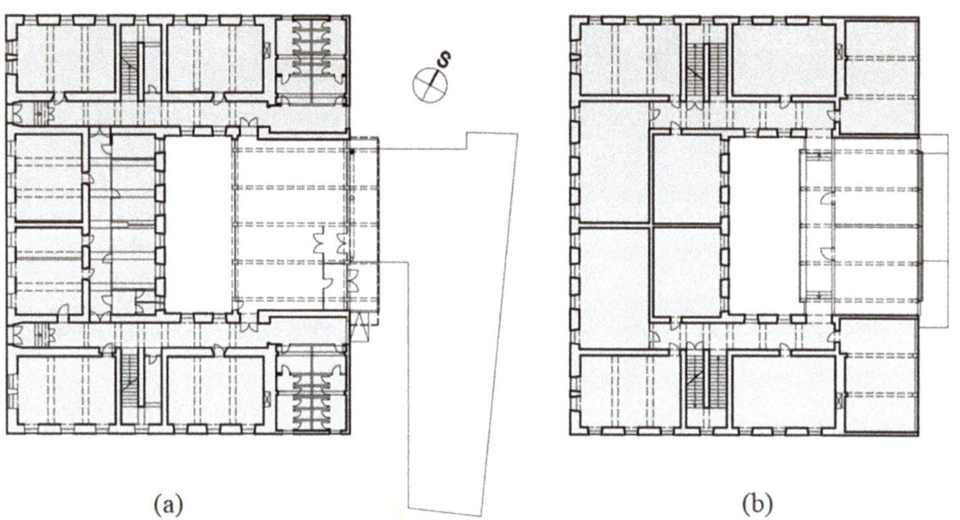

(a) (b)

Figure 7. Floor plans: (**a**) ground floor; (**b**) first floor.

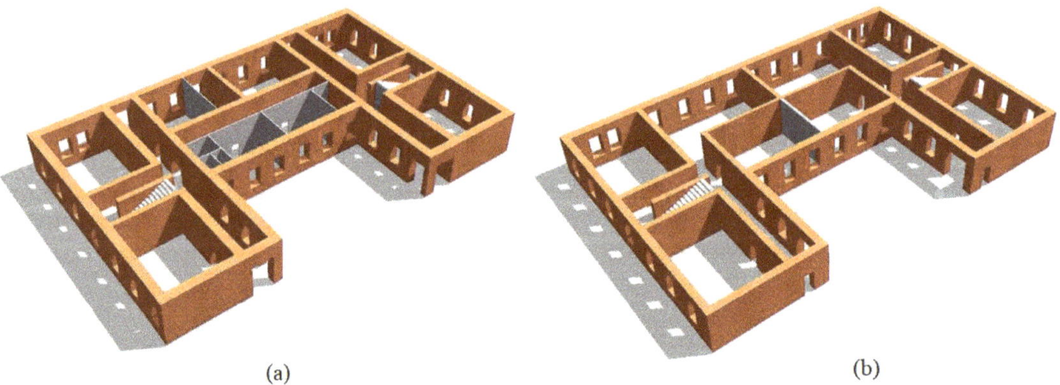

(a) (b)

Figure 8. 3D models of: (**a**) the ground floor; (**b**) the first floor.

2.3. Post-Earthquake Damage Assessment

Following the rapid post-earthquake assessment, the building was assigned the N2 label—not usable due to damage, which means that the structure has reached its load-bearing capacity, and there is a possibility of a collapse of load-bearing and non-load-bearing elements. For a full description of the post-earthquake procedures and damage grading the reader is addressed to [47–50].

A more detailed inspection was carried out in April 2022 within the detailed condition assessment. The following damage was recorded: on the north-western side of the building the roof structure had collapsed, which also caused part of the timber floor structure below it to collapse; the north-western and south-eastern gable walls have collapsed; horizontal cracks are visible on the western gable walls, indicating out-of-plane damage (Figures 9–12). Furthermore, the masonry arches in the vicinity of the staircase on the ground floor were damaged, and part of the barrel vault on the first floor had collapsed. The collapsed vault is near the area of the building where the roof and timber floor structure had collapsed. More detailed failures are explained in Table 1.

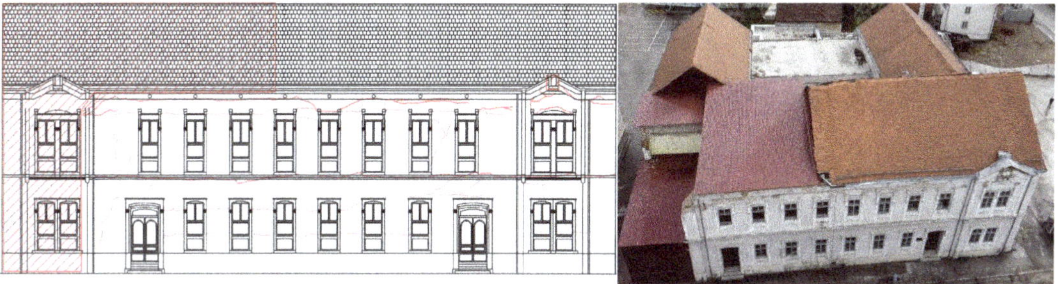

Figure 9. Southwestern façade plan with marked damage.

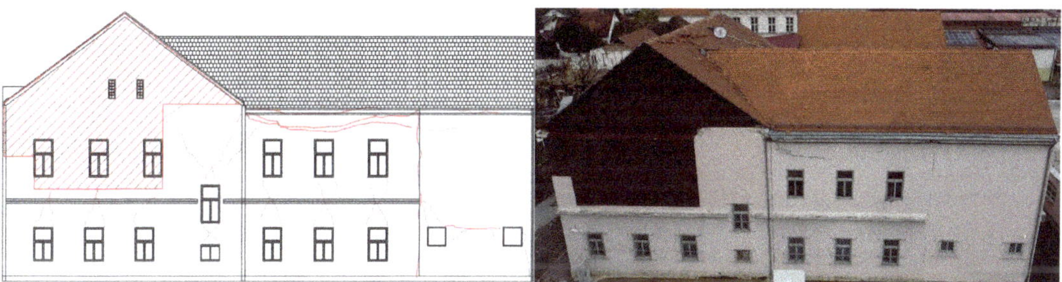

Figure 10. Southeastern façade plan with marked damage.

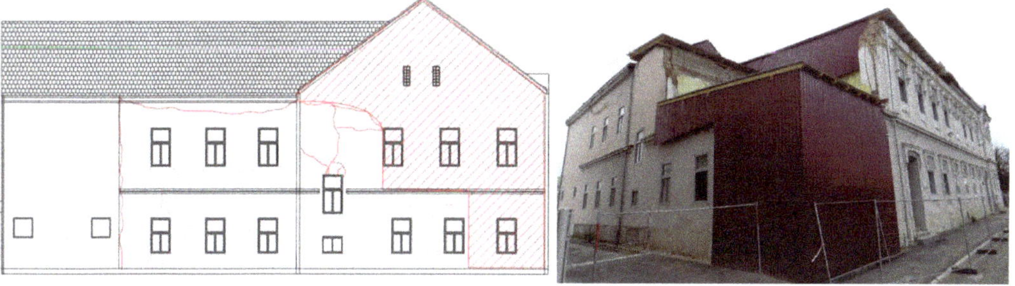

Figure 11. Northwestern façade plan with marked damage.

Figure 12. Northeastern façade plan with marked damage.

Considering the findings of the detailed inspection, the damage to the building is classified as Grade 4: very heavy damage (heavy structural damage, heavy non-structural damage), according to the EMS-98 classification [9] of damage.

Table 1. Typical damages of the case-study building.

Recorded Building Damage

Out-of-plane failure of wall connections

Partial overturning of the gable walls

Out-of-plane wall failure in the attic

Vault damage/collapse

Table 1. *Cont.*

Recorded Building Damage

In-plane shear failure of interior walls

In-plane sliding failure (horizontal cracks)

In-plane shear failure (diagonal cracks)

Tensile failure of masonry and vertical cracks at the corners

Shear failure of spandrels (diagonal cracks)

Partial collapse of the floor structure

In addition to the detailed inspection of the building, basic investigative works were also carried out to determine the material properties. Floor structures were examined to determine the type and thickness of layers and structural timber elements; the wall thickness and the dimensions of the brick elements were measured; mechanical properties of masonry walls (compressive strength and shear strength) were measured at two locations.

One of the hardest things about this building was to decide the exact locations of in-situ testing. Half of the building's roof collapsed and was under influence of external environmental conditions such as wind, rain, and snow. In addition, the facade walls are full of openings, so they were not suitable for testing. The chosen wall should be undisturbed, load bearing, if it is possible without openings, without any damage, etc.

The selected load-bearing walls are not in the vicinity of the collapsed parts of the building; they were not exposed to the weather or damaged, and have as few openings as possible. The compressive strength was tested in the field by directly loading the brick in the wall with a cylindrical press, and brick samples were taken from the building in six different places so that they could be tested in the laboratory for uniaxial compressive strength. Shear strength was measured "in situ" by using a hydraulic press. The mortar was moved horizontally in the vicinity of one brick in order to determine the shear strength. At the same time, the structure of the existing wall was minimally damaged. One brick is isolated in the wall so that it can move on one side, and a hydraulic press is placed on the other side, which applies pressure to the brick until it fails. More explanation of this common procedure in Croatia is given by Lulić et al. [49].

The following values were obtained after in situ testing:

- Shear strength of masonry under zero normal stress, f_{v0} = 0.05–0.06 N/mm^2;
- Compressive strength of the brick elements, f_b = 7.00 N/mm^2.

3. Modelling and Design of Seismic Strengthening Methods

Although, the research on complex and more precise modeling of URM is very active [51–54], in this research the 3D model for the evaluation of the seismic behavior was created in the 3Muri software [55], which is used for the analysis of new or existing buildings, and is especially suitable for performing non-linear static (pushover) analysis, which provides very good results for masonry structures.

Furthermore, the 3Muri software can perform linear static analysis and modal analysis according to Eurocode 6 [56] and Eurocode 8 [26]. The 3Muri uses the FEM—Frame by Macro Element calculation method [57,58]. Once the 3D model of the building is created, the generation of the equivalent frame model transforms every wall element into several deformable pier and spandrel elements, which are connected by rigid nodes. Provisions given by Italian researchers [59,60] about the modeling and seismic response analyses of URM buildings were respected.

Modeling in 3Muri software can be divided up into four phases:

- Phase 1: definition of geometry;
- Phase 2: definition of structural characteristics;
- Phase 3: model analysis;
- Phase 4: structural analysis;
- Optional phase 5: local mesh.

Every object is characterized by its material and geometric properties. Definition of structural characteristics consists of defining material, geometrical, and structural properties. Material properties can be obtained through investigation work, assumed based on Eurocode, MQI method, NDTs, etc. Geometrical properties are obtained through research of archive and existing documentation, site visiting, and assessment. Structural analysis is performed after all structural objects have been defined.

Structural analysis of the model consists of gravity (static) analysis, bending analysis out of plane, linear analysis, modal analysis, and non-linear analysis (pushover analysis). Results of performing non-linear analysis consist of capacity curves, vulnerability indices, damage to the building, periods, activated masses, etc.

The program automatically generates constraints and boundary conditions of the elements; however, additional constraints can be manually defined. For existing structures which have suffered damage during an earthquake (as is the case with First Primary School in Petrinja), one way of verifying the effectiveness of the model is a comparison of actual, recorded damage with the results of the pushover analysis. The result of the pushover analysis is the capacity curve. The deformation of every wall can be shown in every step of the analysis with color-coded damage (for example, red color is bending damage, orange is shear damage). If the building model is accurate, the recorded damage will, in certain measure, correspond with the results of the pushover analysis for any given wall.

The self-weight of the barrel vaults and the timber floors was determined according to the results of the investigation works. The imposed load on the floor structures was determined according to HRN EN 1991-1-1:2012/NA (Category C—areas where people may congregate—areas in schools [61]). The snow load on the roof structure was calculated according to HRN EN 1991-1-3/NA.

For the given location of the building, the ground type C (deep deposits of dense or medium-dense sand, gravel of stiff clay with thickness from several tens to many hundreds of meters) was selected. Considering that the case study has great cultural and historical importance, and is currently being used as a primary school, it is classified as Importance class III, with the corresponding importance factor $\gamma_I = 1.2$. The seismic hazard map for Croatia from the National Annex to Eurocode 8 [62] was used to determine the value of peak ground acceleration (PGA). The design ground acceleration is equal to peak ground acceleration times the importance factor:

For the return period of 95 years:

$$a_{g,95} = \gamma_I \times a_{gR} = 1.2 \times 0.074g = 0.87 \text{ m/s}^2$$

For the return period of 475 years:

$$a_{g,475} = \gamma_I \times a_{gR} = 1.2 \times 0.152g = 1.80 \text{ m/s}^2$$

Given that the building in question is classified as Importance class III, and that the building was heavily damaged, the chosen reconstruction level for the building is level 4: complete reconstruction of the building structure. At reconstruction level 4, the building's seismic resistance should match the conditions given by the Technical Regulations for Building Structures and HRN EN. The value of the vulnerability index for each of the two limit states (Significant Damage and Damage Limitation) must be greater than 1.00. The vulnerability index α is defined as the ratio between the limit capacity acceleration of the building and the reference peak ground acceleration. For Limit State of Significant Damage, the return period of PGA is 475 years, corresponding to the probability of exceedance of 10% in 50 years. For Limit State of Damage Limitation, the return period of PGA is 95 years, corresponding to the probability of exceedance of 10% in 10 years.

The masonry properties used are based on investigative post-earthquake assessments on similar buildings [48,49,63,64] and are shown in Table 2.

Table 2. Masonry properties.

Material	MoE, [N/mm^2]	Shear Modulus G, [N/mm^2]	Specific Weight [kN/m^3]	Characteristic Compression Strength, [N/mm^2]	Shear Strength [N/mm^2]
Masonry	1500.00	500.00	18.00	1.62	0.05

Linear static analysis of the structure was carried out, and it was concluded that all walls meet the load-bearing conditions and the geometric conditions. Furthermore, a modal analysis was carried out; the results are presented in Table 3.

Another important parameter that needs to be defined for pushover analysis is the control node, whose displacement is used to draw the capacity curve. The control node is located on the top floor and as close as possible to the center of mass. The selected node is shown in Figure 13.

Table 3. Modal analysis results.

Mode	T [s]	m_x [kg]	M_x [%]	m_y [kg]	M_y [%]	m_z [kg]	M_z [%]
1	0.29594	1,829,485	70.37	160	0.01	25	0
4	0.21744	108,720	0.42	2,090,799	80.42	1	0

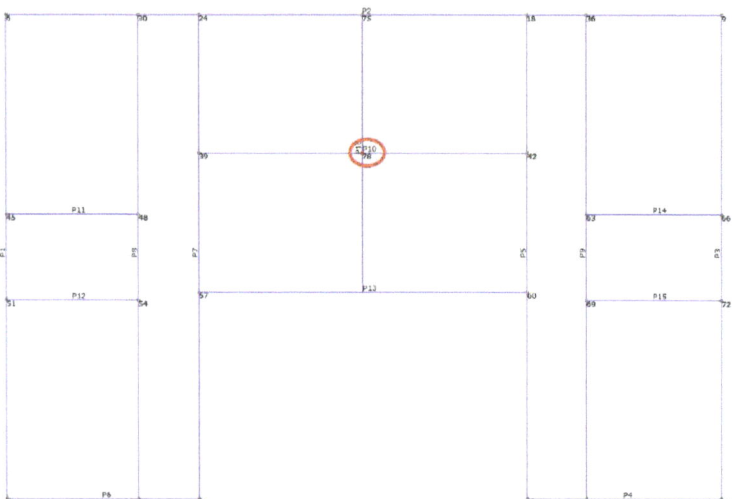

Figure 13. Position of the selected control mode on the top floor.

Local mechanisms were also checked. 3Muri has an optional module that performs local mechanisms analysis. The local mechanisms analysis is performed on those parts of the building where there is a possibility of wall failure by overturning due to an earthquake. To check the local mechanisms, it is necessary to extract from the global numerical model a part of the structure that is assumed to have failed due to overturning; these are usually gable walls or protruding parts of structures. The capacity check is carried out using the kinematic model balance method by comparing the required acceleration that causes the out-of-plane failure of the isolated part of the structure and the minimum required acceleration that depends on the height position of the isolated part of the structure. For the case-study building, gable walls were of course critical.

3.1. Results of the Pushover Analysis

For the case-study building, 24 pushover analyses were performed, depending on the type of load (uniform and modal distribution), direction ($\pm$x, $\pm$y) and the accidental eccentricity value which accounts for inaccuracies in the distribution of masses in the structure. As traditional buildings in Croatia, as well as in all the Mediterranean region, consist of load-bearing masonry walls and timber diaphragms, these diaphragms can be considered flexible or semi-flexible. Lately, a lot of research has been done in the field [65–71].

The results of the performed analyses are presented as 24 corresponding capacity curves (Figure 14).

Figure 15 shows a 3D representation of different types of damage to the building at the last step of the pushover analysis. In *x* direction, most of the damage is bending (pink color) and shear (beige and orange color) damage, but serious failure has not yet occurred. This is not the case for the *y* direction where the wall elements of the southwestern façade are in serious crisis (pink color) and are failing during the elastic phase (blue color). This is in accordance with the actual recorded damage of the building.

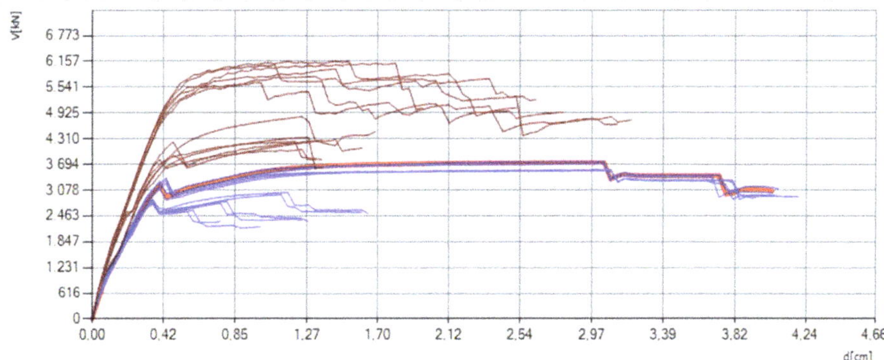

Figure 14. Capacity curves for x (blue) and y (red) direction. The bright red color represents a critical result.

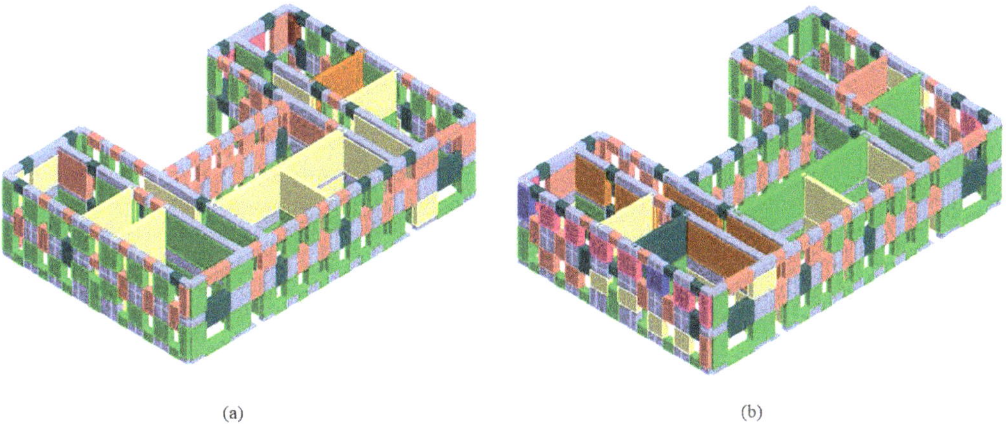

(a)   (b)

Figure 15. 3D view of damage for the most significant pushover analysis: (**a**) in x-direction; (**b**) in y-direction.

Table 4 shows the values of the vulnerability index for the most significant analysis in the x and y directions. Since the values for both Significant Damage and Damage Limitation limit states are lower than 1.00, it can be concluded that the current building structure is deficient and requires seismic strengthening.

Table 4. Values of vulnerability index α for the most significant analysis in the x and y direction.

No.	Direction	α_{SD}	α_{DL}
12	+X	0.333	0.434
19	+Y	0.569	0.697

Figures 16 and 17 show the comparison of obtained failures by 3Muri and observed damage on the real case-study building.

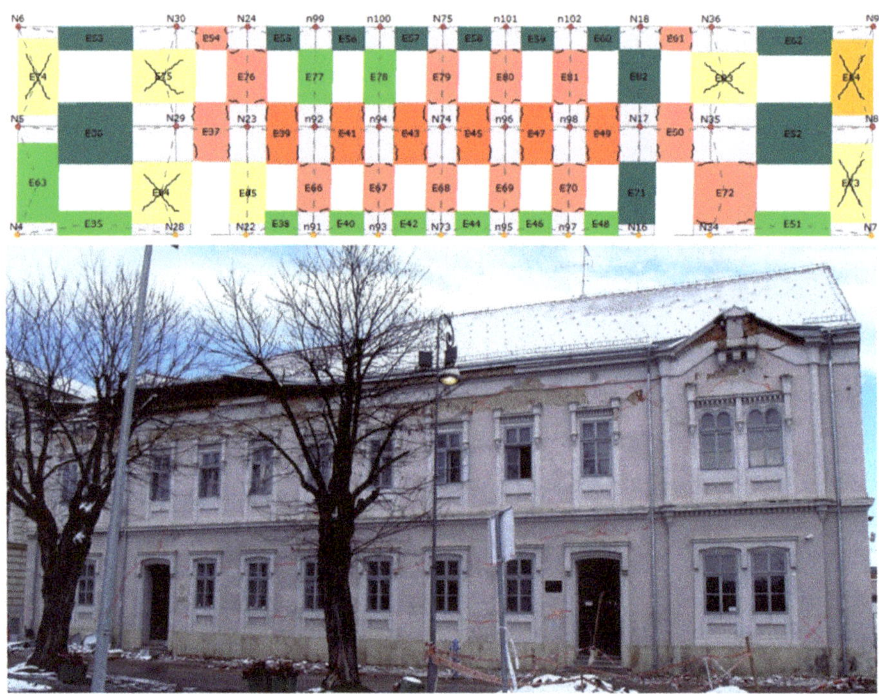

Figure 16. Detailed display of street facade damage in the 3Muri software package and actual damage to the street facade of the building in question.

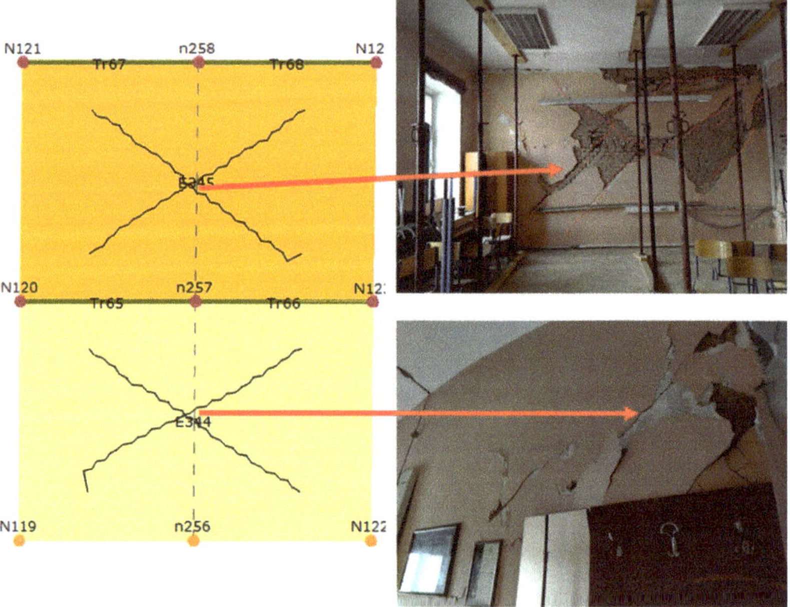

Figure 17. Detailed display of damage in the 3Muri software package and actual damage to the walls of the building in question.

3.2. Seismic Strengthening Strategies

In this paper, the three most commonly applied methods of seismic strengthening in Croatia are presented: reinforcement with concrete shotcrete, with FRP, and with FRCM.

In 3Muri software in the definition of structural characteristics there is a possibility to define retrofitting measurements such as FRCM/FRP, shotcrete, reinforced plaster, reinforced masonry, reinforcement, etc. For FRCM/FRP reinforcement, there is a material library with all the needed data for different providers such as MAPEI, Kerakoll, Ruregold, etc. To apply FRCM or FRP reinforcement, one needs to define which reinforcement and which manufacturer he/she uses. After defining the type of reinforcement and manufacturer, the basic material characteristic should be defined. After that, the software automatically calculates changes in structural characteristics. Applying shotcrete consists of applying reinforcement on masonry walls such as reinforcing mesh and changing the wall MoE so that software can calculate changes in structural characteristics. Before deciding what kind of retrofitting method is the most suited to stabilize the current structure of the building and make the building earthquake-resistant, it is needed to think about how it will affect the building. For example, would it increase the masses of the building, would it modify the structural system, etc. After applying retrofitting measurements, it is necessary to perform model analysis and structural analysis to see if the building has reached the required safety level.

The numerical calculation of all reinforcements was performed for a model with a rigid diaphragm on the first floor (timber–concrete composite floor) because without such a solution it was not possible to meet the limit states of significant damage and limited damage. Although the retrofitting technique for activating timber floors for their energy dissipation is ongoing research [72], still, according to Croatian regulations timber–concrete composite was chosen.

3.2.1. Reinforcement with FRCM

FRCM is a modern and compatible strengthening strategy for existing masonry, which consists in plastering the walls by means of mortar layers with embedded grids or textiles made of long fibers [73]. In this research, the structure is reinforced with five different FRCM systems of varying thicknesses, which are shown in Table 5. Three layers of glass fiber mesh were applied to both faces of the walls to ensure there is no difference in wall stiffness. Glass fiber mesh was chosen due to its availability and cost-effectiveness [74]. The most favorable configuration of different FRCM systems that satisfied the Significant Damage and Limited Damage limit states (Figure 18) was determined by iteration. The results of the pushover analyses are presented as 24 capacity curves (Figure 19).

Table 5. FRCM composite system properties.

FRCM Composite System					
Fiber thickness t_f (mm)	0.18	0.22	0.25	0.28	0.30
Modulus of elasticity E (N/mm^2)			71,000		

3.2.2. Reinforcement with FRP

The structure is reinforced with four different FRP systems of varying thicknesses, which are shown in Table 6. Carbon fiber fabric is applied to both faces of the walls. The most favorable configuration of different FRP systems that satisfied the Significant Damage and Limited Damage limit states (Figure 20) was determined by iteration. The results of the pushover analyses are presented as 24 capacity curves (Figure 21).

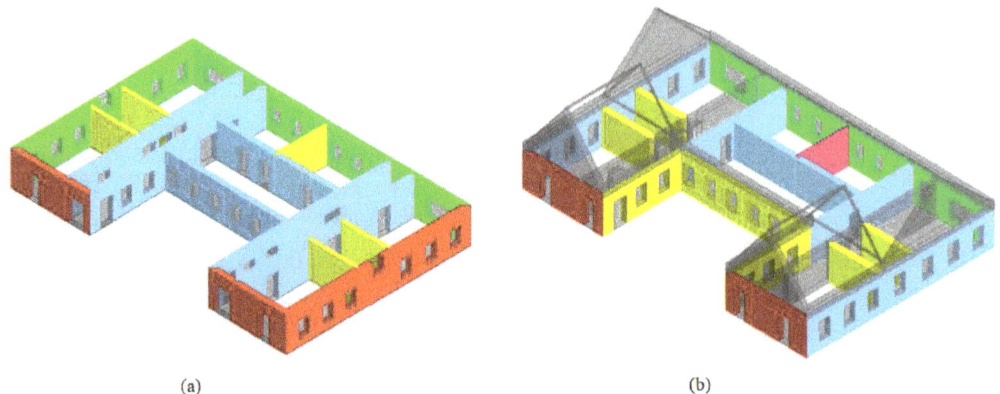

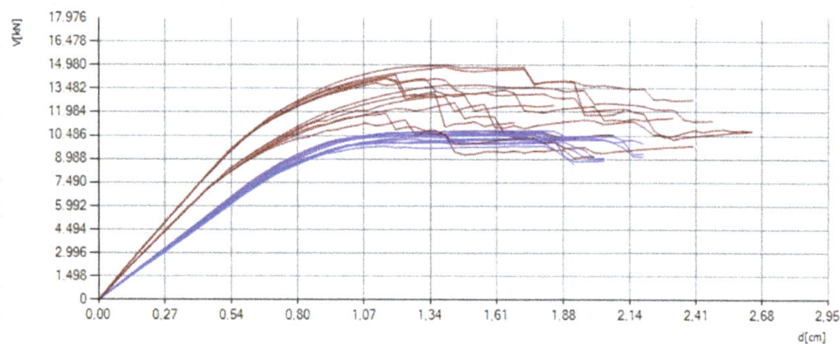

Figure 18. Configuration of different FRCM systems: (**a**) on the ground floor; (**b**) on the first floor.

Figure 19. Capacity curves for the model strengthened by FRCM composite system for x (blue) and y (red) direction.

Figure 20. Configuration of different FRP systems: (**a**) on the ground floor; (**b**) on the first floor.

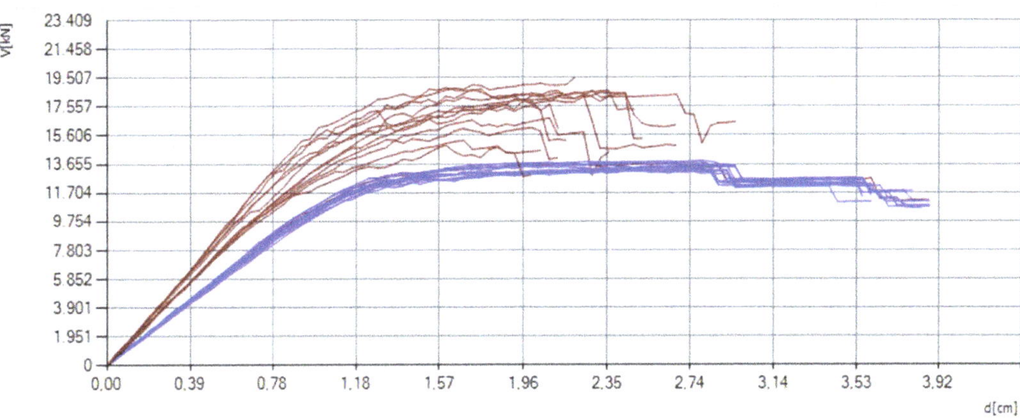

Figure 21. Capacity curves for the model strengthened by FRP composite system for *x* (blue) and *y* (red) direction.

Table 6. FRP composite system properties.

FRP Composite System				
Fiber thickness t_f (mm)	0.15	0.18	0.22	0.25
Modulus of elasticity E (N/mm^2)		390,000		

3.2.3. Reinforcement with Concrete Shotcrete

The third and final analyzed method is reinforcement with shotcrete. The thickness of the concrete layer is 6 cm, the selected reinforcing mesh is Q283 (6 mm diameter bars with 100 mm pitch), and the steel grade is B500B. Unlike the previous two strengthening methods, shotcrete is applied to only one face of the wall, given that such a variant meets the required checks, and results in a lower cost of reconstruction. On the external walls, shotcrete is applied on the inner face of the wall to preserve the appearance of the building, which has great cultural and historic importance. The most favorable configuration of shotcrete reinforcement that satisfied the Significant Damage and Limited Damage limit states (Figure 22) was determined by iteration. The results of the pushover analyses are presented as 24 capacity curves (Figure 23).

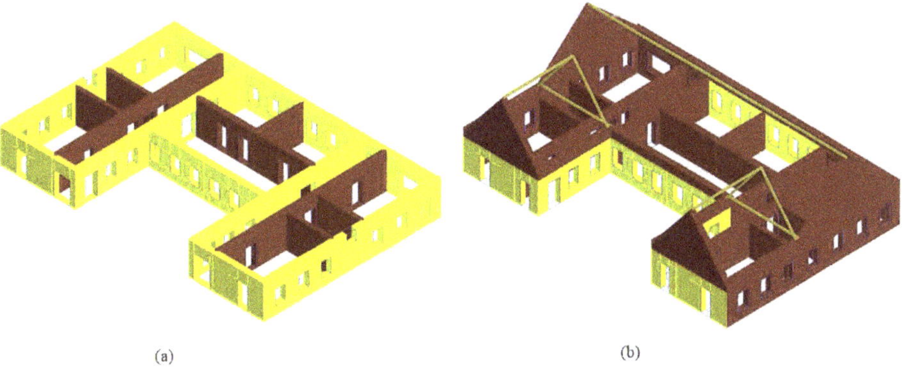

(a) (b)

Figure 22. Configuration of shotcrete reinforcement (yellow color): (**a**) on the ground floor; (**b**) on the first floor.

3.3. Comparison of Strengthening Methods

The risk indices of the most significant pushover analysis for both Significant Damage and Damage Limitation limit state are presented in Table 7. The limit state is considered satisfied if the vulnerability index α is greater than 1.00. The vulnerability index α is the result of the pushover analysis and is defined as the ratio between the limit capacity acceleration of the building and the reference peak ground acceleration. Two vulnerability indices are calculated for every pushover analysis: one for Significant Damage LS and one for Damage Limitation LS. The limit capacity acceleration of the building is the peak ground acceleration for which the structure reaches one of the two limit states. For Limit State of Significant Damage, the return period of PGA is 475 years, corresponding to the probability of exceedance of 10% in 50 years. For Limit State of Damage Limitation, the return period of PGA is 95 years, corresponding to the probability of exceedance of 10% in 10 years. The PGA values are input data for seismic load. The value of the vulnerability index for each of the two limit states must be greater than 1.00. The EN 1998-3 defines the Significant Damage Limit State as follows: the structure is significantly damaged, with some residual lateral strength and stiffness, and vertical elements can sustain vertical loads. Non-structural components are damaged, although partitions and infills have not failed out-of-plane. The structure can sustain after-shocks of moderate intensity. The Damage Limitation Limit State can be defined as follows: the structure is only lightly damaged, with structural elements prevented from significant yielding, and retaining their strength and stiffness properties. Non-structural components, such as partitions and infills, may show cracking, but the damage could be economically repaired. EN1998-3 defines a third limit state: Near Collapse; however, the Croatian National Annex doesn't require checks for that LS. Accidental eccentricity of the center of mass with respect to the rigidity center is computed automatically. According to EN1998, it is calculated as 5% of the floor dimension perpendicular to the direction of the seismic action.

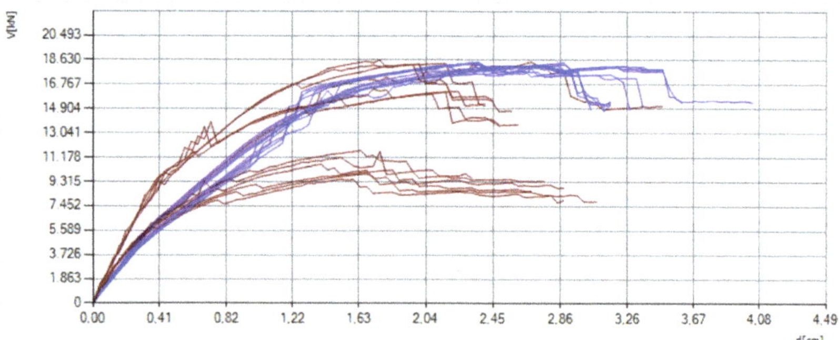

Figure 23. Capacity curves for the model strengthened by shotcrete for the x (blue) and y (red) direction.

Table 7. Vulnerability indices for the most significant analysis in the x and y direction for the strengthened models.

Method	No.	Direction	α_{SD}	α_{DL}
FRCM	13	$-X$	1.002	1.614
FRCM	22	$-Y$	1.175	2.133
FRP	15	$-X$	1.522	1.944
FRP	20	$+Y$	1.257	2.182
Shotcrete	12	$+X$	1.752	2.861
Shotcrete	19	$+Y$	1.377	1.912

All 24 pushover analyses for all strengthening methods satisfied this check. A 3D visualization of damage for the most significant analysis in x and y directions for all strengthening methods is shown in Figure 24.

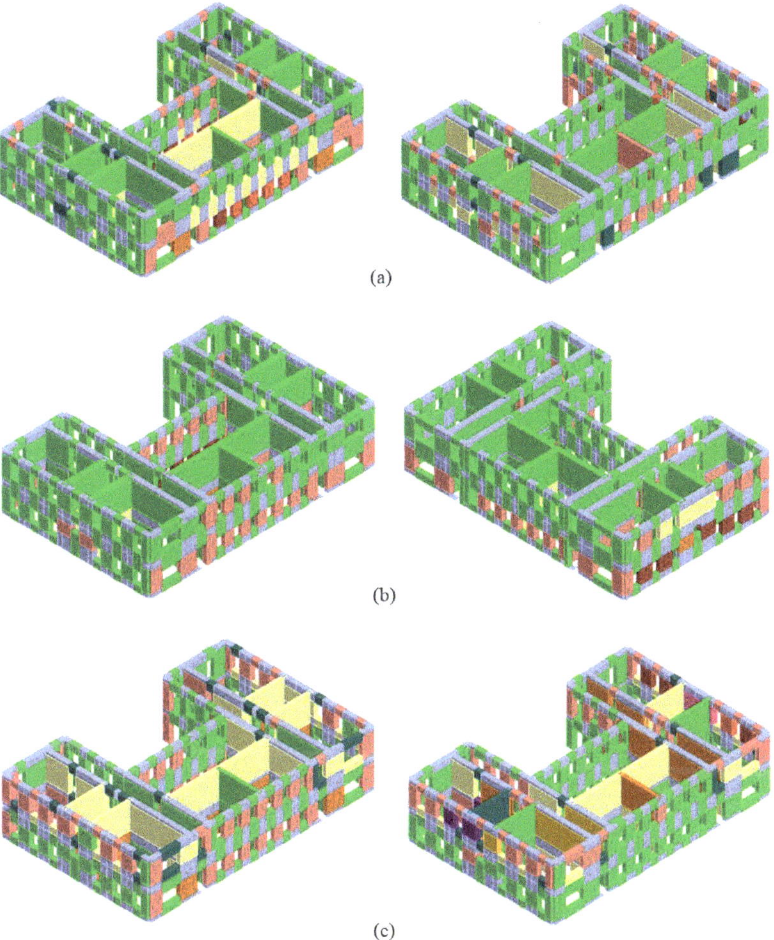

Figure 24. Three-dimensional visualization of damage in the last step of the most significant pushover analyses in x and y direction: (**a**) FRCM strengthening; (**b**) FRP strengthening; (**c**) shotcrete strengthening.

A graphical comparison of vulnerability indices for all 24 pushover analyses is shown in Figure 25. Figure 26 shows the comparison of the risk indices for the unreinforced model and all reinforcement variants. The results show that the highest values of the vulnerability index are mostly obtained for the model reinforced with shotcrete.

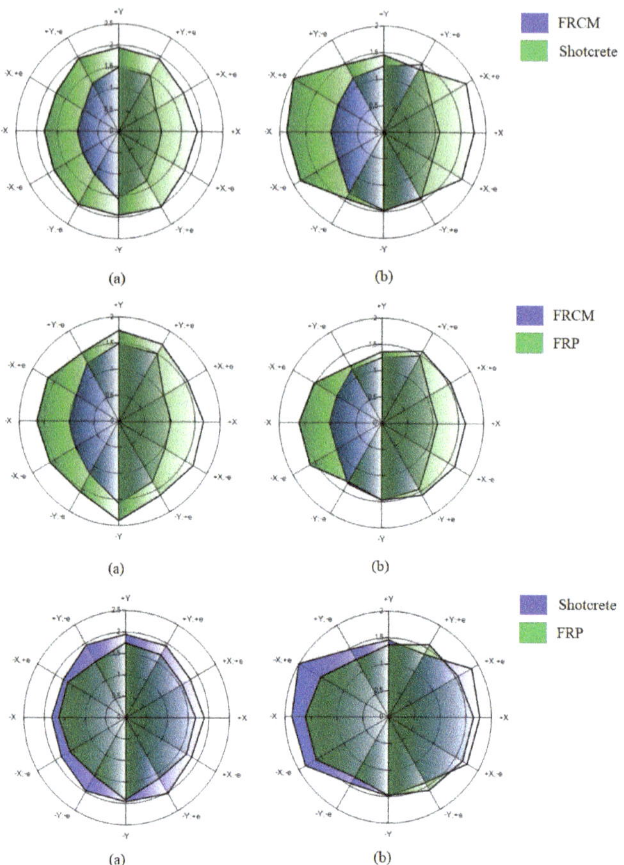

Figure 25. Comparison of vulnerability indices (Significant Damage limit state) of the different strengthening methods: (**a**) uniform load distribution; (**b**) modal load distribution.

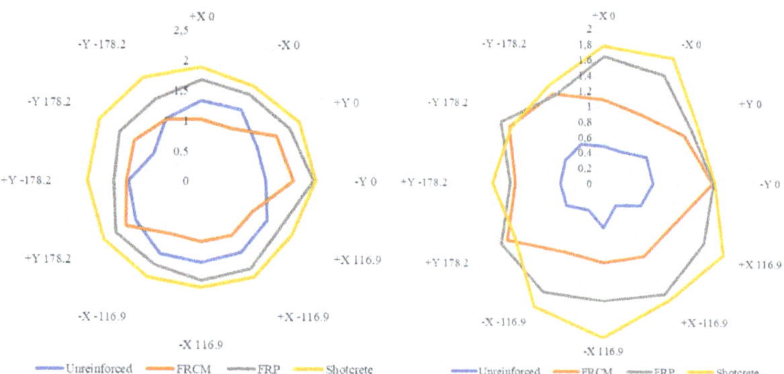

Figure 26. Comparison of vulnerability indices (Significant Damage limit state) of the unreinforced model and all strengthening methods (uniform load distribution—(**left**), modal load distribution—(**right**))

4. Financial Costs and CO$_2$ Impact

Besides achieved earthquake capacity, the methods were compared based on cost-effectiveness. Approximate costs of renovation were calculated for every strengthening method (Tables 8–12) according to [75]. The cost of a new concrete slab on the first floor is added to the renovation costs of each strengthening variant, given that the assumption of a rigid diaphragm was a precondition for the limit state checks to be satisfied.

Current trends in the construction industry indicate a general shift toward sustainability and integrated approaches [76,77]; therefore, the environmental impact of the proposed strengthening methods was also considered. In addition, preserving cultural heritage is not only an obligation to sustain and transmit it to the future generation, but is also a driver of sustainable growth [78]. Using the SimaPro software [79], the environmental impact of each strengthening method was calculated in the form of CO$_2$ greenhouse gas emissions (Table 13). It is important to note that the calculation doesn't take into account CO$_2$ emissions during transportation, construction, usage, and demolition, but only during the material production phase.

Table 8. Calculation of renovation costs—rigid diaphragm on the first floor.

Task Description	Unit of Measure	Unit price (EUR)	Amount	Total Price (EUR)
Dismantling of existing wooden floor layers	m^2	5.98	491.74	2940.61
Loading and unloading the waste in the landfill	m^3	46.48	24.59	1142.80
Filling the space between the joists with expanded polystyrene (EPS)	m^2	7.3	491.74	3589.70
Installation of steel connectors	pcs	1.59	12,300.00	19,557.00
Installation of anchor rods that connect the concrete slab to the walls	pcs	33.2	360.00	11,952.00
Placement of nylon on which the concrete is poured	m^2	0.66	491.74	324.55
Installation of reinforcement mesh	kg	1.33	2203.00	2929.99
Pouring a new concrete slab (6 cm thick)	m^3	112.88	29.05	3279.16
Total (EUR)			45,715.81	

Table 9. Calculation of renovation costs—preparatory works (all strengthening methods).

Task Description	Unit of Measure	Unit Price (EUR)	Amount	Total Price (EUR)
Operating the mobile scaffolding to perform the necessary task	m^2	15.94	1544.69	24,622.30
Complete removal of plaster from the walls	m^2	4.65	3151.16	14,652.89
Waste collection	m^2	1.99	3151.16	6270.81
Loading and unloading the waste in the landfill	m^3	46.48	94.54	4394.42
Total (EUR)			49,940.30	

Table 10. Calculation of renovation costs—FRCM composite system.

Task Description	Unit of Measure	Unit Price (EUR)	Amount	Total Price (EUR)
Preparatory works				49,940.30
Repointing mortar joints	m^2	13.28	884.36	11,744.32
Application of FRCM system	m^2	33.2	2927.24	97,184.27
Connecting the FRCM layers with FRP rope	pcs	15.94	3500.00	55,790.00
Applying new plaster	m^2	11.28	3151.16	35,545.08
Total (EUR)			250,204.09	

Table 11. Calculation of renovation costs—FRP composite system.

Task Description	Unit of Measure	Unit Price (EUR)	Amount	Total Price (EUR)
Preparatory works				49,940.30
Repointing mortar joints	m^2	13.28	884.36	11,744.32
Application of FRP system	m^2	66.4	2927.24	194,368.54
Connecting the FRP layers with FRP rope	pcs	15.94	3500.00	55,790.00
Applying new plaster	m^2	11.29	3151.16	35,576.60
Total (EUR)		347,419.87		

Table 12. Calculation of renovation costs—shotcrete.

Task Description	Unit of Measure	Unit Price (EUR)	Amount	Total Price (EUR)
Preparatory works				49,940.30
Installation of anchors for the reinforcement mesh (4 pcs/m^2)	pcs	2.66	2710	7208.60
Installation of Q283 reinforcement mesh	kg	1.33	3031.08	4031.34
Shotcrete application (6 cm thick)	m^3	112.88	39.57	4466.46
Applying new plaster	m^2	11.29	3151.16	35,576.60
Total (EUR)		101,223.41		

Table 13. CO$_2$ emissions for different materials.

	Unit	CO$_2$ Emissions (kg)	Amount	Total CO$_2$ Emissions (kg)
FRCM				
Concrete	m^3	385.32	43.91	16,919.25
Glass fiber	kg	1.92	1857.86	3570.00
Total				20,489.25
FRP				
Carbon fiber	kg	31.00	1112.35	34,482.85
Total				34,482.85
Shotcrete				
Concrete	m^3	385.32	40.59	15,640.00
Steel	kg	2.28	3031.08	6900.00
Total				22,540.00

The final goal of this paper is to compare the strengthening methods based on the achieved earthquake capacity, cost, and environmental impact (Figure 27). The diameter of the bubbles represents the CO$_2$ emissions. The production of concrete emits the largest amount of CO$_2$ but, from an economic standpoint, this option is the most favorable. Furthermore, glass fibers have a much smaller carbon footprint than concrete and, despite the slightly higher price, the application of the FRCM system should be seriously considered. The FRP system, despite low CO$_2$ emissions, is quite costly compared to the other variants.

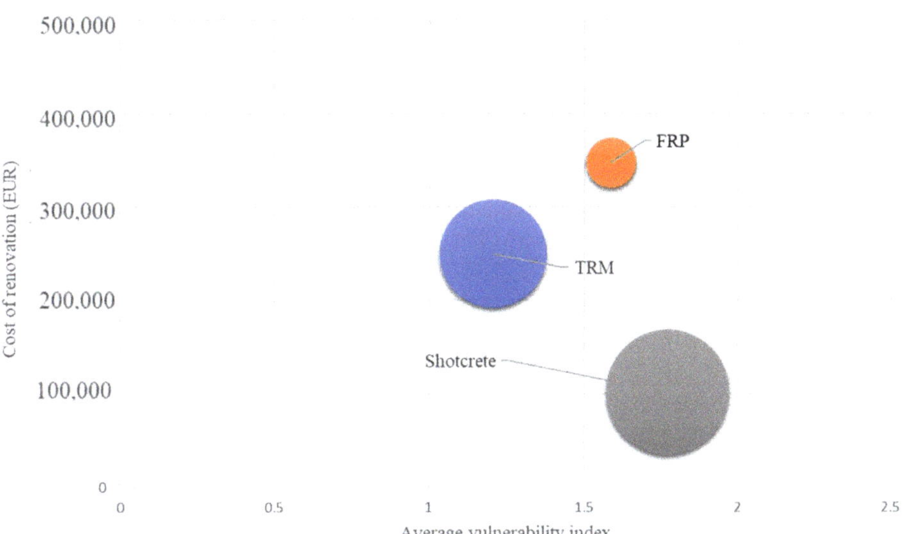

Figure 27. Comparison of methods based on renovation costs, average vulnerability index, and environmental impact.

5. Conclusions

Croatia is one of the most seismically vulnerable countries in Europe. The recent earthquakes in Zagreb and Petrinja exposed many weaknesses of construction in Croatia—dilapidated buildings and materials, numerous mistakes in design and construction, and slow and inconsistent legal framework related to reconstruction are just some of the problems. Experts in the fields of seismology and civil engineering have been warning about the possible catastrophic consequences of earthquakes for years; unfortunately, until the 2020 earthquakes, more attention and resources were directed to increasing the energy efficiency of buildings, while seismic renovation was neglected despite the significant seismic risk.

The main goal of retrofitting is to make buildings earthquake-resistant and to stabilize the current structure. However, in addition, it needs to be in line with the conservation and restoration rules and, of course, with the building owner requests. Some of the basic and commonly used retrofitting methods in Croatia are adding shear walls, jacketing of beams and columns, applying FRCM or FRP, applying shotcrete, and adding timber concrete composite floors. Shotcreting is certainly more widespread because it is much simpler and certainly ensures a sufficient level of load capacity. The problem arises that it is not effective with flexible diaphragms and greatly changes the static and dynamic image of the lateral system. Renovation with FRP allows the freedom to use the flexible floor structure and, as a rule, does not change the dynamic image of the structure. Basic problems in Croatia are non-educated and untrained workers, high costs of retrofitting, and sometimes unrealistic conservation demands.

The selection of the optimal strengthening method will depend on the age and type of the building, required load-bearing capacity and ductility, available means of reconstruction, and many other factors. Shotcrete is the most cost-effective, but large CO_2 emissions and invasiveness are reasons to consider other solutions, especially when a culturally significant building is in question. The FRP composite system achieved satisfactory seismic capacity; the disadvantage of this solution is the high cost of renovation compared to the FRCM system, which meets the requirements for seismic capacity and sustainability while maintaining a slightly lower price.

Author Contributions: Conceptualization, M.S.; methodology, M.S. and I.M.; software, A.S. and M.K.; validation, A.S. and M.S.; formal analysis, A.S. and M.K.; investigation, A.S. and M.S.; resources, M.S. and I.M.; data curation, A.S.; writing—original draft preparation, A.S. and M.S.; writing—review and editing, A.S. and M.S.; visualization, A.S. and M.S.; supervision, M.S. and I.M; project administration, M.S. and I.M.; funding acquisition, M.S. All authors have read and agreed to the published version of the manuscript.

Funding: This research was funded by the Croatian Science Foundation, grant number UIP-2019-04-3749 (ARES project—Assessment and rehabilitation of existing structures—development of contemporary methods for masonry and timber structures), project leader: Mislav Stepinac.

Data Availability Statement: The data presented in this study are available on request from the corresponding author. The data are not publicly available due to privacy reasons.

Acknowledgments: The authors want to thank Boja Čačić Šipoš for helping in the condition assessment and numerical modeling of the case-study building.

Conflicts of Interest: The authors declare no conflict of interest.

References

1. Bilgin, H.; Shkodrani, N.; Hysenlliu, M.; Baytan Ozmen, H.; Isik, E.; Harirchian, E. Damage and performance evaluation of masonry buildings constructed in 1970s during the 2019 Albania earthquakes. *Eng. Fail. Anal.* **2022**, *131*, 105824. [CrossRef]
2. Papadimitriou, P.; Kapetanidis, V.; Karakonstantis, A.; Spingos, I.; Kassaras, I.; Sakkas, V.; Kouskouna, V.; Karatzetzou, A.; Pavlou, K.; Kaviris, G.; et al. First results on the Mw = 6.9 Samos earthquake of 30 October 2020. *Bull. Geol. Soc. Greece* **2020**, *56*, 251–279. [CrossRef]
3. Vlachakis, G.; Vlachaki, E.; Lourenço, P.B. Learning from failure: Damage and failure of masonry structures, after the 2017 Lesvos earthquake (Greece). *Eng. Fail. Anal.* **2020**, *117*, 104803. [CrossRef]
4. Günaydin, M.; Atmaca, B.; Demir, S.; Altunişik, A.C.; Hüsem, M.; Adanur, S.; Ateş, Ş.; Angin, Z. Seismic damage assessment of masonry buildings in Elazığ and Malatya following the 2020 Elazığ-Sivrice earthquake, Turkey. *Bull. Earthq. Eng.* **2021**, *19*, 2421–2456. [CrossRef]
5. Yakut, A.; Sucuoğlu, H.; Binici, B.; Canbay, E.; Donmez, C.; İlki, A.; Caner, A.; Celik, O.C.; Ay, B.Ö. Performance of structures in İzmir after the Samos island earthquake. *Bull. Earthq. Eng.* **2021**, *20*, 7793–7818. [CrossRef]
6. Stepinac, M.; Lourenço, P.B.; Atalić, J.; Kišiček, T.; Uroš, M.; Baniček, M.; Šavor Novak, M. Damage classification of residential buildings in historical downtown after the ML5.5 earthquake in Zagreb, Croatia in 2020. *Int. J. Disaster Risk Reduct.* **2021**, *56*, 102140. [CrossRef]
7. Moretić, A.; Stepinac, M.; Lourenço, P.B. Seismic upgrading of cultural heritage—A case study using an educational building in Croatia from the historicism style. *Case Stud. Constr. Mater.* **2022**, *17*, e01183. [CrossRef]
8. *World Bank Report: Croatia Earthquake—Rapid Damage And Needs Assessment*; June 2020; Government of Croatia: Zagreb, Croatia, 2020.
9. Comisión Sismológica Europea. *Escala Macro Sísmica Europea EMS—98*; Comisión Sismológica Europea: Luxembourg, 1998; Volume 15, ISBN 2879770084.
10. *World Bank Report: Croatia December 2020 Earthquake—Rapid Damage And Needs Assessment*; Government of Croatia: Zagreb, Croatia, 2021.
11. *The Database of Usability Classification, Croatian Centre of Earthquake Engineering (HCPI—Hrvatski Centar za Potresno Inženjerstvo)*; Faculty of Civil Engineering, University of Zagreb: Zagreb, Croatia, 2020.
12. Atalic, J.; Savor Novak, M.; Uros, M. Rizik od potresa za Hrvatsku: Pregled istraživanja i postojećih procjena sa smjernicama za budućnost. *Građevinar* **2019**, *71*, 923–947.
13. Šipoš, T.K.; Hadzima-Nyarko, M. Seismic risk of Croatian cities based on building's vulnerability. *Teh. Vjesn.* **2018**, *25*, 1088–1094. [CrossRef]
14. Perrone, D.; O'Reilly, G.J.; Monteiro, R.; Filiatrault, A. Assessing seismic risk in typical Italian school buildings: From in-situ survey to loss estimation. *Int. J. Disaster Risk Reduct.* **2020**, *44*, 101448. [CrossRef]
15. O'Reilly, G.J.; Perrone, D.; Fox, M.; Monteiro, R.; Filiatrault, A. Seismic assessment and loss estimation of existing school buildings in Italy. *Eng. Struct.* **2018**, *168*, 142–162. [CrossRef]
16. Gaetani d'Aragona, M.; Polese, M.; Prota, A. Stick-IT: A simplified model for rapid estimation of IDR and PFA for existing low-rise symmetric infilled RC building typologies. *Eng. Struct.* **2020**, *223*, 111182. [CrossRef]
17. Domaneschi, M.; Zamani Noori, A.; Pietropinto, M.V.; Cimellaro, G.P. Seismic vulnerability assessment of existing school buildings. *Comput. Struct.* **2021**, *248*, 106522. [CrossRef]
18. Fontana, C.; Cianci, E.; Moscatelli, M. Assessing seismic resilience of school educational sector. An attempt to establish the initial conditions in Calabria Region, southern Italy. *Int. J. Disaster Risk Reduct.* **2020**, *51*, 101936. [CrossRef]
19. Ruggieri, S.; Chatzidaki, A.; Vamvatsikos, D.; Uva, G. Reduced-order models for the seismic assessment of plan-irregular low-rise frame buildings. *Earthq. Eng. Struct. Dyn.* **2022**, *51*, 3327–3346. [CrossRef]

20. Leggieri, V.; Ruggieri, S.; Zagari, G.; Uva, G. Appraising seismic vulnerability of masonry aggregates through an automated mechanical-typological approach. *Autom. Constr.* **2021**, *132*, 103972. [CrossRef]
21. Longobardi, G.; Formisano, A. Seismic vulnerability assessment and consolidation techniques of ancient masonry buildings: The case study of a Neapolitan Masseria. *Eng. Fail. Anal.* **2022**, *138*, 106306. [CrossRef]
22. Brunelli, A.; de Silva, F.; Cattari, S. Site effects and soil-foundation-structure interaction: Derivation of fragility curves and comparison with Codes-conforming approaches for a masonry school. *Soil Dyn. Earthq. Eng.* **2022**, *154*, 107125. [CrossRef]
23. Law on the Reconstruction of Earthquake-Damaged Buildings in the City of Zagreb, Krapina-Zagorje County and Zagreb County (NN 102/2020). Available online: https://narodne-novine.nn.hr/clanci/sluzbeni/2020_09_102_1915.html (accessed on 29 June 2021).
24. Law on Reconstruction of Earthquake-Damaged Buildings in the City of Zagreb, Krapina-Zagorje County, Zagreb County, Sisak-Moslavina County and Karlovac County (NN 102/2020, 10/21). Available online: https://www.zakon.hr/z/2656 /Zakon-o-obnovi-zgrada-o%C5%A1te%C4%87enih-potresom-na-podru%C4%8Dju-Grada-Zagreba%2C-Krapinsko-zagorske-%C5%BEupanije%2C-Zagreba%C4%8Dke-%C5%BEupanije%2C-Sisa%C4%8Dko-moslava%C4%8Dke-%C5%BEupanije-i-Karlova%C4%8Dke-%C5%BEupani (accessed on 29 June 2021).
25. Tehnički Propis o Izmjeni i Dopunama Tehničkog Propisa za Građevinske Konstrukcije. Available online: https://narodne-novine.nn.hr/clanci/sluzbeni/2020_07_75_1448.html (accessed on 29 June 2021).
26. HRN EN 1998-3:2011 Eurocode 8: Design of Structures for Earthquake Resistance—Part 3: Assessment and Retrofitting of Buildings (EN 1998-3:2005+AC:2010). Available online: https://www.phd.eng.br/wp-content/uploads/2014/07/en.1998.3.2005.pdf (accessed on 29 June 2021).
27. Mazzoni, S.; Castori, G.; Galasso, C.; Calvi, P.; Dreyer, R.; Fischer, E.; Fulco, A.; Sorrentino, L.; Wilson, J.; Penna, A.; et al. 2016–2017 central italy earthquake sequence: Seismic retrofit policy and effectiveness. *Earthq. Spectra* **2018**, *34*, 1671–1691. [CrossRef]
28. Calderoni, B.; Cordasco, E.A.; Del Zoppo, M.; Prota, A. Damage assessment of modern masonry buildings after the L'Aquila earthquake. *Bull. Earthq. Eng.* **2020**, *18*, 2275–2301. [CrossRef]
29. Penna, A.; Morandi, P.; Rota, M.; Manzini, C.F.; da Porto, F.; Magenes, G. Performance of masonry buildings during the Emilia 2012 earthquake. *Bull. Earthq. Eng.* **2014**, *12*, 2255–2273. [CrossRef]
30. Bertolini Cestari, C.; Marzi, T. Conservation of historic timber roof structures of Italian architectural heritage: Diagnosis, assessment, and intervention. *Int. J. Archit. Herit.* **2018**, *12*, 632–665. [CrossRef]
31. Thöns, S. Value of Information analyses and decision analyses types. In Proceedings of the COST TU 1402 Training School of Structural Health Monitoring Information, Lake Como, Cadenabbia, Italy, 6–9 November 2017.
32. Scala, S.A.; Del Gaudio, C.; Verderame, G.M. Influence of construction age on seismic vulnerability of masonry buildings damaged after 2009 L'Aquila earthquake. *Soil Dyn. Earthq. Eng.* **2022**, *157*, 107199. [CrossRef]
33. Stepinac, M.; Kisicek, T.; Renić, T.; Hafner, I.; Bedon, C. Methods for the assessment of critical properties in existing masonry structures under seismic loads-the ARES project. *Appl. Sci.* **2020**, *10*, 1576. [CrossRef]
34. Sigmund, Z.; Radujkovic, M.; Lazarevic, D. Decision support model for seismic strengthening technology selection of masonry buildings. *Teh. Vjesn. Technol. Gaz.* **2016**, *23*, 791–800. [CrossRef]
35. Kouris, L.A.S.; Triantafillou, T.C. State-of-the-art on strengthening of masonry structures with textile reinforced mortar (TRM). *Constr. Build. Mater.* **2018**, *188*, 1221–1233. [CrossRef]
36. Kišiček, T.; Stepinac, M.; Renić, T.; Hafner, I.; Lulić, L. Strengthening of masonry walls with FRP or TRM. *Gradjevinar* **2020**, *72*, 937–953. [CrossRef]
37. Ghiassi, B.; Milani, G. *Numerical Modeling of Masonry and Historical Structures: From Theory to Application*; Woodhead Publishing: Sawston, UK, 2019.
38. Bournas, D.A. Concurrent seismic and energy retrofitting of RC and masonry building envelopes using inorganic textile-based composites combined with insulation materials: A new concept. *Compos. Part B Eng.* **2018**, *148*, 166–179. [CrossRef]
39. Al-Lami, K.; D'Antino, T.; Colombi, P. Durability of fabric-reinforced cementitious matrix (FRCM) composites: A review. *Appl. Sci.* **2020**, *10*, 1714. [CrossRef]
40. Rahman, S.; Akib, S.; Khan, M.T.R.; Shirazi, S.M. Experimental study on tsunami risk reduction on coastal building fronted by sea wall. *Sci. World J.* **2014**, *2014*, 729357. [CrossRef]
41. Skejić, D.; Lukačević, I.; Ćurković, I.; Čudina, I. Application of steel in refurbishment of earthquake-prone buildings. *Gradjevinar* **2020**, *72*, 955–966. [CrossRef]
42. Cao, X.Y.; Shen, D.; Feng, D.C.; Wang, C.L.; Qu, Z.; Wu, G. Seismic retrofitting of existing frame buildings through externally attached sub-structures: State of the art review and future perspectives. *J. Build. Eng.* **2022**, *57*, 104904. [CrossRef]
43. Triantafillou, T.C.; Bournas, D.A.; Gkournelos, P. *Novel Technologies for the Seismic Upgrading of Existing European Buildings*; Publications Office of the European Union: Luxembourg, 2022; JRC123314. [CrossRef]
44. Karic, A.; Atalić, J.; Kolbitsch, A. Seismic vulnerability of historic brick masonry buildings in Vienna. *Bull. Earthq. Eng.* **2022**, *20*, 4117–4145. [CrossRef]
45. Blagojević, P.; Bizev, S.; Cvetković, R. Simplified Seismic Assessment of Unreinforced Masonry Residential Buildings in the Balkans: The Case of Serbia. *Buildings* **2021**, *11*, 392. [CrossRef]
46. Available online: http://ss-petrinja.skole.hr/upload/ss-petrinja/newsattach/1020/Dan%20%B9kole-bro%B9ura.pdf (accessed on 2 June 2022).

47. Uroš, M.; Todorić, M.; Crnogorac, M.; Atalić, J.; Šavor Novak, M.; Lakušić, S. (Eds.) *Potresno Inženjerstvo—Obnova Zidanih Zgrada*; Građevinski fakultet, Sveučilišta u Zagrebu: Zagreb, Croatia, 2021; ISBN 978-953-8163-43-7.
48. Pojatina, J.; Barić, D.; Anđić, D.; Bjegović, D. Structural renovation of residential building in Zagreb after the 22 March 2020 earthquake. *Gradjevinar* **2021**, *73*, 633–648. [CrossRef]
49. Lulić, L.; Ožić, K.; Kišiček, T.; Hafner, I.; Stepinac, M. Post-earthquake damage assessment-case study of the educational building after the zagreb earthquake. *Sustainability* **2021**, *13*, 6353. [CrossRef]
50. Stepinac, M.; Rajčić, V.; Barbalić, J. Pregled i ocjena stanja postojećih drvenih konstrukcija. *Gradjevinar* **2017**, *69*, 861–873. [CrossRef]
51. Funari, M.F.; Pulatsu, B.; Szabó, S.; Lourenço, P.B. A solution for the frictional resistance in macro-block limit analysis of non-periodic masonry. *Structures* **2022**, *43*, 847–859. [CrossRef]
52. Santos, F.A.; Caroço, C.; Amendola, A.; Miniaci, M.; Fraternali, F. A concurrent micro/macro fe-model optimized with a limit analysis tool for the assessment of dry-joint masonry structures. *Int. J. Multiscale Comput. Eng.* **2022**, *20*, 53–64. [CrossRef]
53. Tomić, I.; Vanin, F.; Božulić, I.; Beyer, K. Numerical simulation of unreinforced masonry buildings with timber diaphragms. *Buildings* **2021**, *11*, 205. [CrossRef]
54. Asıkoğlu, A.; Vasconcelos, G.; Lourenço, P.B. Overview on the nonlinear static procedures and performance-based approach on modern unreinforced masonry buildings with structural irregularity. *Buildings* **2021**, *11*, 147. [CrossRef]
55. 3muri User Manual 12.2.1. Available online: https://www.3muri.com/en/brochures-and-manuals/ (accessed on 3 July 2022).
56. HRN EN 1996-1-1:2012 Eurocode 6: Design of Masonry Structures—Part 1-1: General Rules for Reinforced and Unreinforced Masonry Structures (EN 1996-1-1:2005+A1:2012). Available online: https://www.phd.eng.br/wp-content/uploads/2015/02/en.1996.1.1.2005.pdf (accessed on 2 June 2022).
57. Lagomarsino, S.; Penna, A.; Galasco, A.; Cattari, S. TREMURI program: An equivalent frame model for the nonlinear seismic analysis of masonry buildings. *Eng. Struct.* **2013**, *56*, 1787–1799. [CrossRef]
58. Penna, A.; Bracchi, S.; Salvatori, C.; Morandini, C.; Rota, M. Extending Analysis Capabilities of Equivalent Frame Models for Masonry Structures. In *European Conference on Earthquake Engineering and Seismology*; Springer: Cham, Switzerland, 2022; pp. 473–485. [CrossRef]
59. Lagomarsino, S.; Cattari, S.; Angiolilli, M.; Bracchi, S.; Rota, M.; Penna, A. Modelling and seismic response analysis of existing URM structures. Part 2: Archetypes of Italian historical buildings. *J. Earthq. Eng.* **2022**. [CrossRef]
60. Penna, A.; Rota, M.; Bracchi, S.; Angiolilli, M.; Cattari, S.; Lagomarsino, S. Modelling and seismic response analysis of existing URM structures. Part 1: Archetypes of Italian modern buildings. *J. Earthq. Eng.* **2022**. [CrossRef]
61. HRN EN 1991-1-1:2012 Eurocode 1: Actions on Structures—Part 1-1: General Actions—Densities, Self-Weight, Imposed Loads for Building (EN 1991-1-1:2002+AC:2009). Available online: https://www.phd.eng.br/wp-content/uploads/2015/12/en.1991.1.1.2002.pdf (accessed on 2 June 2022).
62. Herak, M.; Allegretti, I.; Herak, D.; Kuk, V.; Marić, K.; Markušić, S.; Sović, I. Maps of Seismic Areas of the Republic of Croatia. Available online: https://www.bib.irb.hr/615698 (accessed on 2 June 2022).
63. Krolo, J.; Damjanović, D.; Duvnjak, I.; Smrkić, M.F.; Bartolac, M.; Košćak, J. Methods for determining mechanical properties of walls. *Gradjevinar* **2021**, *73*, 127–140. [CrossRef]
64. Milić, M.; Stepinac, M.; Lulić, L.; Ivanišević, N.; Matorić, I.; Šipoš, B.Č.; Endo, Y. Assessment and Rehabilitation of Culturally Protected Prince Rudolf Infantry Barracks in Zagreb after Major Earthquake. *Buildings* **2021**, *11*, 508. [CrossRef]
65. Nakamura, Y.; Magenes, G.; Griffith, M. Comparison of pushover methods for simple building systems with flexible diaphragms. In Proceedings of the Australian Earthquake Engineering Society 2014 Conference, Lorne, Victoria, 21–23 November 2014.
66. Mirra, M.; Ravenshorst, G.; de Vries, P.; van de Kuilen, J.W. An analytical model describing the in-plane behaviour of timber diaphragms strengthened with plywood panels. *Eng. Struct.* **2021**, *235*, 112128. [CrossRef]
67. Mirra, M.; Ravenshorst, G.; van de Kuilen, J.-W. Comparing in-plane equivalent shear stiffness of timber diaphragms retrofitted with light and reversible wood-based techniques. *Pract. Period. Struct. Des. Constr.* **2021**, *26*, 04021031. [CrossRef]
68. Brignola, A.; Pampanin, S.; Podestà, S. Experimental evaluation of the in-plane stiffness of timber diaphragms. *Earthq. Spectra* **2012**, *28*, 1687–1709. [CrossRef]
69. Peralta, D.F.; Bracci, J.M.; Hueste, M.B.D. Seismic Behavior of Wood Diaphragms in Pre-1950s Unreinforced Masonry Buildings. *J. Struct. Eng.* **2004**, *130*, 2040–2050. [CrossRef]
70. Ciocci, M.P.; Marques, R.F.P.; Lourenço, P.B. Applicability of FEM and Pushover Analysis to Simulate the Shaking-Table Response of a Masonry Building Model with Timber Diaphragms. 2021. Available online: https://scholar.archive.org/work/7w3zeqlhv5g3lbqbwk7rlh6lm4/access/wayback/https://www.scipedia.com/wd/images/d/df/Draft_Content_751443795p1164.pdf (accessed on 15 July 2022).
71. Adhikari, R.K.; D'Ayala, D.F. Applied element modelling and pushover analysis of unreinforced masonry buildings with flexible roof diaphragm. *COMPDYN Proc.* **2019**, *2*, 3836–3851. [CrossRef]
72. Mirra, M.; Ravenshorst, G. A Seismic Retrofitting Design Approach for Activating Dissipative Behavior of Timber Diaphragms in Existing Unreinforced Masonry Buildings. In *Current Perspectives and New Directions in Mechanics, Modelling and Design of Structural Systems*; CRC Press: Boca Raton, FL, USA, 2022; pp. 1901–1907. [CrossRef]
73. Pantò, B.; Boem, I. Masonry elements strengthened with TRM: A review of experimental, design and numerical methods. *Buildings* **2022**, *12*, 1307. [CrossRef]
74. Galić, J.; Vukić, H.; Andrić, D.; Stepinac, L. *Tehnike Popravaka i Pojačanja Zidanih Zgrada*; Arhitektonski Fakultet: Zagreb, Croatia, 2020.

75. Galić, J.; Vukić, H.; Andrić, D.; Stepinac, L. *Priručnik za Protupotresnu Obnovu Postojećih Zidanih Zgrada*; Arhitektonski Fakultet: Zagreb, Croatia, 2020.
76. Milovanovic, B.; Bagaric, M.; Gaši, M.; Stepinac, M. Energy renovation of the multi-residential historic building after the Zagreb earthquake—Case study. *Case Stud. Therm. Eng.* **2022**, *38*, 102300. [CrossRef]
77. Milovanović, B.; Bagarić, M. How to achieve nearly zero-energy buildings standard. *Gradjevinar* **2020**, *72*, 703–720. [CrossRef]
78. Lourenço, P.B.; Barontini, A.; Oliveira, D.V.; Ortega, J. Rethinking Preventive Conservation: Recent Examples. In *Geotechnical Engineering for the Preservation of Monuments and Historic Sites III.*; CRC Press: Boca Raton, FL, USA, 2022; pp. 70–86. [CrossRef]
79. LCA Software for Informed Change-Makers. Available online: https://simapro.com/ (accessed on 23 June 2022).

Article

A Design Methodology for the Seismic Retrofitting of Existing Frame Structures Post-Earthquake Incident Using Nonlinear Control Systems

Assaf Shmerling [1,*] and Matthias Gerdts [2]

1   Department of Civil and Environmental Engineering, Ben-Gurion University of the Negev, Beer-Sheva 84105, Israel
2   Institute of Applied Mathematics and Scientific Computing, Universität der Bundeswehr München, 85577 Neubiberg, Germany
*   Correspondence: assafs@bgu.ac.il

Abstract: A structural design methodology for retrofitting weakened frame systems following earthquakes is developed and presented. The design procedure refers to frame systems in their degraded strength and stiffness states and restores their dynamic performance using nonlinear control systems. The control law associated with the employed systems regards the gains between the negative state feedback and the control force, which consists of linear, nonlinear, and hysteretic portions. Structural optimization is introduced in designing the nonlinear control systems, and the controller gains are optimized using the fixed-point iteration to improve the frame system's dynamic performance. The fixed-point iteration method relates to first-order PDE equations; hence, a new state-space formulation for weakened inelastic frame systems is developed and presented using the frame system's lateral force equilibrium equation. The design scheme and optimization strategy differ from designing passive control systems, given that the nonlinear control system's force consists of linear, nonlinear, and hysteretic portions. The utilization of the fixed-point iteration in the structural design area is by itself a novel application due to its robustness in addressing the gains of any type of nonlinear control system. This paper's nonlinear control system chosen to exhibit the application is Buckling Restrained Braces (BRBs) since force consists of linear and hysteretic portions. The implementation of hysteretic control force is rare in structural control applications. In the case of BRBs, the fixed-point iteration optimizes the cross-sectional areas. Two system optimization examples of 3-story and 15-story inelastic frames are provided and described. The examples demonstrate the fixed-point iteration's applicability and robustness in optimizing control gains of nonlinear systems and regulating the dynamic response of weakened frame structures.

Keywords: Inelastic frame systems; post-earthquake design; nonlinear control; fixed-point iteration; control gain optimization

Citation: Shmerling, A.; Gerdts, M. A Design Methodology for the Seismic Retrofitting of Existing Frame Structures Post-Earthquake Incident Using Nonlinear Control Systems. *Buildings* **2022**, *12*, 1886. https://doi.org/10.3390/buildings12111886

Academic Editors: Mislav Stepinac, Chiara Bedon, Marco Francesco Funari, Tomislav Kišiček and Ufuk Hancilar

Received: 16 September 2022
Accepted: 1 November 2022
Published: 4 November 2022

1. Introduction

Seismic retrofitting inelastic lateral load resisting systems is challenging primarily due to the unpredictable earthquake response and final damage state. It becomes even more challenging when retrofitting a system that is already damaged, following an earthquake incident, since the system is characterized by stiffness and strength degradations. Nevertheless, the design philosophy and application remain the same when utilizing control systems for reinstating and improving the existing system. According to well-accepted global damage indices, the weakened state of the structure after an earthquake can be considered by its maximum fundamental vibrations period, related to the most significant stiffness degradation.

The global damage indices indicate the damage state of the whole system to determine its functionality level—unlike local damage indices, which look at a particular member (see, for example, [1–4]). Global damage equations compare the parameters of the structure (e.g., modal frequencies) before and after the earthquake excitation to quantify the system's weakening. For example, DiPasquale et al. [5] developed the "plastic softening" and "final softening" global damage indices for reinforced concrete structures. The "plastic softening" damage index provides a good measurement of plastic deformation and soil-structure interaction during seismic excitation (regarding the final and maximum period ratio). In contrast, the "final softening" indicates the state of the structure after dynamic excitation (regarding the initial and final period ratio). DiPasquale and Cakmak [6] followed suit and developed the "maximum softening" global damage index, indicating the ultimate stiffness degradation related to the maximum fundamental vibration period under seismic excitation. In recent years, a few researchers followed suit in developing new terms for the global damage index (e.g., [7,8]). According to the review paper of Williams and Sexsmith [9], it is considered the best indicator of the global damage state. In this paper, we adopt the maximum softening philosophy and address the maximum fundamental vibrations period of the frame system. At this point, we employ nonlinear control systems.

Various design methodologies have been developed and exemplified their effectiveness in regulating the earthquake response of frame systems by utilizing control systems. The procedures for designing control systems usually aim at optimizing their parameters to minimize a particular objective function as a part of an optimization problem. See, for example, the applications of Lagrange multipliers in [10–13], gradient-based optimization in [14–20], the linear quadratic regulator in [21–26], and the direct probability integral approach in [27–29]. In such cases, when optimizing static parameters (i.e., static gains), most optimization algorithms combine a technique to assess/estimate/address the inelastic response by an equivalent linear analysis and, if necessary, recalculate the gains corresponding to the inelastic system. One example is the procedure by Shmerling and Levy [18], which utilizes the direct-displacement-based approach for simplifying inelastic frame systems into an equivalent linear system in interstory drift deformations.

Researchers with unique optimization typologies usually regard the system's equilibrium equation, induced force, or stress. For example, Potra and Simiu [30] express the column's stress under extreme loads (e.g., earthquakes) and introduce it to a nonlinear dynamic programming algorithm. Shmerling and Levy [31] use the general system interconnection paradigm to represent the dynamic equilibrium of rigid frame systems in the Laplace domain. Smarra et al. [32] represent the LQR objective function and state-space in a discrete manner suitable for the predictive horizon approach implementation. This paper proposes a new inelastic state-space formulation for nonlinear static gains control systems and later refers to BRBs as a nonlinear control system following [33,34].

The effectiveness of the BRBs application in upgrading and improving the dynamic performance of civil structures is well acknowledged. Besides being considered for upgrading rigid frame systems, BRBs have also been proposed to enhance the performance of different infrastructures, such as nuclear plant turbine buildings [35] and steel arch bridges [36]. There are various optimization techniques for designing BRBs. Balling et al. [37] present a nine-step algorithm that performs nonlinear time history and reconfigures the BRBs according to the redesign equations. Hoffman and Richards [38] introduce four different genetic algorithms (baseline, forced diversity, adaptive mutation, and noncrossover adaptive mutation). Abedini et al. [39] solve an optimization problem in which the objective function is the BRBs' weight and the dissipated energy using the metaheuristic salp swarm and colliding bodies algorithms. Pan et al. [40] design procedure that calculates the minimal weight of BRBs subject to the global buckling prevention criterion and the stiffness–strength relationship curve. Rezazadeha and Talatahari [41] address the seismic input energy to the structure and the absorbed yielding energy and utilize the vibrating particles system algorithm to achieve the optimal BRBs configuration. Tu et al. [42] optimally allocate BRBs to frame structures while referring to the deformation and damage constraints.

In this paper, the motivation for employing BRBs stems from its control law, which consists of linear and hysteretic control forces corresponding to the resisting force's elastic and inelastic portions. Control force of hysteretic nature is more challenging to implement in control theory due to the gains being subject to integration terms. Nevertheless, the fixed-point iteration method is suitable for such a control system due to addressing its state trajectory.

2. Inelastic State-Space

The inelastic state-space formulation presented herein enhances the equations scheme in [26] for frame structures, which refers to the lateral force equilibrium of an inelastic rigid frame system under lateral loads. In this paper, the frame structure is equipped with a nonlinear control system of static gains that induces either nonlinear, hysteretic, or linear portions—depending on the system type. The nonlinear control system chosen in this paper is BRB, whose applied force consists of hysteretic and linear parts. In this case, the static gains are the BRBs' cross-sectional areas.

The addressed lateral force equilibrium considers the inelastic behavior of the structure since even while we use a control device to keep the system close to its elastic range, this may not always be the case. The expression of the condensate equation governing the lateral force equilibrium is:

$$
\mathbf{m}\,\ddot{\mathbf{x}}(t) + \mathbf{c}\,\dot{\mathbf{x}}(t) + \mathbf{T}'_{x \to d}\mathbf{f}^F\left(\dot{\mathbf{f}}^F,\mathbf{d},\dot{\mathbf{d}}\right) + \mathbf{u}\left(\Sigma \mathbf{A},\dot{\mathbf{u}},\mathbf{x},\dot{\mathbf{x}}\right) = \mathbf{p}(t)
$$
$$
\text{and}:
$$
$$
\mathbf{x}(0) = 0
$$
$$
\dot{\mathbf{x}}(0) = 0
$$

(1)

where $(\ )'$ denotes the conjugate-transpose, $\mathbf{x}(t)$ is the N-dimensional ceilings' relative-to-ground displacement vector, $\dot{\mathbf{x}}(t)$ is the N-dimensional ceilings' relative-to-ground velocities vector, $\ddot{\mathbf{x}}(t)$ is the N-dimensional ceilings' relative-to-ground accelerations vector, $\mathbf{m}$ is the $N \times N$ static-condensate diagonal mass matrix, $\mathbf{c}$ is the $N \times N$ inherent damping matrix, $\mathbf{p}(t)$ is the N-dimensional lateral dynamic load vector, $\mathbf{f}^F\left(\dot{\mathbf{f}}^F,\mathbf{d},\dot{\mathbf{d}}\right)$ is the N-dimensional structural rigid frame system's lateral resisting force vector, and $\mathbf{u}\left(\Sigma \mathbf{A},\mathbf{u},\mathbf{x},\dot{\mathbf{x}}\right)$ is the N-dimensional nonlinear static gains control force vector. The vector function $\mathbf{d}(t)$, which is referred by $\mathbf{f}^F\left(\dot{\mathbf{f}}^F,\mathbf{d},\dot{\mathbf{d}}\right)$, is the N-dimensional interstory drifts vector—given by the transformation from displacement coordinates into drift coordinates:

$$
\mathbf{d}(t) = \mathbf{T}_{x \to d}\mathbf{x}(t) \ \leftrightarrow \ \mathbf{T}_{x \to d} = \begin{bmatrix} 1 & & & \\ -1 & 1 & & \\ & \ddots & \ddots & \\ & & -1 & 1 \end{bmatrix}
$$

(2)

where $\mathbf{T}_{x \to d}$ denotes the transformation matrix, and its reverse form is:

$$
\mathbf{x}(t) = \mathbf{T}_{d \to x}\mathbf{d}(t) \ \leftrightarrow \ \mathbf{T}_{d \to x} = \begin{bmatrix} 1 & & \\ \vdots & \ddots & \\ 1 & \cdots & 1 \end{bmatrix}
$$

(3)

Figure 1 depicts the structural deformations in terms of $\mathbf{x}(t)$ and $\mathbf{d}(t)$ to exemplify the difference between the two.

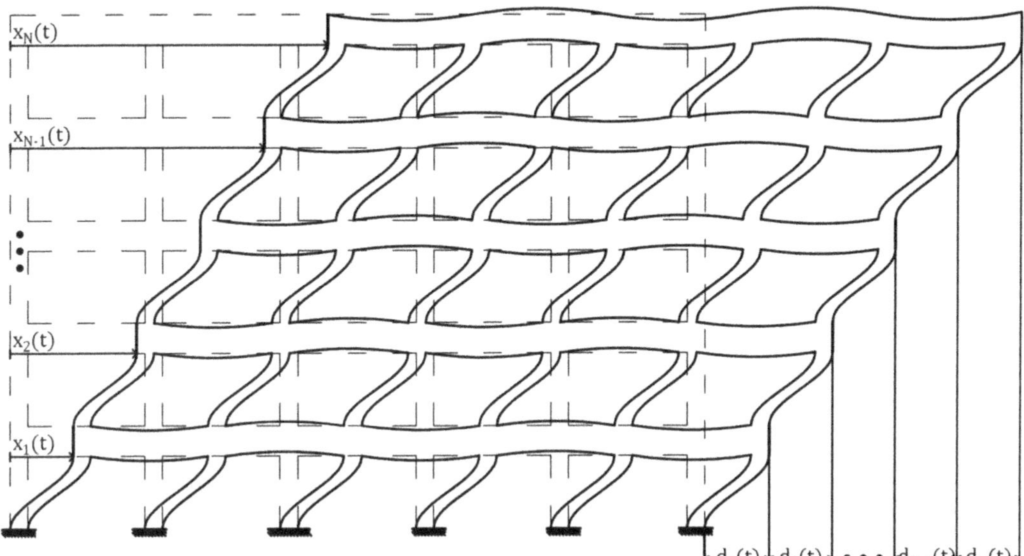

Figure 1. Rigid frame system deformations in terms of relative-to-ground displacements $(x_1, x_2, \ldots, x_{N-1}, x_N)$ and interstory drifts $(d_1, d_2, \ldots, d_{N-1}, d_N)$.

The hysteretic model of $\mathbf{f}^F$ is expressed as a combination of its elastic and hysteretic portions. Considering that:

$$\mathbf{f}^F\left(\dot{\mathbf{f}}^F, \mathbf{d}, \dot{\mathbf{d}}\right) = \int_0^t \dot{\mathbf{f}}^F\left(\mathbf{f}^F, \mathbf{d}, \dot{\mathbf{d}}\right) d\tau = \mathbf{k}^{F,el}\,\mathbf{d}(t) + \int_0^t \dot{\mathbf{f}}^{F,hys}\left(\mathbf{f}^F, \mathbf{d}, \dot{\mathbf{d}}\right) d\tau \tag{4}$$

where $\mathbf{k}^{F,el}$ is the $N \times N$ static-condensate matrix elastic stiffness portion about $\mathbf{d}(t)$, and $\mathbf{f}^{F,hys}$ is the N-dimensional hysteretic portion of $\mathbf{f}^F$, and give that:

$$\dot{\mathbf{f}}^{F,hys}(t) = \mathbf{k}^{F,hys}\left(\mathbf{f}^F, \mathbf{d}, \dot{\mathbf{d}}\right)\dot{\mathbf{d}}(t) \tag{5}$$

where $\mathbf{k}^{F,hys}$ is the $N \times N$ static-condensate hysteretic stiffness matrix, the force $\mathbf{f}^F$ is formulated as:

$$\mathbf{f}^F\left(\dot{\mathbf{f}}^F, \mathbf{d}, \dot{\mathbf{d}}\right) = \mathbf{k}^{F,el}\,\mathbf{d}(t) + \int_0^t \mathbf{k}^{F,hys}\left(\mathbf{f}^F, \mathbf{d}, \dot{\mathbf{d}}\right)\dot{\mathbf{d}}(\tau) d\tau \tag{6}$$

The control force $\mathbf{u}$ is defined by linear, hysteretic, and nonlinear control force portions. That is:

$$\mathbf{u}\left(\Sigma\mathbf{A}, \dot{\mathbf{u}}, \mathbf{x}, \dot{\mathbf{x}}\right) = \mathbf{k}^{u,el}(\Sigma\mathbf{A})\mathbf{x}(t) + \mathbf{c}^{u,el}(\Sigma\mathbf{A})\dot{\mathbf{x}}(t) + \int_0^t \mathbf{k}^{u,hys}(\Sigma\mathbf{A}, \mathbf{u}, \mathbf{x}, \dot{\mathbf{x}})\dot{\mathbf{x}}(\tau) d\tau + \mathbf{f}^{u,NL}(\Sigma\mathbf{A}, \mathbf{x}, \dot{\mathbf{x}}) \tag{7}$$

where $\Sigma\mathbf{A}$ is an N-dimensional vector consisting of the static gain variables, $\mathbf{k}^{u,el}$ is the $N \times N$ linear stiffness matrix, $\mathbf{c}^{u,el}$ is the $N \times N$ linear damping matrix, $\mathbf{k}^{u,hys}$ is the $N \times N$ hysteretic stiffness matrix, and $\mathbf{f}^{u,NL}$ denotes the N-dimensional nonlinear portion of $\mathbf{u}$.

The applied dynamic load vector $\mathbf{p}(t)$ is modeled as the quasi-static cyclic loading whose maximum amplitude reaches twice the total yielding force. As the numerical examples exemplify, adhering to such load supports the system remaining close to its elastic state under a significant earthquake to maintain our weakened frame structure.

Figure 2 illustrates the normalized nth story load $\mathbf{p}_n(t)$ about the total yielding force so that $\mathbf{f}_n^{F,yld}$ is the nth story columns' shear force at first yield versus the normalized load duration about the highest modal period T_1.

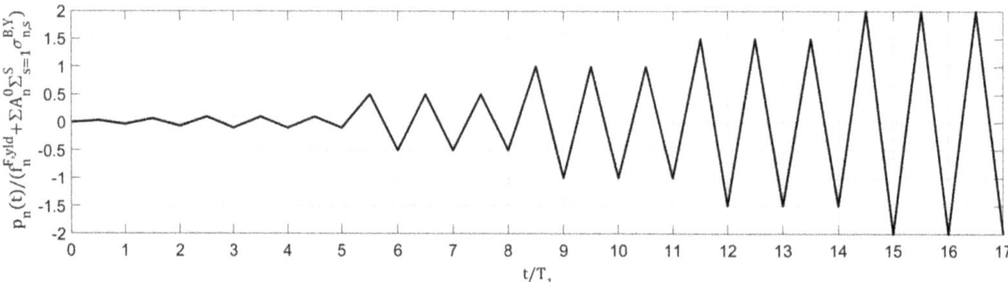

Figure 2. Quasi-static cyclic loading.

The inelastic state-space formulation refers to the term $\dot{\mathbf{f}}^{F,hys} = \mathbf{k}^{F,hys}\dot{\mathbf{d}}$ of Equation (5) as a separate entity within the following *4N*-dimensional state-vector $\mathbf{z}(t)$:

$$\mathbf{z}(t) = \begin{bmatrix} \mathbf{x}(t) \\ \int_0^t \mathbf{f}^{F,hys}(\mathbf{z}(t))d\tau \\ \dot{\mathbf{x}}(t) \\ \mathbf{f}^{F,hys}(\mathbf{z}(t)) \end{bmatrix} \tag{8}$$

Consequently, the corresponding *4N* × *4N* state matrix $\mathbf{A}(\mathbf{z})$ is:

$$\mathbf{A}(\mathbf{z}(t)) = \begin{bmatrix} 0 & 0 & \mathbf{I} & 0 \\ 0 & 0 & 0 & \mathbf{I} \\ -\mathbf{m}^{-1}\mathbf{T}'_{x \to d}\mathbf{k}^{F,el}\mathbf{T}_{x \to d} & 0 & -\mathbf{m}^{-1}\mathbf{c} & -\mathbf{m}^{-1} \\ 0 & 0 & \mathbf{k}^{F,hys}(\mathbf{z}(t))\mathbf{T}_{x \to d} & 0 \end{bmatrix} \tag{9}$$

and the inelastic state-space equation is:

$$\dot{\mathbf{z}}(t) = \mathbf{A}(\mathbf{z}(t))\mathbf{z}(t) + \mathbf{B}\big(\mathbf{p}(t) + \mathbf{u}(\Sigma\mathbf{A}, \dot{\mathbf{u}}, \mathbf{x}, \dot{\mathbf{x}})\big) \tag{10}$$

where the *4N* × *N* input-to-state matrix $\mathbf{B}$ is:

$$\mathbf{B} = \begin{bmatrix} 0 \\ 0 \\ \mathbf{m}^{-1} \\ 0 \end{bmatrix} \tag{11}$$

Since $\mathbf{z}(t)$ consists of $\mathbf{x}(t)$ and $\dot{\mathbf{x}}(t)$ the nonlinear static gains control force is henceforth denoted as $\mathbf{u}(\Sigma\mathbf{A}, \mathbf{z}(t))$.

Figure 3 depicts the closed-loop control process. The closed-loop paradigm comprises the inelastic state-space, defined above, and the control law regarding the various control force portions (linear, hysteretic, and nonlinear). The matrices $\mathbf{T}^{hys}$, $\mathbf{T}^{lin}$, and $\mathbf{T}^{NL}$ are portion transformations matrices applied to the negative feedback of $\mathbf{z}$ to yield the respected portions of the control force, and $\mathbf{G}(\Sigma\mathbf{A}, \mathbf{z}(t))$ is the $N \times 4N$ gain matrix. The BRBs system is chosen to regulate the frame structure and to exemplify the developed methodology. The BRB's control law is composed of linear force, relating to the BRB elastic portion, and hysteretic force, relating to the material inelasticity.

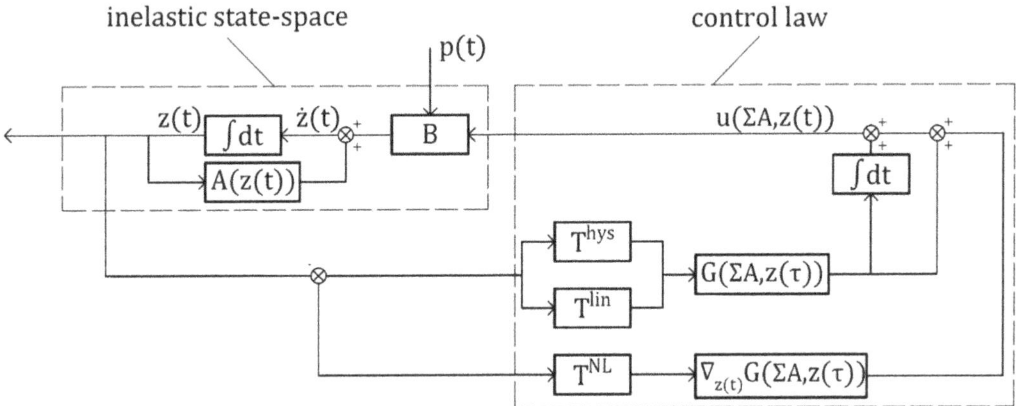

Figure 3. Proposed state-space paradigm.

3. BRB Control Law

The BRB system is chosen herein since it consists of an inelastic force portion. Inelastic behavior is more challenging to implement in control theory because of its hysteretic nature. Various research projects have examined BRB behavior over the last decades. Recently, Tremblay et al. [43] tested six BRBs segments and examined different brace cores, mechanisms, and profiles with/without unbinding material. The research shows that certain BRB specimens exhibited a predictable elastic response and a ductile and stable inelastic response, without fracture, under the cyclic quasi-static loading plus four additional large-amplitude tension cycles. This paper's analytical model considers such idealized cyclic behavior. Figure 4 depicts a rigid frame system supplied with BRBs—showcasing its installation.

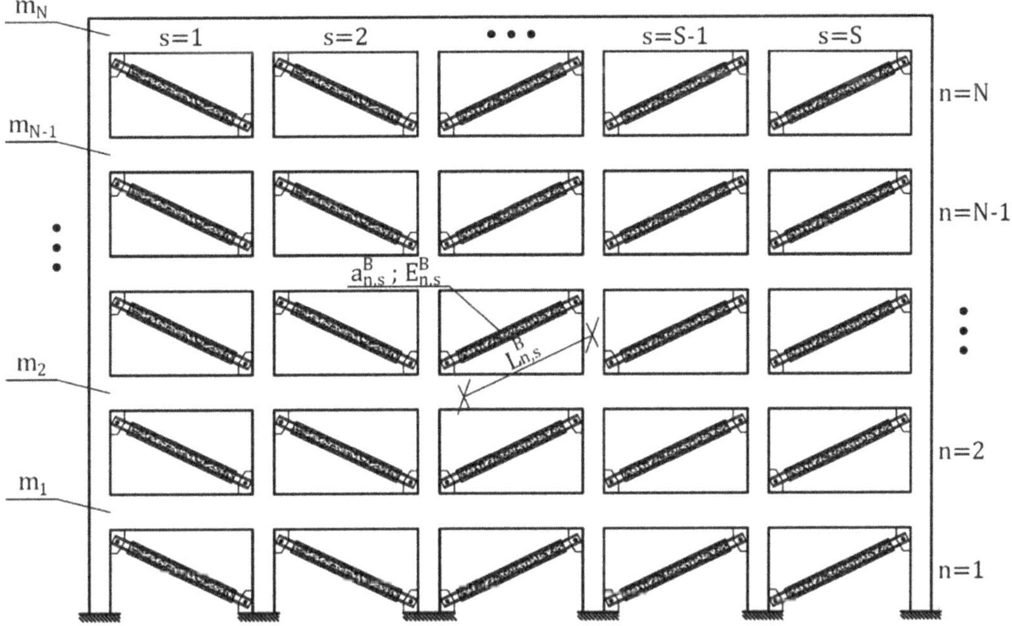

Figure 4. Rigid frame system elevation scheme equipped with BRBs.

The BRB cyclic model is composed of linearly elastic and hysteretic stiffness portions and refers to each span of the rigid frame system individually. Therefore, the control force of Equation (7) is reduced to:

$$\mathbf{u}(\Sigma\mathbf{A},\mathbf{u},\mathbf{x},\dot{\mathbf{x}}) = \mathbf{k}^{u,el}(\Sigma\mathbf{A})\mathbf{x}(t) + \int_0^t \mathbf{k}^{u,hys}(\Sigma\mathbf{A},\mathbf{u},\mathbf{x},\dot{\mathbf{x}})\dot{\mathbf{x}}(\tau)d\tau \tag{12}$$

and the matrices $\mathbf{k}^{u,el}$ and $\mathbf{k}^{u,hys}$ are expressed as:

$$\mathbf{k}^{u,el}(\Sigma\mathbf{A}) = (\mathbf{T}_{x\to\Delta})'\mathrm{diag}\left\{\left[\left(\mathbf{a}^B\odot\mathbf{E}^B\oslash\mathbf{L}^B\right)\mathbf{1}^S\right]\odot\Sigma\mathbf{A}\right\}\mathbf{T}_{x\to\Delta} \tag{13}$$

$$\mathbf{k}^{u,hys}(\Sigma\mathbf{A},\mathbf{u},\mathbf{x},\dot{\mathbf{x}}) = (\mathbf{T}_{x\to\Delta})'\mathrm{diag}\left\{\left[\left(1-\mathbf{a}^B\right)\odot\left(\rho^B(\Sigma\mathbf{A},\mathbf{u},\mathbf{x},\dot{\mathbf{x}})\odot\mathbf{E}^B\oslash\mathbf{L}^B\right)\mathbf{1}^S\right]\odot\Sigma\mathbf{A}\right\}\mathbf{T}_{x\to\Delta} \tag{14}$$

where $\odot$ denotes element-wise multiplication, $\oslash$ marks element-wise division, $\mathbf{a}^B$ is an $N \times S$ matrix consisting of the ratios between the BRBs' axial plastic and elastic stiffnesses, $\mathbf{E}^B$ is an $N \times S$ matrix composed of the BRBs' elasticity modulus, $\mathbf{A}^B$ is an $N \times S$ matrix containing the BRBs' effective cross-sectional area, $\mathbf{L}^B$ is an $N \times S$ matrix comprised of the BRBs' length, ρ^B is an $N \times S$ matrix that corresponds to the material's elastic/plastic/unloading stages, $\mathbf{1}^S$ an S-dimensional vector of ones, $\mathbf{T}_{x\to\Delta}$ is the geometric transformation matrix from $\mathbf{x}(t)$ into the BRB's axial deformation $\Delta(t)$, and $\Sigma\mathbf{A}$ is an N-dimensional vector composed of the total cross-sectional area quantities:

$$\Sigma\mathbf{A} = \begin{bmatrix} \Sigma_{s=1}^S A_{1,s}^B \\ \vdots \\ \Sigma_{s=1}^S A_{N,s}^B \end{bmatrix} \tag{15}$$

The parameter $A_{n,s}^B$ denotes the cross-sectional area of the BRB installed at the sth span of the nth story.

The matrix ρ^B is defined using the Bouc–Wen model equations. Assuming the BRB does not experience stiffness and strength degradations, the unloading curve is parallel to the elastic stiffness—which gives:

$$\rho^B(\Sigma\mathbf{A},\mathbf{u},\mathbf{x},\dot{\mathbf{x}}) = 1 - 0.5\left|\sigma^{in}(\sigma^B,\mathbf{x})\oslash(1-\mathbf{a}^B)\oslash\sigma^{B,Y}\right|^\nu\odot\left[1+\mathrm{sign}\left(\dot{\Delta}^B(\dot{\mathbf{x}})\odot\sigma^{in}(\sigma^B,\mathbf{x})\right)\right]$$

and :
$$\sigma^B(\Sigma\mathbf{A},\mathbf{u}) = \left[\left(\mathbf{u}(\Sigma\mathbf{A},\dot{\mathbf{u}},\mathbf{x},\dot{\mathbf{x}})\oslash\Sigma\mathbf{A}\right)\mathbf{1}^{S'}\right]\odot\left[\mathbf{A}^B\oslash\left(\Sigma\mathbf{A}\mathbf{1}^{S'}\right)\right] \tag{16}$$

where $\sigma^{B,Y}$ is an $N \times S$ matrix of the BRBs' yield stress, $\dot{\Delta}^B(t)$ is the BRBs' axial deformation rate, and $\sigma^{in}(t)$ is an $N \times S$ matrix referring to the inelastic portion of the BRBs' axial stress. The matrices $\dot{\Delta}^B(t)$ and $\sigma^{in}(t)$ are defined as:

$$\dot{\Delta}^B(\dot{\mathbf{x}}) = \mathbf{T}_{x\to\Delta}\dot{\mathbf{x}}(t)\mathbf{1}^{S'} \tag{17}$$

$$\sigma^{in}\left(\sigma^B,\mathbf{x}\right) = \sigma^B\left(\mathbf{f}^B\right) - \mathbf{a}^B\odot\mathbf{E}^B\oslash\mathbf{L}^B\odot\left(\mathbf{T}_{x\to\Delta}\mathbf{x}(t)\mathbf{1}^{S'}\right) \tag{18}$$

The BRB cyclic model involves a control law with a linear force portion and a hysteretic force portion. That is:

$$\mathbf{u}(\Sigma\mathbf{A},\dot{\mathbf{u}},\mathbf{z}(t)) = -\mathbf{G}(\Sigma\mathbf{A},\mathbf{z}(\tau))\mathbf{T}^{lin}\mathbf{z}(\tau) - \int_0^t \mathbf{G}(\Sigma\mathbf{A},\mathbf{u},\mathbf{z}(t))\mathbf{T}^{hys}\mathbf{z}(\tau)d\tau$$
and :
$$\mathbf{G}(\Sigma\mathbf{A},\mathbf{u},\mathbf{z}(t)) = \left[\mathbf{k}^{B,el}(\Sigma\mathbf{A})\;0\;\mathbf{k}^{u,hys}(\Sigma\mathbf{A},\mathbf{u},\mathbf{z})\;0\right]$$

$$\mathbf{T}^{lin} = \begin{bmatrix} \mathbf{I} & & \\ & 0 & \\ & & 0 \\ & & 0 \end{bmatrix} ;\; \mathbf{T}^{hys} = \begin{bmatrix} 0 & & \\ & 0 & \\ & & \mathbf{I} \\ & & 0 \end{bmatrix} \tag{19}$$

where $\mathbf{k}^{B,el}(\mathbf{\Sigma A})$ and $\mathbf{k}^{u,hys}(\mathbf{\Sigma A}, \mathbf{u}, \mathbf{z})$ are defined by Equations (13) and (14), respectively. The implicit nature of $\mathbf{u}$ requires an iterative optimization approach to finding the optimal $\mathbf{\Sigma A}$. Here, the fixed-point iteration is introduced since it is capable of optimizing $\mathbf{\Sigma A}$ regardless of $\mathbf{u}$ form.

4. Fixed-Point Iteration

The intended utilization of the fixed-point iteration optimizes the static gain variables of nonlinear nonautonomous control systems. This paper's fixed-point iteration minimizes the rigid frame system's interstory drift and the drift velocity trajectories while referring to BRBs. The objective function is subject to the inelastic state-space, initial conditions, and control variable limitations. The complete optimization problem is given by:

$$
\begin{aligned}
\underset{\mathbf{\Sigma A}}{\text{minimize}} \quad & \left\{ J = \int_0^{t_f} \mathbf{z}^T(t)\mathbf{Q}\mathbf{z}(t)dt = \int_0^{t_f} \mathbf{d}^T(t)\mathbf{Q}_1\mathbf{d}(t) + \dot{\mathbf{d}}^T(t)\mathbf{Q}_2\dot{\mathbf{d}}(t)dt \right\} \\
\text{subject to} \quad & \dot{\mathbf{z}}(t) = \mathbf{A}(\mathbf{z}(t))\mathbf{z}(t) + \mathbf{B}(\mathbf{p}(t) + \mathbf{u}(\mathbf{\Sigma A}, \mathbf{z})) \\
& \mathbf{z}(0) = \mathbf{0} \\
& \mathbf{\Sigma A}^{max} - \mathbf{\Sigma A} \geq \mathbf{0} \\
& \mathbf{\Sigma A} \geq \mathbf{0}
\end{aligned}
\tag{20}
$$

Since dealing with BRBs, $\mathbf{\Sigma A}$ stands for the total cross-sectional areas, which are limited by 0 and $\mathbf{\Sigma A}^{max}$. The two objective function forms in Equation (20) imply that the weighting matrix $\mathbf{Q}$ transforms $\mathbf{z}(t)$ into $\mathbf{d}(t)$ and $\dot{\mathbf{d}}(t)$. That is:

$$
\mathbf{Q} = \begin{bmatrix} (\mathbf{T}_{x \to d})'\mathbf{Q}_1\mathbf{T}_{x \to d} & & \\ & \mathbf{0} & \\ & & (\mathbf{T}_{x \to d})'\mathbf{Q}_2\mathbf{T}_{x \to d} \\ & & & \mathbf{0} \end{bmatrix}
\tag{21}
$$

where $\mathbf{Q}_1$ and $\mathbf{Q}_2$ are diagonal weighting matrices whose components govern the minimization priority between $\mathbf{d}(t)$ and $\dot{\mathbf{d}}(t)$, respectively. The hysteretic components of $\mathbf{z}$ do not participate in the objective function.

The fixed-point iteration stems from the Lagrange function expression added by the initial condition $\mathbf{z}(0) = \mathbf{0}$ and the Karush–Kuhn–Tucker (KKT) conditions for $\mathbf{\Sigma A} - \mathbf{\Sigma A}^{max} \leq 0$ and $\mathbf{\Sigma A} \geq 0$. That is:

$$
\begin{aligned}
\mathcal{L}(\mathbf{z}, \mathbf{\Sigma A}, \lambda, \sigma, \mu) = & \int_0^{t_f} H(t, \mathbf{z}(t), \mathbf{\Sigma A}, \lambda(t)) - \lambda'(t)\dot{\mathbf{z}}(t)dt + \dots \\
& \mu_1'(\mathbf{\Sigma A} - \mathbf{A}^{max}) + \mu_2'(-\mathbf{\Sigma A}) + \sigma'(\mathbf{z}(0) - \mathbf{0})
\end{aligned}
\tag{22}
$$

where $\lambda(t)$ is the $4N$-dimensional time-varying Lagrange multipliers vector, σ is a $4N$-dimensional multipliers vector that governs, $\mathbf{z}(0) = \mathbf{0}$, μ_1 and μ_2 are the KKT multipliers vector governing the design limitation inequality constraints, and H is the Hamilton function, which is given by:

$$
H(t, \mathbf{z}(t), \mathbf{\Sigma A}, \lambda(t)) = \mathbf{z}(t)'\mathbf{Q}\mathbf{z}(t) + \lambda'(t)[\mathbf{A}(t)\mathbf{z}(t) + \mathbf{B}(\mathbf{p}(t) + \mathbf{u}(t))]
\tag{23}
$$

Appendix A elaborates on the Lagrange function and develops the conditions for optimality. It ends with the following set of requirements:

Adjoint conditions :
$$\dot{\lambda}(t) = -\nabla_{\mathbf{z}(t)} H(t, \mathbf{z}(t), \mathbf{\Sigma A}, \lambda(t))$$
$$\dot{\eta}(t) = -\nabla_{\mathbf{\Sigma A}} H(t, \mathbf{z}(t), \mathbf{\Sigma A}, \lambda(t))$$
Transversality conditions :
$$\lambda(t_f) = \mathbf{0}$$
$$\eta(t_f) = \mathbf{0} \tag{24}$$
Stationary condition :
$$\mu_1 - \mu_2 + \eta(0) = 0$$
Complementarity condition :
$$\mu = \begin{bmatrix} \mu_1 \\ \mu_2 \end{bmatrix} \geq \mathbf{0}, \ \mu_1'(\mathbf{\Sigma A} - \mathbf{\Sigma A}^{max}) - \mu_2'\mathbf{\Sigma A} = 0, \quad \begin{bmatrix} \mathbf{\Sigma A} - \mathbf{A}^{max} \\ -\mathbf{\Sigma A} \end{bmatrix} \leq \mathbf{0}$$

where the function $\eta(t)$ is defined as:

$$\eta(t) = \int_t^{t_f} \nabla_{\mathbf{\Sigma A}} H(t) d\tau \tag{25}$$

and the gradients of H in $\mathbf{z}(t)$ and $\mathbf{\Sigma A}$ are:

$$\nabla_{\mathbf{z}(t)} H(t, \mathbf{z}(t), \mathbf{\Sigma A}, \lambda(t)) = 2\mathbf{Q}\,\mathbf{z}(t) + \left(\mathbf{A}(\mathbf{z}(t)) + \nabla_{\mathbf{z}(t)}\mathbf{A}(\mathbf{z}(t))'\mathbf{z}'(t) + \mathbf{u}'_{\mathbf{z}(t)}(\mathbf{\Sigma A}, \dot{\mathbf{u}}, \mathbf{z}(t))\mathbf{B}' \right)\lambda(t) \tag{26}$$

$$\nabla_{\mathbf{\Sigma A}} H(t, \mathbf{z}(t), \mathbf{\Sigma A}, \lambda(t)) = \mathbf{u}'_{\mathbf{\Sigma A}}(\mathbf{\Sigma A}, \dot{\mathbf{u}}, \mathbf{z}(t))\mathbf{B}'\lambda(t) \tag{27}$$

Regarding the gradient expressions, the control force derivatives $\mathbf{u}_{\mathbf{z}(t)}(\mathbf{\Sigma A}, \dot{\mathbf{u}}, \mathbf{z}(t))$ and $\mathbf{u}_{\mathbf{\Sigma A}}(\mathbf{\Sigma A}, \dot{\mathbf{u}}, \mathbf{z}(t))$ are written as:

$$\mathbf{u}_{\mathbf{z}(t)}(\mathbf{\Sigma A}, \dot{\mathbf{u}}, \mathbf{z}(t)) = -\mathbf{k}^{B,el}(\mathbf{\Sigma A}) - \int_0^t \mathbf{k}^{B,hys}{}_{\mathbf{z}(\tau)}(\mathbf{\Sigma A}, \mathbf{u}, \mathbf{z}(t))\mathbf{z}(\tau) + \mathbf{k}^{u,hys}(\mathbf{\Sigma A}, \mathbf{u}, \mathbf{z})d\tau \tag{28}$$

$$\mathbf{u}_{\mathbf{\Sigma A}}(\mathbf{\Sigma A}, \dot{\mathbf{u}}, \mathbf{z}(t)) = -\mathbf{k}^{B,el}{}_{\mathbf{\Sigma A}}(\mathbf{\Sigma A})\mathbf{z}(t) - \int_0^t \mathbf{k}^{B,hys}{}_{\mathbf{\Sigma A}}(\mathbf{\Sigma A}, \mathbf{u}, \mathbf{z}(t))\mathbf{z}(\tau)d\tau$$
and :
$$\mathbf{k}^{B,el}{}_{\mathbf{\Sigma A}}(\mathbf{\Sigma A}) = (\mathbf{T}_{x\to\Delta})'\text{diag}\left\{ \left[(\mathbf{a}^B \odot \mathbf{E}^B \oslash \mathbf{L}^B)\mathbf{1}^S \right] \right\} \mathbf{T}_{x\to\Delta} \tag{29}$$
$$\mathbf{k}^{B,hys}{}_{\mathbf{\Sigma A}}(\mathbf{\Sigma A}, \mathbf{z}(t)) = (\mathbf{T}_{x\to\Delta})'\text{diag}\left\{ \left[(\mathbf{1} - \mathbf{a}^B) \odot (\mathbf{\rho}^B(\mathbf{\sigma}^B, \mathbf{x}, \dot{\mathbf{x}}) \odot \mathbf{E}^B \oslash \mathbf{L}^B)\mathbf{1}^S \right] \right\} \mathbf{T}_{x\to\Delta}$$

In Equation (28), the expression $\mathbf{G}_{\mathbf{\Sigma A}}(\mathbf{\Sigma A}, \mathbf{z}(\tau))$ is determined numerically.

The fixed-point iteration implementation is in discrete time. Accordingly, the satisfaction of the adjoint and transversality conditions is by solving $\dot{\mathbf{z}}(t) = \mathbf{A}(\mathbf{z}(t))\mathbf{z}(t) + \mathbf{B}(\mathbf{p}(t) + \mathbf{u}(\mathbf{\Sigma A}, \dot{\mathbf{u}}, \mathbf{z}(t)))$, with the initial value $\mathbf{z}(0) = \mathbf{0}$, followed by solving in backward-time $\dot{\lambda}(t) = -\nabla_{\mathbf{z}(t)} H(t, \mathbf{z}(t), \mathbf{\Sigma A}, \lambda(t))$ and $\dot{\eta}(t) = -\nabla_{\mathbf{\Sigma A}} H(t, \mathbf{z}(t), \mathbf{\Sigma A}, \lambda(t))$ with $\lambda(t_f) = \mathbf{0}$ and $\eta(t_f) = \mathbf{0}$. The complementarity term of Equation (24) is satisfied by defining the components of μ_1 and μ_2 as:

$$\eta_{1,n} = \begin{cases} 0 & \leftrightarrow \ \mathbf{\Sigma A}_n \neq \mathbf{\Sigma A}_n^{max} \\ -\eta_n(0) & \leftrightarrow \ \mathbf{\Sigma A}_n = \mathbf{\Sigma A}_n^{max} \end{cases} \quad \forall \ n = 1, \ldots, N \tag{30}$$

$$\eta_{2,n} = \begin{cases} 0 & \leftrightarrow \ \mathbf{\Sigma A}_n \neq 0 \\ \eta_n(0) & \leftrightarrow \ \mathbf{\Sigma A}_n = 0 \end{cases} \quad \forall \ n = 1, \ldots, N \tag{31}$$

That leaves us with the stationary term.

The introduction of the fixed-point iteration modifies $\mathbf{\Sigma A}$ and leads $\dot{\eta}$ to result in $\eta(0)$ that satisfies the stationary term—reaching an optimal point. Substituting Equations (30) and (31) into the stationarity condition of Equation (24) gives:

$$-\eta(0)'(\mathbf{\Sigma A} - \mathbf{\Sigma A}^{max}) = 0 \tag{32}$$

The form of Equation (32) is ready for the fixed-point iteration applications so that the consequent iterative term is:

$$\mathbf{\Sigma A}_n^{k+1} = \Pi_{\mathbf{\Sigma A}}\left(\mathbf{\Sigma A}_n^k + \gamma\eta_n^k(0) \right) \quad \forall \ n = 1, \ldots, N \tag{33}$$

where γ is the convergence parameter, ΣA_n^k denotes the nth component of ΣA^k at the kth iteration, and $\eta_n^k(0)$ denotes the nth component of $\eta^k(0)$ that stems from ΣA^k. The fixed-point iteration application is by the following procedure consisting of four initial steps and four iterative steps.

(i)   Set the iteration index to $k = 0$ and choose an initial ΣA^0
(ii)   Define the maximum control gains boundary vector ΣA^{max}
(iii)   Decide the weighting sub-matrices Q_1 and Q_2
(iv)   Choose γ and define the number of maximum iterations k^{max}
(v)   Calculate $z^k(t)$ by solving the initial value problem:

$$\dot{z}(t) = A(z(t))z(t) + B\left(p(t) + u\left(\Sigma A^k, \dot{u}, z(t)\right)\right) \quad ; \quad z(0) = 0$$

(vi)   Calculate $\lambda^k(t)$ and $\eta^k(t)$ backward in time by solving:

$$\dot{\lambda}^k(t) = -\nabla_{z(t)}H\left(t, z(t), \Sigma A^k, \lambda^k(t)\right) \quad ; \quad \lambda^k(t_f) = 0$$

$$\dot{\eta}^k(t) = -\nabla_{\Sigma A}H\left(t, z(t), \Sigma A^k, \lambda^k(t)\right) \quad ; \quad \eta^k(t_f) = 0$$

(vii)   Update the components of ΣA^k using the fixed-point iteration method:

$$\Sigma A_n^{k+1} = \Pi_{\Sigma A}\left(\Sigma A_n^k + \gamma\eta_n^k(0)\right) \quad \forall \quad n = 1,\ldots,N$$

(viii)   Check if $-\eta^k(0)'\left(\Sigma A^{k+1} - \Sigma A^{max}\right) = 0$ is satisfied or $k = k^{max}$. If yes, then finish. If not, increase k by 1 and go back to step (v).

The eight-step procedure is a straightforward approach to solving Equation (20). Notice that step (vi) addresses the adjoint and transversality conditions, while step (vii) addresses the complementarity and stationary terms. The procedure stops when the stationary condition is satisfied.

5. Optimization Examples

Two optimization examples examine the fixed-point iteration application for inelastic rigid frame systems in a weakened state following an earthquake incident. The first example optimizes a low-rise three-story structure, and the second deals with a high-rise 15-story structure. The employed BRBs are IPN profiles of modulus of elasticity $E_{n,s}^B = 205,000$ MPa, axial yield stress of $\sigma_{n,s}^{B,Y} = 225$ MPa, and $a_{n,s}^B = 0.1$ ratios between the plastic and the elastic stiffness.

The numerical evaluation of the dynamic response is conducted using the Newmark-β of the average acceleration scheme. The Newton–Raphson method is implemented in addressing the implicit hysteretic resisting and control forces. Equation (24) adjoint terms are calculated backward in discrete time based on the extended-mean-value theorem employed by Newmark-β.

5.1. Example 1: Three-Story Rigid Frame System

The three-story and two-spans rigid frame system's elevation scheme is depicted in Figure 5. The entire story's height is 4.5 m, the columns' effective length is 3.5 m, and the span's length is 4.0 m. The story stiffness quantities $k_{n=1,\ldots,3}^F$ in Figure 5 are calculated for clumped columns with squared cross-sections of 0.6 m $\times$ 0.6 m, 0.5 m $\times$ 0.5 m, and 0.4 m $\times$ 0.4 m dimensions in stories 1 to 3, respectively. The frame members' modulus of elasticity is set as 27,000 MPa, and the story masses are $m_{n=1,\ldots,3} = 40$ ton. The stories' yield force $f_n^{F,yld}$ specified in Figure 5 regard interstory drift of 0.5% of the columns' effective length, and the ratio between the plastic and the elastic stiffness is $a_n^F = 0.5$. The inherent damping ratio is assumed to be 5%, and the inherent damping matrix is calculated by Caughey damping.

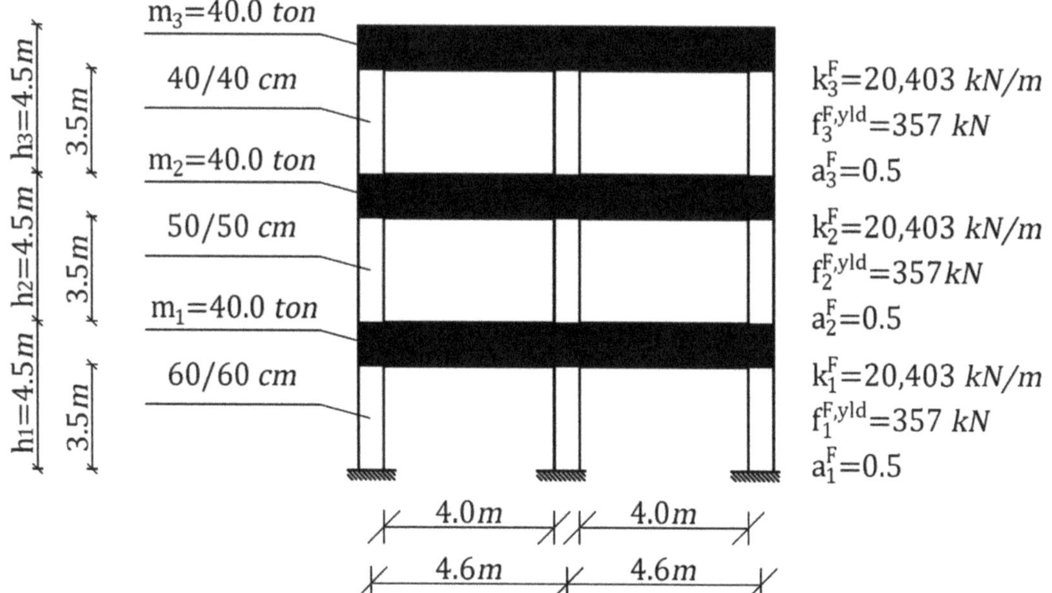

Figure 5. The bare three-story rigid frame system.

The highest modal period of the frame structure (without BRBs) is 0.87 s, a quantity that is considerably high for a three-story system—implying a weakened system. The initial BRBs allocation to start the fixed-point iteration consists of uniformly distributed IPN180 elements of cross-sectional area $A_{n,s}^0 = 27.9$ cm^2 and effective length $L_{n,s}^B \cong 5.3$ m. After the initial BRBs allocation, the highest modal period is reduced to $T_1 \cong 0.25$ s. This example's quasi-static cyclic loading is defined accordingly and is depicted in Figure 6.

The total cross-sectional areas corresponding to the IPN180 profiles and the static gains to be optimized are $\Sigma A_1^0 = \Sigma A_2^0 = \Sigma A_3^0 \cong 55.8$ cm^{2}s. The static gains maximum corresponds to the IPN600 profile of cross-sectional area 254 cm^2, hence $\Sigma A_1^{max} = \Sigma A_2^{max} = \Sigma A_3^{max} = 508$ cm^2. The weighting matrices are defined by default as $Q_1 = I$ and $Q_2 = I$. The iterative convergence parameters are set for each story individually so that of $\gamma_1 = 0.007$, $\gamma_2 = 0.005$, $\gamma_3 = 0.002$, and $k^{max} = 50$. That concludes the algorithm's preparation steps (i)–(iv).

Following the preparation steps, the iterative process is initiated in steps (v)–(viii). Figure 7 shows the iterative process of ΣA_n^k versus the iteration index $k = 0, 1, \ldots, 50$, which converged to static gains of $\Sigma A_1^{50} \cong 177.0$ cm^2, $\Sigma A_2^{50} \cong 191.0$ cm^2, and $\Sigma A_1^{50} \cong 161.0$ cm^2. The IPN allocation corresponding to the optimal static gains consists of IPN340 in the first story (i.e., $A_{s=1,2,n=1} = 86.7$ cm^2), IPN360 in the second story (i.e., $A_{s=1,2,n=2} = 97.0$ cm^2), and IPN320 in the third story (i.e., $A_{s=1,2,n=3} = 77.7$ cm^2). The elevation scheme of the optimal BRBs allocation is depicted in Figure 8b.

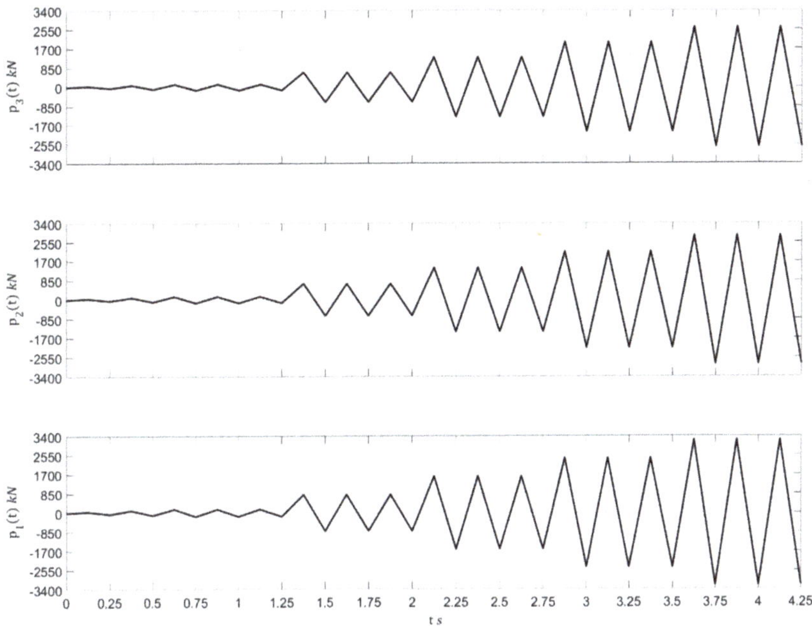

Figure 6. Example 1 quasi-static cyclic loading.

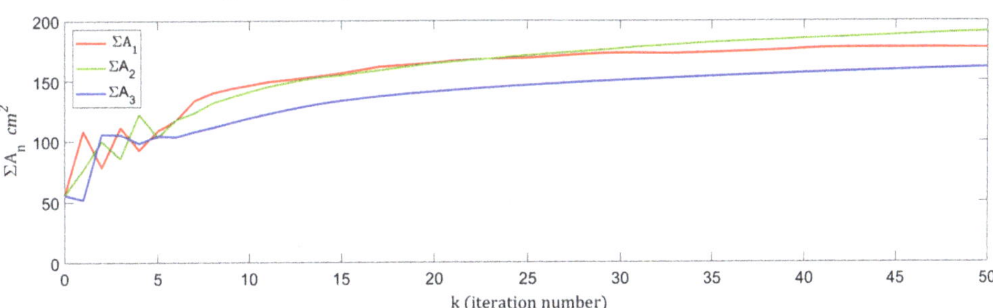

Figure 7. Fixed-point iterative process for the three-story rigid frame system.

The optimal static gains quantity is also examined for equivalent uniform IPN profile distribution. The equivalency is in terms of total cross-sectional area. In a case where modifying the fixed-point iteration's distribution with uniform distribution increases the objective function, it indicates that the solution is indeed optimal. Accordingly, the members' cross-sectional area is calculated by:

$$\Sigma A_1^{50} + \Sigma A_2^{50} + \Sigma A_1^{50} \cong 529.0 \text{ cm}^2 \quad \rightarrow \quad A_{s,n} = \frac{529}{6} \cong 88.0 \text{ cm}^2 \quad \forall \quad n = 1,\ldots,3 \quad \& \quad s = 1,2$$

The quantity $A_{s,n} = 88.0 \text{ cm}^2$ correlates to the IPN340 profile and corresponds to the uniform BRBs distribution depicted in Figure 8c.

Figure 9 illustrates the portions of Equation (20) objective function $\int_0^{t_f} \mathbf{d}^T(t)\mathbf{d}(t)dt$ and $\int_0^{t_f} \dot{\mathbf{d}}^T(t)\dot{\mathbf{d}}(t)dt$, in t, given that $\mathbf{Q}_1 = \mathbf{I}$ and $\mathbf{Q}_2 = \mathbf{I}$. Figure 8 depicts the initial IPN distribution, the optimal IPN distribution, and the equivalent uniform IPN distribution under Figure 6 loading. The plot exemplifies that the objective function is significantly reduced and that changing the optimal distribution with a uniform distribution across all stories slightly increases the objective function, which suggests the fixed-point iteration design is indeed optimal.

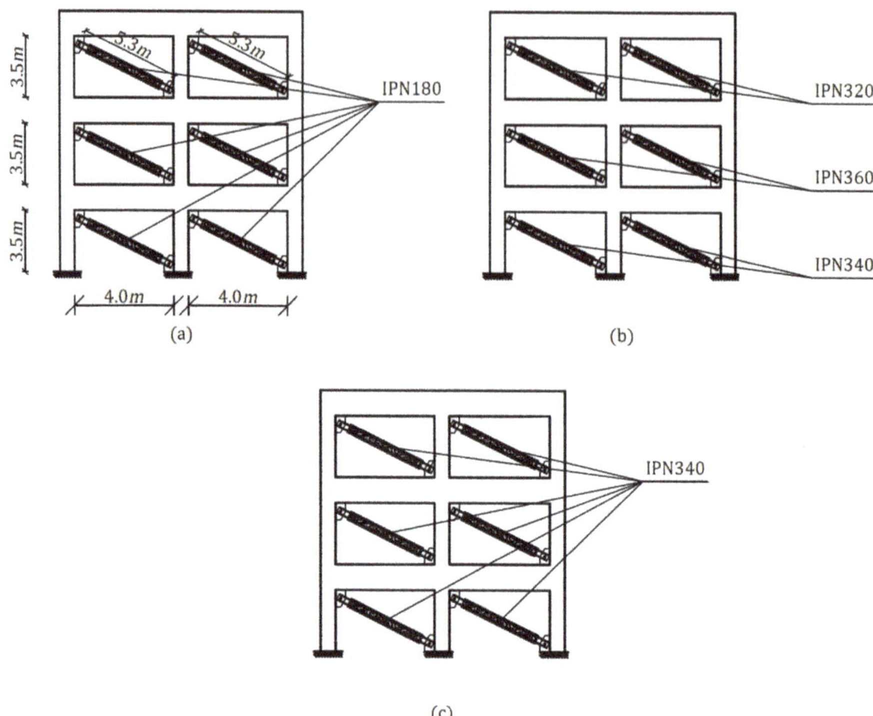

Figure 8. Three-story inelastic rigid frame system configurations: (**a**) initial BRBs allocation, (**b**) optimal BRBs allocation, and (**c**) equivalent uniform BRBs distribution.

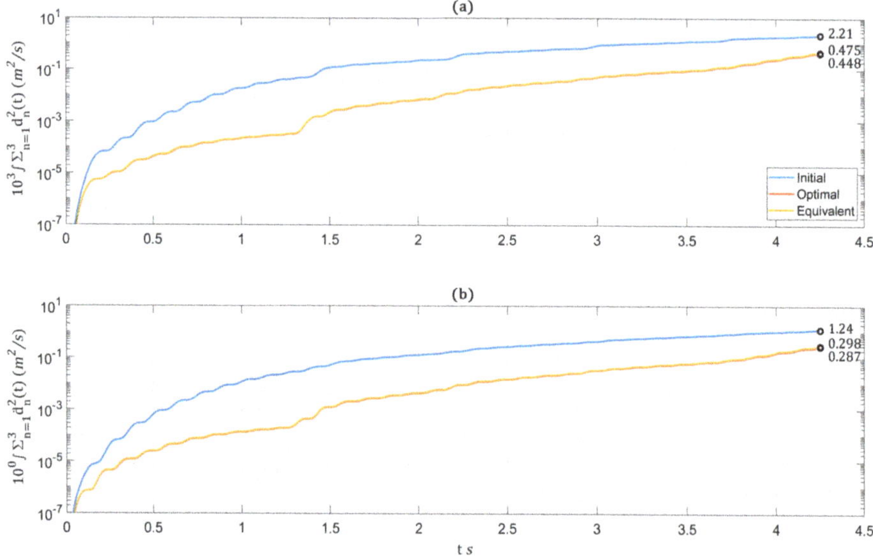

Figure 9. Trajectories of the three-story inelastic rigid frame system: (**a**) interstory drifts (**b**) interstory drift velocities.

The earthquake response of the three IPN allocation possibilities is examined under the Valparaiso 2017 ground acceleration—recorded by the Torpederas station (east–west component) of 0.91 g's peak ground acceleration. Figure 10 depicts the stories' hysteretic resisting forces versus the relative interstory drift. The employment of BRBs maintained the columns by remaining in their elastic regime while the BRBs slightly met the plastic range. These substantial results are thanks to addressing the quasi-static cyclic loading of twice the total yielding force maximum amplitude.

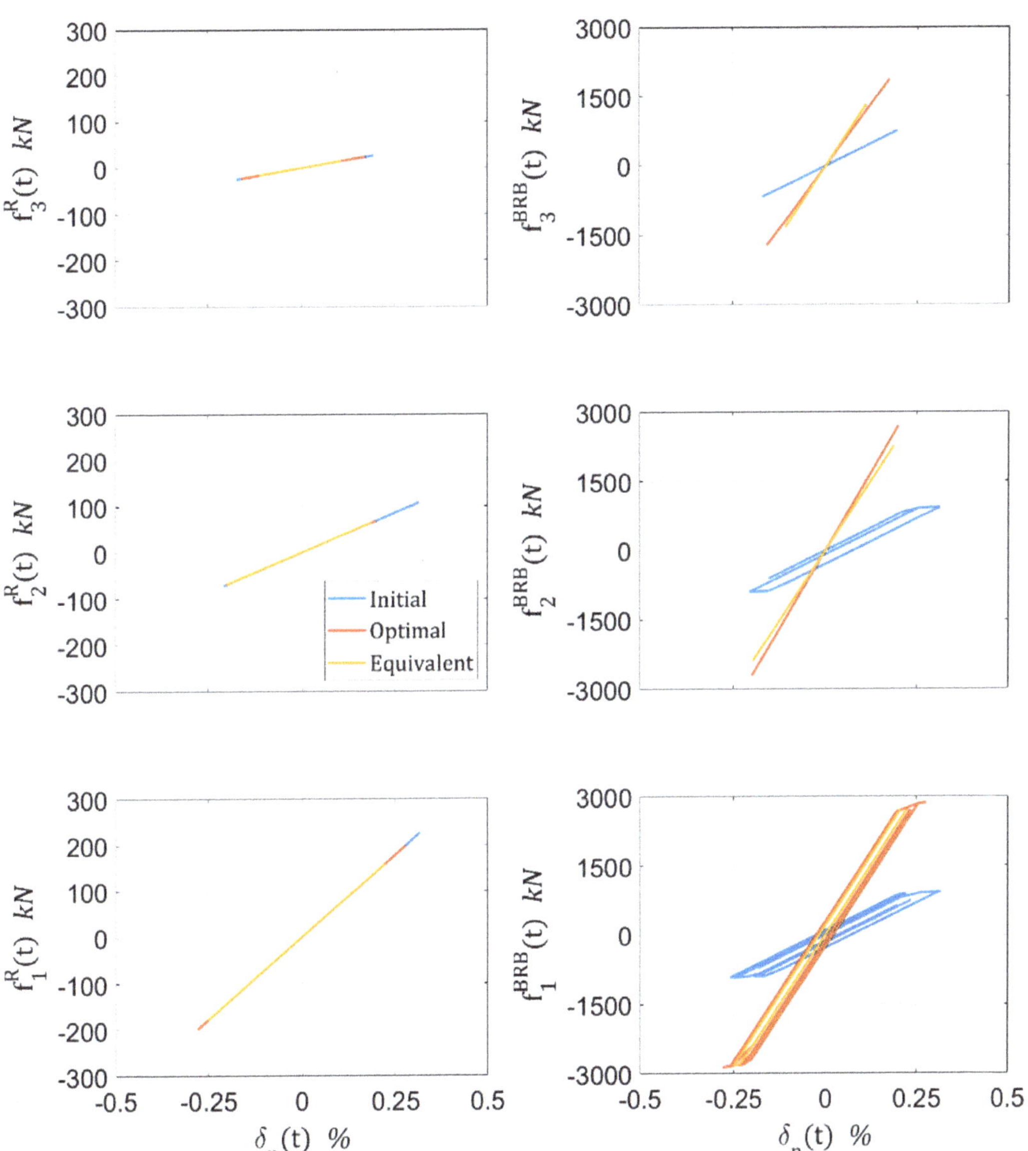

Figure 10. Hysteretic resisting forces of the three-story frame system under the Valparaiso 2017 earthquake.

5.2. Example 2: 15-Story Rigid Frame System

The second example deals with regulating the dynamic response of the 75.0 m high 15-story frame system depicted in Figure 11. The stories' height is 5.0 m, and the spans' length is 7.0 m. The clumped columns are of 4.0 m effective length, and their variation in column cross-sectional dimensions is specified in Figure 11. The material properties of the BRBs and Columns are the same as in Example 1. The BRBs' effective length is $L_{n,s}^B \cong 8.6$—considering the lateral angle of approximately $45.0°$. The quantities of the ceilings' mass m_n, the lateral story stiffnesses k_n^F, the columns' shear force at first yield $f_n^{F,yld}$, and the ratio between the plastic and the elastic stiffness a_n^F are specified in Figure 11 as well.

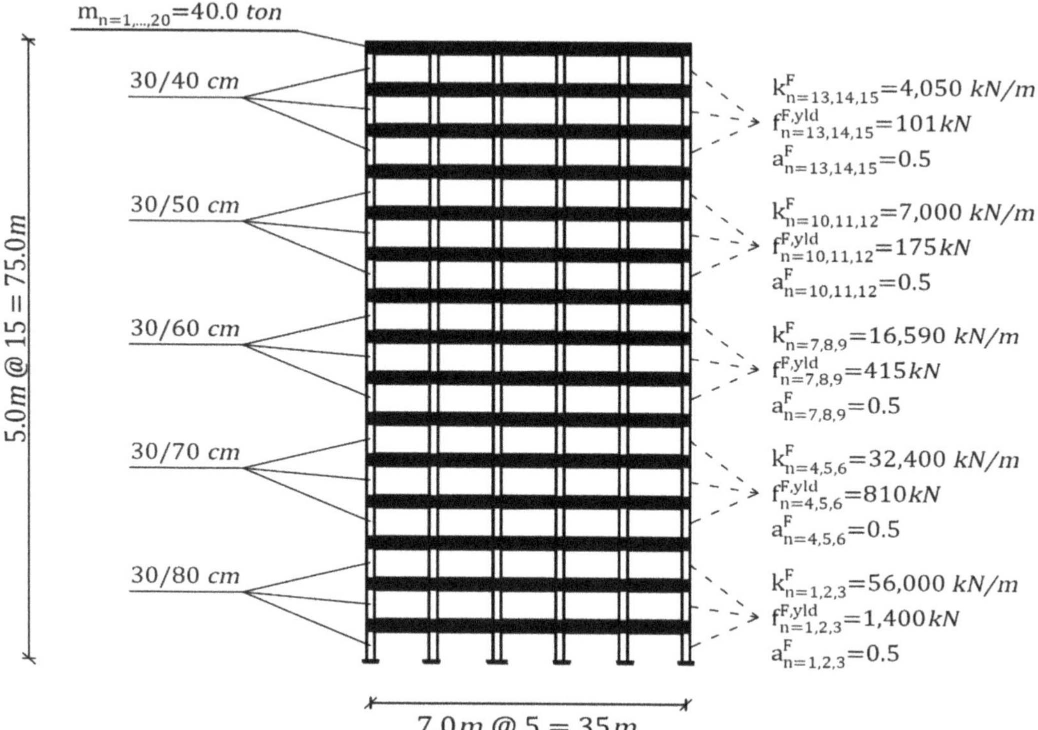

Figure 11. Fifteen-story inelastic rigid frame system.

The dominant undamped free-vibrations period is approximately $T_1 = 0.92$ s. In this example, at the initial iteration, IPN180 profiles are assigned to all spans and stories, which provides $\Sigma A_1^0 = \ldots = \Sigma A_{15}^0 \cong 139.5$ cm^{2}s. Considering the IPN600 profile as the maximum limitation on the gain variables, we have $\Sigma A_1^{max} = \ldots = \Sigma A_{15}^{max} = 1270$ cm^2. The initial frame system equipped with BRBs is depicted in Figure 12a, and the related quasi-static cyclic loadings are illustrated in Figure 13.

The algorithm convergence parameter is decided by $\gamma = 10^{-2}$. The fixed-point iteration method goes under the iterative process shown in Figure 14. At the $k = 50$ iteration, the algorithm converged into the optimal solution whose resultant optimal gain variables (i.e., total cross sectional IPN areas) correspond to the BRBs allocation scheme depicted in Figure 12b. Additionally, the total sum of optimal gains is calculated and divided uniformly to all frame spans, yielding the distribution in Figure 12c to showcase that Figure 12b is indeed optimal.

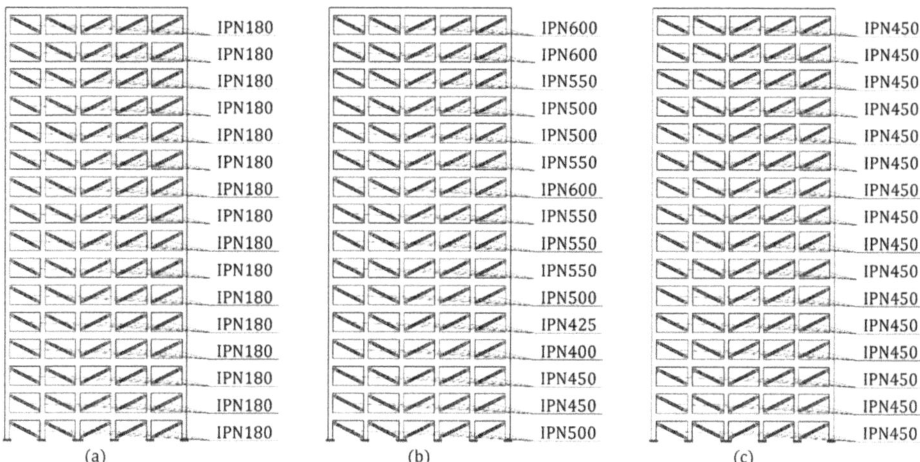

Figure 12. Fifteen-story rigid frame system configurations: (**a**) initial BRBs allocation (**b**) optimal BRBs allocation (**c**) equivalent uniform BRBs distribution.

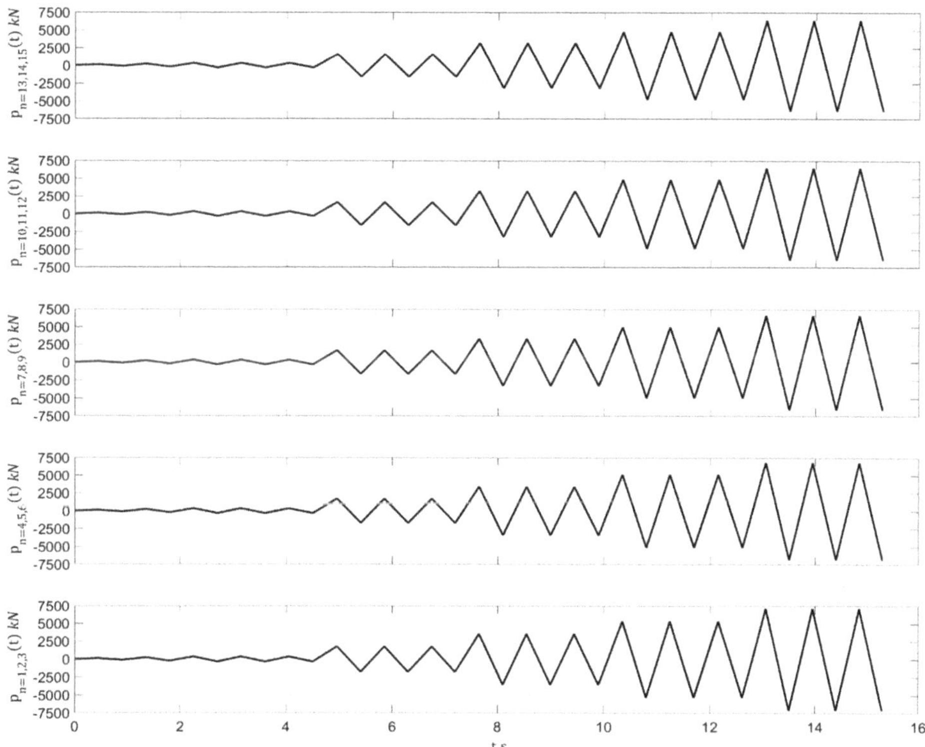

Figure 13. Example 2 quasi-static cyclic loadings.

Figure 15 shows the trajectories of $\int_0^{t_f} \mathbf{d}^{\mathrm{T}}(t)\mathbf{d}(t)dt$ and $\int_0^{t_f} \dot{\mathbf{d}}^{\mathrm{T}}(t)\dot{\mathbf{d}}(t)dt$, as portions of the objective function, for the initial, optimal, and equivalent uniform BRB distributions. It is shown that the optimal distribution provides the minimum objective function—indicating the solution integrity. The three distributions are also examined for the Valparaiso 2017 earthquake of 0.91 g's PGA. Figure 16 shows the hysteretic behavior of stories 1,4,7,10, and 13 regarding the columns' and BRBs' hysteretic forces. The forces exemplify that the frame system remains in its elastic range regardless of the strong earthquake.

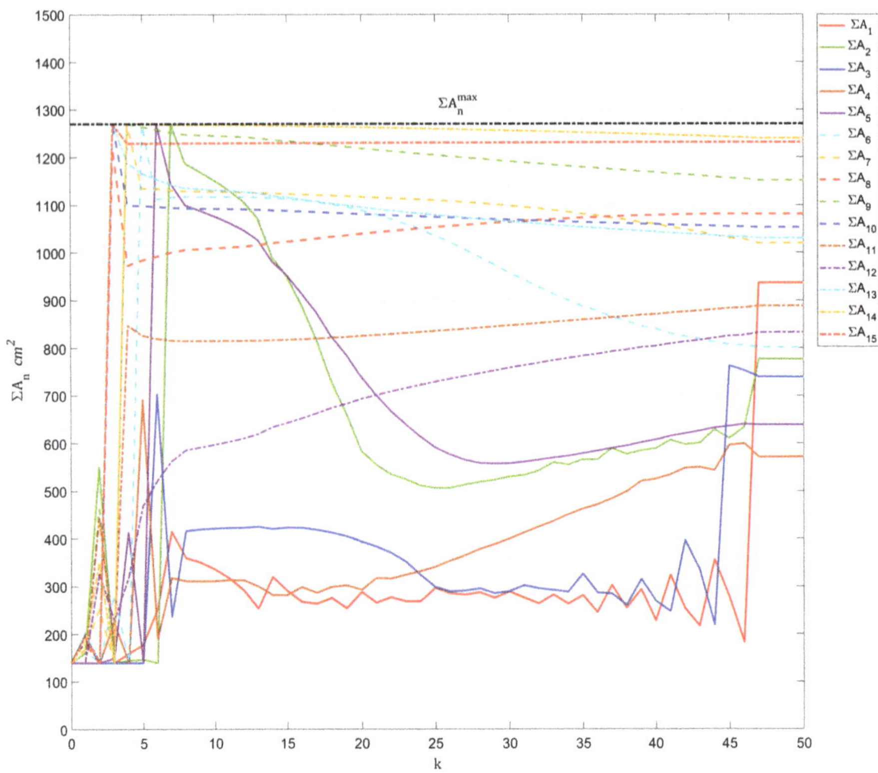

Figure 14. Fixed-point iterative process for the 15-story rigid frame system.

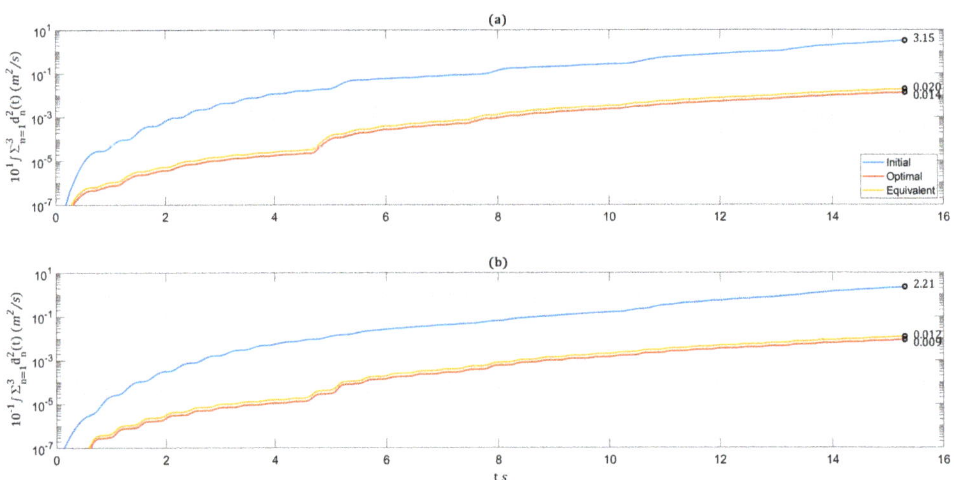

Figure 15. Trajectories of the 15-story inelastic rigid frame system: (**a**) interstory drift (**b**) interstory drift velocities.

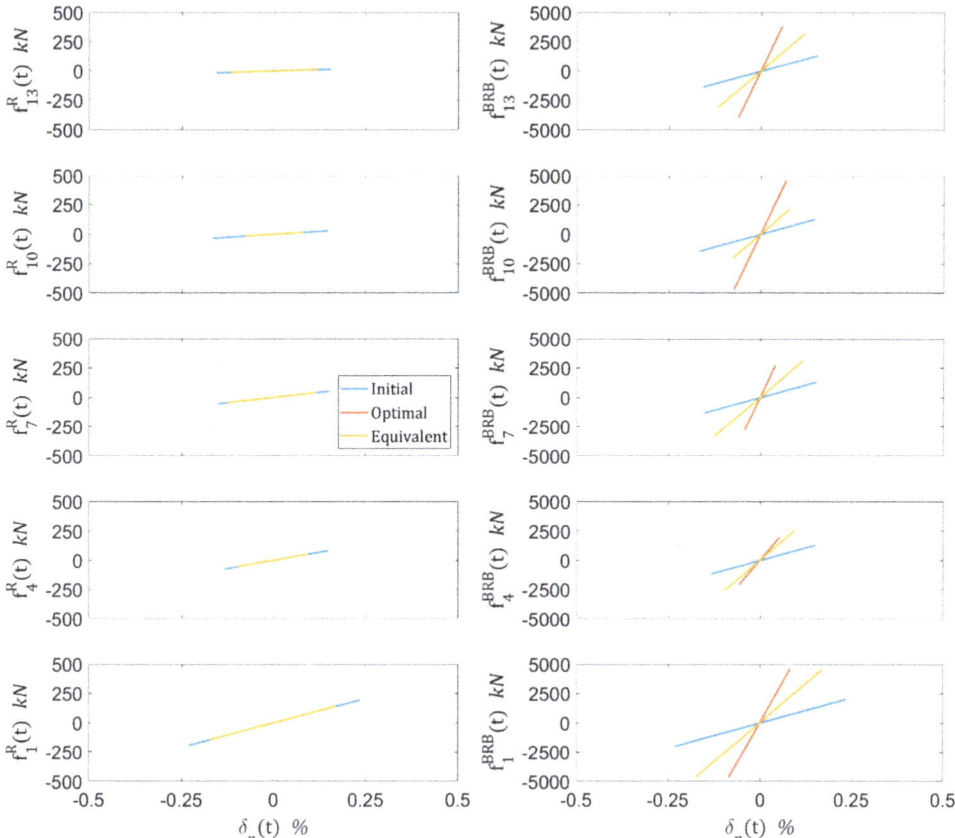

Figure 16. Hysteretic resisting forces of the 15-story frame system under the Valparaiso 2017 earthquake.

6. Conclusions

The paper presents a practical optimization procedure for retrofitting frame structure post-earthquake event. The optimization addresses nonlinear control systems whose mechanism consists of linear, nonlinear, and hysteretic portions. Such control systems are characterized by static parameters (static gains)—referring to the control system's geometric or material characteristics. This paper's optimization procedure is the first to utilize the fixed-point iteration method for controlling and regulating the dynamic response of frame structures.

The state-space equation, associated with the fixed-point iteration, defines the dynamic equilibrium. It is derived from the frame structure's lateral force equilibrium and regards a closed-loop paradigm with negative state feedback and the controller—consisting of linear, nonlinear, and hysteretic portions. The nonlinear control system minimizes the cumulation sum of squared interstory drifts deformations and velocities by calculating the optimal gains while subject to design boundaries. The solution procedure comprises four initial and four iterative steps that are repeated until all optimality conditions are satisfied or the maximum number of iterations has been reached.

The fixed-point iteration scheme presented in this paper differs from other control algorithms in being suitable to address linear, nonlinear, and hysteretic control law. The BRB system is employed in showcasing the application of the developed methodology. Choosing the BRB system is due to having linear and hysteretic portions. Two optimization examples address weakened frame systems (following earthquake incidents) and demonstrate the design procedure practicability and illustrate the fixed-point iterative capability in optimizing multiple control gain variables. It should be noted that the fixed-point iteration converges into local optima. Thus, each of the first four steps of the solution procedure (i.e., deciding on the weighting matrices, choosing the convergence parameter, defining the maximum control gains, and setting the initial control gain variables) significantly influences the final solution.

In closing, the methodology of this paper help to optimize the static specifications of control systems that produce either linear, nonlinear, or hysteretic forces to regulate the seismic vibrations of inelastic systems. The static specifications relate to geometrical, strength, or material parameters. While this paper addresses weakened frame structures, the methodology is relevant to any lateral-load resisting force system whose state vector is calculated.

Author Contributions: Conceptualization, A.S. and M.G.; methodology, A.S.; software, A.S.; validation, A.S.; formal analysis, A.S.; investigation, A.S.; resources, A.S.; data curation, A.S.; writing—original draft preparation, A.S. and M.G; writing—review and editing, A.S.; visualization, A.S.; supervision, A.S.; project administration, M.G.; funding acquisition, M.G. All authors have read and agreed to the published version of the manuscript.

Funding: This research received no external funding.

Data Availability Statement: The data presented in this study are available on request from the corresponding author.

Conflicts of Interest: The authors declare no conflict of interest.

Abbreviations

$\mathbf{a}^{\mathrm{B}}$	$N \times S$ matrix consisting of the ratios between the BRBs' axial plastic and elastic stiffnesses
$\mathbf{c}$	$N \times N$ inherent damping matrix
$\mathbf{c}^{\mathrm{u,el}}$	$N \times N$ linear damping matrix of $\mathbf{u}$
$\mathbf{d}$	N-dimensional interstory drifts vector
$\mathbf{f}^{\mathrm{u,NL}}$	N-dimensional nonlinear portion of $\mathbf{u}$
$\mathbf{f}^{\mathrm{F}}$	N-dimensional structural rigid frame system's lateral resisting force vector
$\mathbf{f}^{\mathrm{F,hys}}$	N-dimensional hysteretic portion of $\mathbf{f}^{\mathrm{F}}$
$f_{\mathrm{n}}^{\mathrm{F,yld}}$	nth story columns' shear force at first yield
k	Fixed-point iteration number
k^{max}	Fixed-point maximum iteration number
$\mathbf{k}^{\mathrm{u,el}}$	$N \times N$ linear stiffness matrix of $\mathbf{u}$
$\mathbf{k}^{\mathrm{u,hys}}$	$N \times N$ hysteretic stiffness matrix of $\mathbf{u}$
$\mathbf{k}^{\mathrm{F,el}}$	$N \times N$ static-condensate matrix elastic stiffness portion about $\mathbf{d}$
$\mathbf{k}^{\mathrm{F,hys}}$	$N \times N$ static-condensate hysteretic stiffness matrix
$\mathbf{m}$	$N \times N$ static-condensate diagonal mass matrix
$\mathbf{p}$	N-dimensional lateral applied dynamic load vector
$\mathbf{u}$	N-dimensional control force vector
$\mathbf{x}$	N-dimensional ceilings' relative-to-ground displacement vector
$\dot{\mathbf{x}}$	N-dimensional ceilings' relative-to-ground velocities vector
$\ddot{\mathbf{x}}$	N-dimensional ceilings' relative-to-ground accelerations vector
$\mathbf{z}(t)$	$4N$-dimensional state-vector
$\mathbf{A}$	$4N \times 4N$ state matrix
$\mathbf{A}^{\mathrm{B}}$	$N \times S$ matrix containing the BRBs' effective cross-sectional area
$A_{n,s}^{B}$	cross-sectional area of the BRB installed at the sth span of the nth story
$\mathbf{B}$	$4N \times N$ input-to-state matrix
$\mathbf{E}^{\mathrm{B}}$	$N \times S$ matrix composed of the BRBs' elasticity modulus
$\mathbf{G}$	$N \times 4N$ gain matrix
H	Hamilton function

$\mathbf{L}^B$	$N \times S$ matrix comprised of the BRBs' length
$\mathbf{Q}_1$	diagonal weighting matrix whose components govern the minimization priority of $\mathbf{d}(t)$
$\mathbf{Q}_2$	diagonal weighting matrix whose components govern the minimization priority of $\dot{\mathbf{d}}(t)$
$\mathbf{T}_{d \to x}$	transformation matrix from drift coordinates into displacement coordinates
$\mathbf{T}_{x \to d}$	transformation matrix from displacement coordinates into drift coordinates
$\mathbf{T}_{x \to \Delta}$	geometric transformation matrix from $\mathbf{x}$ into the BRB's axial deformation $\mathbf{\Delta}$
T_1	highest modal period
$\mathbf{T}^{hys}$	transformation matrix applied to the negative feedback of $\mathbf{z}$ to yield the hysteretic portion of $\mathbf{u}$
$\mathbf{T}^{lin}$	transformation matrix applied to the negative feedback of $\mathbf{z}$ to yield the linea portion of $\mathbf{u}$
$\mathbf{T}^{NL}$	transformation matrix applied to the negative feedback of $\mathbf{z}$ to yield the nonlinear portion of $\mathbf{u}$
$\boldsymbol{\lambda}$	$4N$-dimensional Lagrange multipliers vector
$\boldsymbol{\mu}_1$	KKT multipliers vector governing the design limitation inequality constraints
$\boldsymbol{\mu}_2$	KKT multipliers vector governing the design limitation inequality constraints
$\boldsymbol{\rho}^B$	$N \times S$ matrix that corresponds to the material's elastic/plastic/unloading stages
$\boldsymbol{\sigma}$	$4N$-dimensional Lagrange multipliers vector that governs initial conditions
$\boldsymbol{\sigma}^{in}$	$N \times S$ matrix of the inelastic portion of the BRBs' axial stress
$\boldsymbol{\sigma}^{B,Y}$	$N \times S$ matrix of the BRBs' yield stress
$\dot{\mathbf{\Delta}}^B$	$N \times S$ matrix of BRBs' axial deformation rate vector
$\boldsymbol{\Sigma}\mathbf{A}$	N-dimensional vector consisting of the static gain variables
$\boldsymbol{\Sigma}\mathbf{A}^{max}$	N-dimensional vector consisting of the maximum allowable static gains
$\mathcal{L}$	Lagrange function
$\mathbf{0}$	N-dimensional vector of zeros
$\mathbf{1}^S$	S-dimensional vector of ones

Appendix A

1. Given the optimization problem:

$$\underset{\boldsymbol{\Sigma}\mathbf{A}}{\text{minimize}} \quad \left\{ J = \int_0^{t_f} \mathbf{z}^T(t)\mathbf{Q}\mathbf{z}(t)dt \right\}$$

2.
$$\text{subject to} \quad \dot{\mathbf{z}}(t) = \mathbf{A}(\mathbf{z}(t))\mathbf{z}(t) + \mathbf{B}(\mathbf{p}^{max}\sin\omega_1 t + \mathbf{u}(\boldsymbol{\Sigma}\mathbf{A}, \mathbf{z}))$$
$$\mathbf{z}(0) = \mathbf{0} \tag{A1}$$
$$\boldsymbol{\Sigma}\mathbf{A}^{max} - \boldsymbol{\Sigma}\mathbf{A} \geq \mathbf{0}$$
$$\boldsymbol{\Sigma}\mathbf{A} \geq \mathbf{0}$$

3. In reference to Theorem 2.3.24 in Chapter 2 in the book of Gerdts [44]. Assume $\mathbf{z}^*(t)$ is the optimal trajectory of the state vector and $\boldsymbol{\Sigma}\mathbf{A}^*$ composed of the optimal control gains, then there exists $\boldsymbol{\lambda}(t)$, $\boldsymbol{\sigma}$, and $\boldsymbol{\mu}$ such that:

4. $\nabla_{\mathbf{z},\boldsymbol{\Sigma}\mathbf{A}}\mathcal{L}(\mathbf{z}^*, \boldsymbol{\Sigma}\mathbf{A}^*, \boldsymbol{\lambda}, \boldsymbol{\sigma}, \boldsymbol{\mu}) = \mathcal{L}_{\mathbf{z}}(\mathbf{z}^*, \boldsymbol{\Sigma}\mathbf{A}^*, \boldsymbol{\lambda}, \boldsymbol{\sigma}, \boldsymbol{\mu})\Delta\mathbf{z} + \mathcal{L}_{\boldsymbol{\Sigma}\mathbf{A}}(\mathbf{z}^*, \boldsymbol{\Sigma}\mathbf{A}^*, \boldsymbol{\lambda}, \boldsymbol{\sigma}, \boldsymbol{\mu})\Delta\boldsymbol{\Sigma}\mathbf{A} = 0 \quad$ (A2)

5. Where $\Delta\mathbf{z}(t)$ and $\Delta\boldsymbol{\Sigma}\mathbf{A}$ Denote small changes to the optimal solution of $\mathbf{z}^*(t)$ and $\boldsymbol{\Sigma}\mathbf{A}^*$:

6. $\Delta\mathbf{z}(t) = \mathbf{z}(t) - \mathbf{z}^*(t) \approx \mathbf{0} \quad$ (A3)

7. $\Delta\boldsymbol{\Sigma}\mathbf{A} = \boldsymbol{\Sigma}\mathbf{A} - \boldsymbol{\Sigma}\mathbf{A}^* \approx \mathbf{0} \quad$ (A4)

8. Then, Equation (A2) implies that the small changes result in:

9. $\mathcal{L}_{\mathbf{z}}\Delta\mathbf{z} = \boldsymbol{\sigma}'\Delta\mathbf{z}(0) + \int_0^{t_f}\left(\nabla_{\mathbf{z}(t)}\mathcal{H}(t)\right)\Delta\mathbf{z}(t)dt - \int_0^{t_f}\boldsymbol{\lambda}'(t)\Delta\dot{\mathbf{z}}(t)dt = 0 \quad$ (A5)

10. $\mathcal{L}_{\boldsymbol{\Sigma}\mathbf{A}}\Delta\boldsymbol{\Sigma}\mathbf{A} = \left(\boldsymbol{\mu}_1 - \boldsymbol{\mu}_2 + \int_0^{t_f}\nabla_{\boldsymbol{\Sigma}\mathbf{A}}\mathcal{H}(t)dt\right)\Delta\boldsymbol{\Sigma}\mathbf{A} = 0 \quad$ (A6)

11. Equation (A5) is further elaborated by using integration by parts to replace $\int_0^{t_f}\boldsymbol{\lambda}'(t)\Delta\dot{\mathbf{z}}(t)dt$ and deriving:

12. $\mathcal{L}_{\mathbf{z}}d\mathbf{z} = \boldsymbol{\lambda}(t_f) - (\boldsymbol{\lambda}(0) - \boldsymbol{\sigma})^T\Delta\mathbf{z}(0) + \int_0^{t_f}\left(\nabla_{\mathbf{z}}\mathcal{H}(t) + \dot{\boldsymbol{\lambda}}(t)\right)\Delta\mathbf{z}(t)dt = 0 \quad$ (A7)

13. Accordingly, the following adjoint conditions and transversality conditions have to be satisfied to comply with Equation (A7):

14. $\dot{\boldsymbol{\lambda}}(t) = -\nabla_{\mathbf{z}}\mathcal{H}(t, \mathbf{z}^*(t), \boldsymbol{\Sigma}\mathbf{A}^*, \boldsymbol{\lambda}(t)) \quad$ (A8)

15. $\boldsymbol{\lambda}(t_f) = \mathbf{0} \quad$ (A9)

16. Also, the definition for $\boldsymbol{\sigma}$ is:

17. $\boldsymbol{\sigma} = \boldsymbol{\lambda}(0) \quad$ (A10)

18. Address the equality of Equation (A6) by defining the time function $\boldsymbol{\eta}(t)$:

19. $\boldsymbol{\eta}(t) = \int_t^{t_f}\nabla_{\boldsymbol{\Sigma}\mathbf{A}}\mathcal{H}(\tau)d\tau \quad$ (A11)

20. Thus, having additional adjoint and transversality conditions:
21. $\dot{\boldsymbol{\eta}}(t) = -\nabla_{\boldsymbol{\Sigma A}}\mathcal{H}(t, \mathbf{z}^*(t), \boldsymbol{\Sigma A}^*, \boldsymbol{\lambda}(t))$ (A12)
22. $\boldsymbol{\eta}(t_f) = 0$ (A13)
23. Consequently, Equation (A6) provides the stationary condition:
24. $\boldsymbol{\mu}_1 - \boldsymbol{\mu}_2 + \boldsymbol{\eta}(0) = 0$ (A14)
25. Thus, having the necessary conditions of Equation (24) regarding Equation (20) while requiring the KKT complementarity conditions:
26. $\boldsymbol{\mu} = \begin{bmatrix} \boldsymbol{\mu}_1 \\ \boldsymbol{\mu}_2 \end{bmatrix} \geq 0 : \boldsymbol{\mu}_1^{\mathrm{T}}(\boldsymbol{\Sigma A} - \boldsymbol{\Sigma A}^{\max}) - \boldsymbol{\mu}_2^{\mathrm{T}}(\boldsymbol{\Sigma A}) = 0 \;\leftrightarrow\; \begin{bmatrix} \boldsymbol{\Sigma A} - \boldsymbol{\Sigma A}^{\max} \\ -\boldsymbol{\Sigma A} \end{bmatrix} \leq 0$ (A15)

References

1. Park, Y.-J.; Ang, A.H.-S. Mechanistic Seismic Damage Model for Reinforced Concrete. *J. Struct. Eng.* **1985**, *111*, 722–739. [CrossRef]
2. Bozorgnia, Y.; Bertero, V. Improved Shaking and Damage Parameters for Post-Earthquake Applications. In Proceedings of the SMIP01 Seminar on Utilization of Strong-Motion Data, Los Angeles, CA, USA, 12 September 2001; Citeseer: Princeton, NJ, USA, 2001; pp. 1–22.
3. Mehanny, S.S.F.; Deierlein, G.G. Seismic Damage and Collapse Assessment of Composite Moment Frames. *J. Struct. Eng.* **2001**, *127*, 1045–1053. [CrossRef]
4. Jeong, S.-H.; Elnashai, A.S. New Three-Dimensional Damage Index for RC Buildings with Planar Irregularities. *J. Struct. Eng.* **2006**, *132*, 1482–1490. [CrossRef]
5. DiPasquale, E.; Ju, J.; Askar, A.; Çakmak, A.S. Relation between Global Damage Indices and Local Stiffness Degradation. *J. Struct. Eng.* **1990**, *116*, 1440–1456. [CrossRef]
6. DiPasquale, E.; Cakmak, A.S. Seismic Damage Assessment Using Linear Models. *Soil Dyn. Earthq. Eng.* **1990**, *9*, 194–215. [CrossRef]
7. He, H.; Cong, M.; Lv, Y. Earthquake Damage Assessment for RC Structures Based on Fuzzy Sets. *Math. Probl. Eng.* **2013**, *2013*, 254865. [CrossRef]
8. Rodriguez, M.E. Evaluation of a Proposed Damage Index for a Set of Earthquakes. *Earthq. Eng. Struct. Dyn.* **2015**, *44*, 1255–1270. [CrossRef]
9. Williams, M.S.; Sexsmith, R.G. Seismic Damage Indices for Concrete Structures: A State-of-the-Art Review. *Earthq. Spectra* **1995**, *11*, 319–349. [CrossRef]
10. Aydin, E.; Boduroglu, M.H.; Guney, D. Optimal Damper Distribution for Seismic Rehabilitation of Planar Building Structures. *Eng. Struct.* **2007**, *29*, 176–185. [CrossRef]
11. Fujita, K.; Moustafa, A.; Takewaki, I. Optimal Placement of Viscoelastic Dampers and Supporting Members under Variable Critical Excitations. *Earthq. Struct.* **2010**, *1*, 43–67. [CrossRef]
12. Beskhyroun, S.; Wegner, L.D.; Sparling, B.F. Integral Resonant Control Scheme for Cancelling Human-Induced Vibrations in Light-Weight Pedestrian Structures. *Struct. Control Health Monit.* **2011**, *19*, 55–69. [CrossRef]
13. Apostolakis, G. Optimal Evolutionary Seismic Design of Three-Dimensional Multistory Structures with Damping Devices. *J. Struct. Eng.* **2020**, *146*, 1–16. [CrossRef]
14. Liu, Q.; Zhang, J.; Yan, L. An Optimal Method for Seismic Drift Design of Concrete Buildings Using Gradient and Hessian Matrix Calculations. *Arch. Appl. Mech.* **2010**, *80*, 1225–1242. [CrossRef]
15. Daniel, Y.; Lavan, O. Gradient Based Optimal Seismic Retrofitting of 3D Irregular Buildings Using Multiple Tuned Mass Dampers. *Comput. Struct.* **2014**, *139*, 84–97. [CrossRef]
16. Wang, S.; Mahin, S.A. High-Performance Computer-Aided Optimization of Viscous Dampers for Improving the Seismic Performance of a Tall Building. *Soil Dyn. Earthq. Eng.* **2018**, *113*, 454–461. [CrossRef]
17. Franchin, P.; Petrini, F.; Mollaioli, F. Improved Risk-Targeted Performance-Based Seismic Design of Reinforced Concrete Frame Structures. *Earthq. Eng. Struct. Dyn.* **2018**, *47*, 49–67. [CrossRef]
18. Shmerling, A.; Levy, R. Seismic Structural Design Methodology for Inelastic Shear Buildings That Regulates Floor Accelerations. *Eng. Struct.* **2019**, *187*, 428–443. [CrossRef]
19. Wen, Y.; Hui, B. Stochastic Optimization of Multiple Tuned Inerter Dampers for Mitigating Seismic Responses of Bridges with Friction Pendulum Systems. *Int. J. Struct. Stab. Dyn.* **2022**, *22*, 2250137. [CrossRef]
20. Su, C.; Xian, J. Topology Optimization of Non-Linear Viscous Dampers for Energy-Dissipating Structures Subjected to Non-Stationary Random Seismic Excitation. *Struct. Multidiscip. Optim.* **2022**, *65*, 200. [CrossRef]
21. Gluck, N.; Reinhorn, A.M.; Gluck, J.; Levy, R. Design of Supplemental Dampers for Control of Structures. *J. Struct. Eng.* **1996**, *122*, 1394–1399. [CrossRef]
22. Yamada, K.; Kobori, T. On-Line Predicted Future Seismic Excitation. *Earthq. Eng. Struct. Dyn.* **1996**, *25*, 631–644. [CrossRef]
23. Jalili-Kharaajoo, M.; Nikouseresht, Y.; Mohebbi, A.; Moshiri, B.; Ashari, A.E.; Nagashima, I.; Engineering, E.; Dynamics, S. Notice of Violation of IEEE Publication Principles: Application of linear quadratic regulator (LQR) in displacement control of an active mass damper. In Proceedings of the IEEE Conference on Control Applications, Istanbul, Turkey, 25 June 2003.

24. Chang, C.M.; Shia, S.; Yang, C.Y. Design of Buildings with Seismic Isolation Using Linear Quadratic Algorithm. *Procedia Eng.* **2017**, *199*, 1610–1615. [CrossRef]
25. Shmerling, A.; Levy, R.; Reinhorn, A.M. Seismic Retrofit of Frame Structures Using Passive Systems Based on Optimal Control. *Struct. Control Health Monit.* **2018**, *25*, e2038. [CrossRef]
26. Shmerling, A. Reversed Optimal Control Approach for Seismic Retrofitting of Inelastic Lateral Load Resisting Systems. *Int. J. Dyn. Control* **2022**, *10*, 2034–2052. [CrossRef]
27. Chen, G.; Yang, D. Direct Probability Integral Method for Stochastic Response Analysis of Static and Dynamic Structural Systems. *Comput. Methods Appl. Mech. Eng.* **2019**, *357*, 112612. [CrossRef]
28. Li, B.; Wang, J.; Yang, J.; Wang, Y. Probabilistic Seismic Performance Evaluation of Composite Frames with Concrete-Filled Steel Tube Columns and Buckling-Restrained Braces. *Arch. Civ. Mech. Eng.* **2021**, *21*, 73. [CrossRef]
29. Zhou, Y.; Jing, M.; Pang, R.; Xu, B.; Yao, H. A Novel Method for the Dynamic Reliability Analysis of Slopes Considering Dependent Random Parameters via the Direct Probability Integral Method. *Structures* **2022**, *43*, 1732–1749. [CrossRef]
30. Potra, F.A.; Simiu, E. Multihazard design: Structural optimization approach. *J. Optim. Theory Appl.* **2010**, *144*, 120–136. [CrossRef]
31. Shmerling, A.; Levy, R. Seismic upgrade of structures using the H∞ control problem for a general system interconnection paradigm. *Struct. Control Health Monit.* **2018**, *25*, e2162. [CrossRef]
32. Smarra, F.; Girolamo GDDi Gattulli, V.; Graziosi, F.; D'Innocenzo, A. Learning Models for Seismic-Induced Vibrations Optimal Control in Structures via Random Forests. *J. Optim. Theory Appl.* **2020**, *187*, 855–874. [CrossRef]
33. Hamoda, A.; Emara, M.; Mansour, W. Behavior of Steel I-Beam Embedded in Normal and Steel Fiber Reinforced Concrete Incorporating Demountable Bolted Connectors. *Compos. Part B Eng.* **2019**, *174*, 106996. [CrossRef]
34. Mansour, W.; Tayeh, B.A.; Tam, L. Finite Element Analysis of Shear Performance of UHPFRC-Encased Steel Composite Beams: Parametric Study. *Eng. Struct.* **2022**, *271*, 114940. [CrossRef]
35. Liou, D.D. Synopsis of buckling-restrained braced frame design. In *Structures Congress 2014*; ASCE: Reston, VI, USA, 2014; pp. 1551–1562. [CrossRef]
36. Usami, T.; Lu, Z.; Ge, H. A seismic upgrading method for steel arch bridges using buckling-restrained braces. *Earthq. Eng. Struct. Dyn.* **2005**, *34*, 471–496. [CrossRef]
37. Balling, R.J.; Balling, L.J.; Richards, P.W. Design of Buckling-Restrained Braced Frames Using Nonlinear Time History Analysis and Optimization. *J. Struct. Eng.* **2009**, *135*, 461–468. [CrossRef]
38. Hoffman, E.W.; Richards, P.W. Efficiently Implementing Genetic Optimization with Nonlinear Response History Analysis of Taller Buildings. *J. Struct. Eng.* **2014**, *140*, A4014011. [CrossRef]
39. Abedini, H.; Hoseini Vaez, S.R.; Zarrineghbal, A. Optimum design of buckling-restrained braced frames. *Structures* **2020**, *25*, 99–112. [CrossRef]
40. Pan, W.H.; Tong, J.Z.; Guo, Y.L.; Wang, C.M. Optimal design of steel buckling-restrained braces considering stiffness and strength requirements. *Eng. Struct.* **2020**, *211*, 110437. [CrossRef]
41. Rezazadeh, F.; Talatahari, S. Seismic energy-based design of BRB frames using multi-objective vibrating particles system optimization. *Structures* **2020**, *24*, 227–239. [CrossRef]
42. Tu, X.; He, Z.; Huang, G. Seismic multi-objective optimization of vertically irregular steel frames with setbacks upgraded by buckling-restrained braces. *Structures* **2022**, *39*, 470–481. [CrossRef]
43. Tremblay, R.; Bolduc, P.; Neville, R.; DeVall, R. Seismic testing and performance of buckling-restrained bracing systems. *Can. J. Civ. Eng.* **2006**, *33*, 183–198. [CrossRef]
44. Gerdts, M. *Optimal Control of ODEs and DAEs*; Walter de Gruyter: Berlin, Germany, 2012. [CrossRef]

Article

The Role of Disaster Risk Governance for Effective Post-Disaster Risk Management—Case of Croatia

Zvonko Sigmund [1,*], Mladen Radujković [2] and Josip Atalić [1]

1. Faculty of Civil Engineering, University of Zagreb, 10000 Zagreb, Croatia; josip.atalic@grad.unizg.hr
2. Alma Mater Europaea ECM, 2000 Maribor, Slovenia; mladen.radujkovic@almamater.si
* Correspondence: zvonko.sigmund@grad.unizg.hr

Abstract: Risk governance is mostly viewed through the lens of disaster or emergency management departments, agencies, or organizations. Visible in times of crises, risk governance is rarely seen as part of everyday public or private functions such as planning, social welfare, investments, or fiscal responsibilities. This paper emphasizes the importance of disaster risk governance in disaster risk management activities on the example of the post-disaster recovery of Croatia after a series of strong seismic events in mainland Croatia. The analysis is made based on a thorough review of national documents of Croatia and other selected countries overlapped with the national journals reporting on the situation from the affected areas. In accordance with the authors' opinion, the necessary elements of disaster risk governance are clearly stated through the four Sendai framework priorities, and this statement is supported by the facts from the case study. Without either the political will or the enabling surrounding the disaster, risk management is next to impossible. The Croatian case study emphasizes the importance of disaster risk governance, showcasing the adaptation process for the post-disaster recovery process to start.

Keywords: disaster risk governance; disaster risk management; Sendai framework; Croatia; case study

Citation: Sigmund, Z.; Radujković, M.; Atalić, J. The Role of Disaster Risk Governance for Effective Post-Disaster Risk Management—Case of Croatia. *Buildings* **2022**, *12*, 420. https://doi.org/10.3390/buildings12040420

Academic Editor: Pierfrancesco De Paola

Received: 14 February 2022
Accepted: 25 March 2022
Published: 31 March 2022

Publisher's Note: MDPI stays neutral with regard to jurisdictional claims in published maps and institutional affiliations.

1. Introduction

Natural disasters, alongside climate change, cause ever increasing losses, with a $3\times$ increase in losses only in the last 20 years [1]. In order to improve the rate of implementation of scientific advances effectively in disaster risk reduction, it is important to understand what the major barriers for effective disaster risk management are.

Disaster risk governance has traditionally been fragmented between local, state, and national entities and between sectors, and compartmentalized in highly variable bureaucratic structures [2], which is the case in Croatia as well. Risk governance is mostly viewed through the lens of disaster or emergency management departments, agencies, or organizations, which often have little interaction among other governmental, civil society, or corporate entities. Visible in times of crises, risk governance is rarely seen as part of everyday public or private functions such as planning, social welfare, investments, or fiscal responsibilities [2,3].

Building on a premises published in [4], where, after the capacity for disaster risk governance needed to be enabled through a broad list of planned actions, ranging from material resources—access to equipment and technology; human resources—skills, knowledge, awareness; structures—organizations and policies; processes—decision making, coordination, delivery; and enabling mechanisms—political support, advocacy, staff incentives [4], the authors showcased the Croatian disaster risk reduction system prior and after the earthquake series in the year 2020 while building the case around the Sendai framework for disaster risk reduction, and some other important cases identified through a wider scope of research conducted in [5].

This paper aims to emphasize the importance of disaster risk governance in the implementation of disaster risk management in the example of Croatia, mainly concentrating on the construction industry. The research area is focused on the implementation of DRR and DRM principles in the area of seismic disaster risk management. Disaster risk governance principles, as were defined and planned through the regulatory framework, as well as the changes that were introduced after the earthquake series that struck mainland Croatia during the year of 2020, are reviewed in this paper.

Seismicity of Croatia

The grounds for a more holistic approach to managing disaster risk, and thereby the DRM capacity, have been expressed within the critical literature in this field for some time [6,7]. This includes moving beyond a focus on a DRM of preparedness and emergency management to building capacity in disaster prevention, mitigation, and long-term recovery [8]. This need, to advance the DRM, becomes a necessity as soon as a disaster happens, as it did in Croatia in the year 2020.

For this paper, the UNDRR terminology glossary [9] is used for the terms "disaster risk governance" and "disaster risk management". Here, disaster risk governance is defined as "The system of institutions, mechanisms, policy and legal frameworks and other arrangements to guide, coordinate and oversee disaster risk reduction and related areas of policy", and disaster risk management is "the application of disaster risk reduction policies and strategies to prevent new disaster risk, reduce existing disaster risk and manage residual risk, contributing to the strengthening of resilience and reduction of disaster losses" [9].

The seismicity of the territory is unevenly distributed, with the most seismic activity happening at the coastal areas of the country and in a small part of north-west mainland Croatia (Figure 1). Croatia, due to its geographical shape, spreads out through a wide variety of seismically active regions. The territory of Croatia is a part of the Alpine–Mediterranean seismic region, which comprises of several geotectonic units. The dominant geotechnic units are the Pannonian Basin to the north, the Eastern Alps, the Dinarides, the Dinarides–Adriatic Platform transition zone and the Adriatic Platform itself [10].

The seismicity of north-west Croatia can be characterized as moderate with rare occurrences of strong events, both features typical for regions of intraplate seismicity. Although not the most earthquake-prone region, Croatia is extremely seismically vulnerable due to its economic and political positions. Mainland Croatia, and more precisely the north-west part of the Croatian mainland, is inhabited by 45% of the Croatian population with 55% of the Croatian national product [11].

The history of strong earthquakes in the area near the fault is marked by a major earthquake in Zagreb in 1880, which is considered to have been M6.3, in 1909 in the Pokupsko region (M6.0), and in 1969 in Banja Luka (M6.6) [12].

Recently, Croatia was struck with two major earthquakes: the Zagreb earthquake that struck in March 2020 (M5.0), just after the Croatian government had issued a complete lockdown due to the COVID-19 pandemic; and the Petrinja (about 50 km from Zagreb) earthquake (M6.4) in December 2020.

On 22 March 2020, Zagreb was struck by an M5.5 earthquake that had been expected for more than 100 years and revealed all the deficiencies in the construction of buildings in the Croatian capital, especially those built in the first half of the 20th century [13]. A pronounced issue that arose was the damaging of many historical buildings which were, in many cases, used for various public purposes: hospitals, schools, theaters, local or state administration, etc. The earthquake was followed by 10 aftershocks of M3+ during the next 4 months [14]. One person succumbed to injuries caused by the earthquake, about 24,000 buildings were reported to have damages, of which about 5000 buildings were heavily damaged [15]. The total damages and losses, according to the rapid damage and needs assessment, were 11.3 billion euros [16].

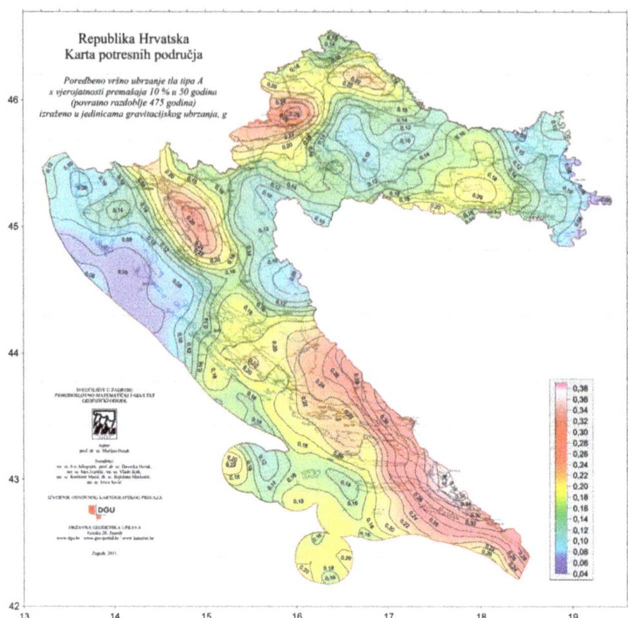

Figure 1. Map of seismic areas of the Republic of Croatia [17].

The Petrinja earthquake begun with an earthquake of M5.0, followed by M4.5 and M3.8 in the same day on the 28 December 2020, [12]. The behavior was considered to be a sign of calming down; this, however, was not the case. On 29 December the main shock struck Petrinja with M6.4 [18,19]. In less than three days after the main earthquake, almost 2000 aftershocks followed. Until 15 January 2021, there were nine M4+ aftershocks, of which the strongest was of a magnitude of 5.0. During the aftermath of the Petrinja earthquake [12], 7 persons were confirmed dead, and about 45,000 buildings were reported to have damages, of which about 11,000 buildings were assessed by engineers to be unusable due to the damages [20]. The total damages and losses, according to the rapid damage and needs assessment, were assessed to 4.8 billion euros [21]. The Petrinja area is still seismically active even after a one-year period.

2. Materials and Methods

Here it is important to point out that in the context of the construction industry, seismic risk can simply be presented as the product of probability of seismic activities' occurrence and the exposure of assets to the unwanted activity; exposure of the assets to the unwanted result is presented by the existing buildings that are insufficiently resistant to seismic activities and people residing in the threatened areas [22]. When dealing with earthquakes and existing built environment, one can only increase the resistance of existing buildings to seismic activities. Here, the importance of disaster risk governance and the related policies have a major role to play. Thus, in the paper, the authors are mainly reviewing the national documents (legislative framework) enabling post-disaster recovery.

The authors of the paper are building the case on the premise that disaster risk reduction should be enabled through a broad list of planned actions involving resources, laws and policies, political will, and implementation skills, whereby the disaster risk management activities are enabled. The argument is supported by the case of Croatian post-disaster situation, where the post-disaster recovery regulatory framework did not exist, but was developed at the time of writing of this paper. For the analysis of the national disaster risk reduction's state of the art, the existing regulatory framework was compared to the 4 priorities of the Sendai framework for disaster risk reduction. As there are almost no relevant publications explaining the Croatian disaster risk governance principles, the state of the art relevant for the topic of governance principles in Croatia was researched based on available literature collected from official national publishers and the national journals as Official gazette, existing and available laws and bylaws, governmental publications on the topic, and other similar sources. The missing links between different priorities and identified gaps in the disaster risk management structure were discussed with the Croatian Sendai Focal point, in which research was conducted for the publication of [23].

The statements on the possible approaches were formed based on the previously collected data sets of regulatory framework and review of a selected sample of approaches used in different countries. To gain a better insight in the functionality of an integrated seismic disaster risk reduction strategy, the already existing seismic DRR strategies with their legal framework were compared. In addition, the reader is presented with the comparison of existing seismic DRR strategies.

Hereafter, the authors commented on the challenges that were faced in the process of creating what now is a fully functional disaster risk reduction management system in Croatia. The development of the post disaster recovery regulatory framework was showcased to identify the possible improvements in creating and implementing the national disaster risk management system.

The development of the regulatory framework that was developed using a trial-and-error approach rather than a planned and thought out approach was used to highlight the importance of particular Sendai framework priorities. Here, to support the case, a selected list of cases from different countries was used as a positive example. The review and the complete research on the listed cases can be found in [5].

3. Seismic Disaster Risk Management—Case of Croatia

Major DRR oriented organizations, such as FEMA [24], OECD [25], UNDRR [23], IFRC [8] and others, agree that in order to ensure that the DRR strategies can be carried out effectively, stimulative measures need to be provided. In this section, the reader is presented with regulations and the legal framework stimulating the effective use of DRR policies.

So far, Croatian disaster risk governance was mainly oriented towards disaster response (a military approach), which is based on a decades-old regulatory framework, as was elaborated thoroughly in the previous work [26]. Nevertheless, Croatia has just recently (within the last few years) started switching its focus from disaster risk preparedness to disaster risk management with the introduction of the Homeland Security System Act [27].

While mainly oriented towards disaster response, in general, the Croatian disaster risk management system (regulatory framework) recognizes only two areas of disaster risk management: prevention and response. Therefore, the Croatian disaster risk management system can hardly be fully valorized through the objectives of the Sendai framework for disaster risk reduction. The previous system and the new developments are going to be presented in the next subchapters.

3.1. Croatian Disaster Risk Prevention Regulatory Framework

So far (prior to the earthquake series), the Croatian government had focused most policies and regulations only in the preparedness and the immediate disaster recovery phases of disaster risk management [28], which had left prevention and recovery unattended by laws or policies.

As the main publicly available platform, there is the Croatian platform for disaster risk reduction. It is organized within the Ministry of the Interior of the Republic of Croatia as an activity task of the Civil Protection Directorate. The main task of the Croatian platform for disaster risk reduction is to facilitate disaster risk reduction [29] activities, so as to integrate and facilitate the interface for communication and decision making by involving the political, operational, and scientific communities. The work of the platform is regulated mainly with the Homeland Security System Act [27] and the Civil Protection

System Act [30]. The Homeland Defense Act regulates the involvement of military forces in immediate post-disaster relief and recovery activities and the integration of military forces with the civil protection teams after the post-disaster activities. Therefore, the work of the platform for disaster risk reduction is indirectly, but still closely connected to the Homeland Defense Act, where crisis management activities are regulated [31]. These laws are also the main regulatory framework, regulating the activities and responsibilities of the Civil Protection Directorate and other involved parties (as shown in Figure 2).

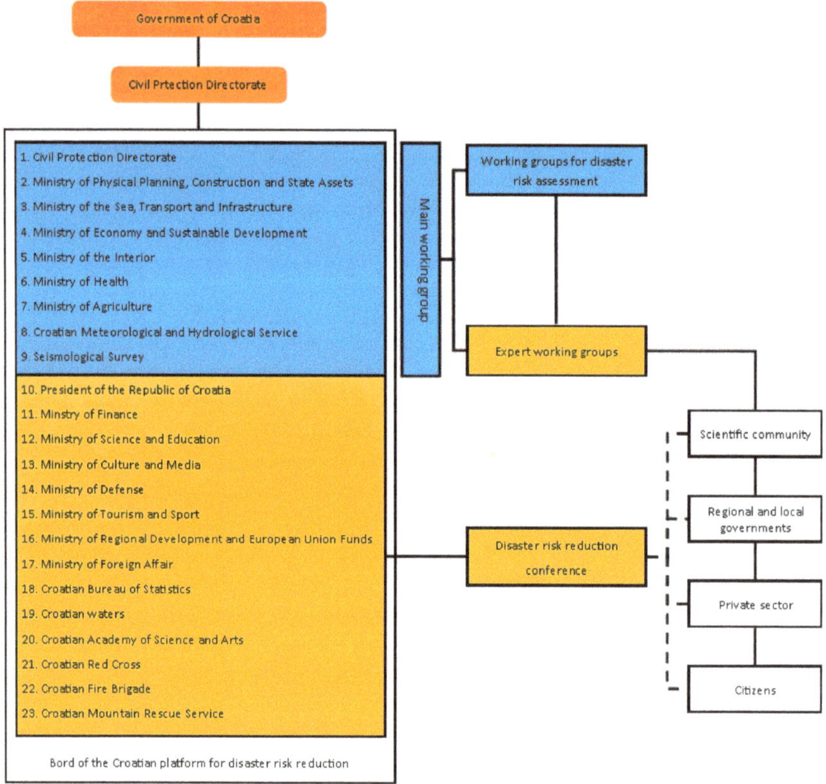

Figure 2. Schematic display of the organization of the Croatian platform for disaster risk reduction [32].

The Homeland Security System Act regulates and enables the integration of the work of governmental and nongovernmental bodies with the aim of increasing national safety. On the other hand, the Civil Protection System Act is the main regulatory basis for all civil protection activities. The Defense Act regulates the involvement of military forces in case of a crisis. Hereafter, military forces can be requested for supporting the humanitarian and disaster stress relief activities.

The Civil Protection System Act regulates the obligations of public authorities and operational capacities, from the local to the state level. It develops a special capacities-headquarters for units and civil protection teams whose activities are needed in a state of emergency, and thus creates a new organizational framework [30] for the country or a region during the emergency state.

The national disaster risk assessment document comments on the available structure of national civil protection: national civil protection is standardized well enough; however, the standardization is achieved through local level strategic documents and regulations which results in general organizational inconsistency. In addition, a major problem, as commented by the national DR assessment document, is the supervision of the regulation's implementation and the organizational structure inconsistency of executive bodies at all levels of the country. Thus, obligations in the field of risk management are either insufficiently recognized or their implementation is not supported to the necessary extent [28].

When considering the question of understanding disaster risk, so far, on the governmental level, the priority of "understanding disaster risk" has been covered only superficially by the publication of the national risk assessment document [28]. Here, no continuous activities have been conducted to enable the rise of awareness to risk exposure at a national level. Further on, the national risk assessment document clearly states that the awareness of risks is still unsatisfactory, and that particular attention should be paid to communicate the disaster risk and possible necessary actions in case of an emergency to citizens effectively, in order to increase the resilience of the citizens themselves and prepare them for an effective interaction with organized parts of the operational capacities of the civil protection system [28]. As the Civil Protection Directorate is the main responsible governmental organization for awareness raising, just recently, the directorate has started with a number of awareness raising projects such as, for instance, an educational awareness raising project for elementary schools which has resulted in the introduction of disaster preparedness training in the elementary school curriculum [30].

So far, Croatia has not developed a disaster risk reduction strategy. Hereafter, even though the only disaster-related activity of the Ministry of Defense is to support civil protection activities in cases of crisis, in its main organizational assessment report, the Ministry of Defense identified a natural related crisis as one of the main risk sources. Hereafter, the Ministry of Defense has prepared a national security strategy which stresses the importance of improving and strengthening the disaster response and short-term recovery capabilities [33,34], which is in line with the activities conducted by the military forces, but not in line with the goals of sustainable long-term recovery planning a government should have.

3.2. Croatian Laws and Regulations in the Construction Industry Prior the Earthquake Series in Year 2020

Approximately 40–60% of residential units in the region of the Croatian mainland were built prior to the first seismic design codes, based on the analysis conducted using the data presented in the assessment of the vulnerability of the Republic of Croatia to natural and technical technological disasters and major accidents [35].

As Croatia, in general terms, doesn't have an active seismic disaster risk reduction plan, the only Croatian regulatory framework regulating activities in the area of reducing seismic risks would be the Construction Law, which is mainly oriented around regulating any types of activities concerning the built environment [36]. In a built environment, one could reduce the disaster risk posed by an earthquake by changing the use of a building, reducing the risk by moving the threatened to other, safer locations, or by strengthening the existing buildings. Both measures affect the "basic requirements" of the building defined by the Construction Law, which requires obtaining a building permit [36]. In some cases, this can be an exhausting and time-taking process. Furthermore, in a case in which one would need to strengthen a historically protected building, one would act considering the Law on the Protection and Preservation of Cultural Heritage [37]. In this case, as defined by the Law on the Protection and Preservation of Cultural Heritage, the permittance process would be even more complicated and would include even more interested parties in the process [37].

As a part of the European Union, Croatia has adopted Eurocodes as the main construction guidelines and norms. Eurocode 1998-3 does not propose any type of active seismic risk mitigation procedure. The choice of whether to manage seismic threats passively or actively for existing structures is made through the definition of Eurocode 8-3 [38], and left to be defined in national addendums. The passive approach considers the seismic assessment of existing buildings only in cases of activities or events that, for instance, relate to the use of the building and its continuity, whereas the active approach may require owners of certain buildings to consider taking action in terms of the seismic protection of their property.

The Croatian national addendum, the Eurocode 8-3/NA [39], makes no mention of preventive seismic protection, thus the passive approach to seismic risk reduction is used, as defined by the Croatian building law.

It can safely be concluded that in terms of seismic disaster risk reduction, the Croatian construction regulation is rather incomplete. The required actions prior to strengthening or even repair works of a larger scale could present a problem even in the case of a disaster.

3.3. Croatian Disaster Recovery Framework after Earthquake Series in Year 2020

Prior to the earthquake, the only law to regulate recovery was the *Law on the mitigation and elimination of the consequences of natural disasters*. This law regulates governmental financial responsibility towards all those affected by disasters and the operationalization of the activities of the Ministry of Finances in cases of disasters. The responsibility is instrumentalized through financial support, but includes an assessment of the effects of disastrous events and the allocation of partial financial relief to affected areas [40]. Other institutionalized measures for disaster recovery were so far regulated only after the occurrence of the disaster, as was the case of the area destructed by the flooding in 2014 [41].

As soon as the first earthquake struck Zagreb, on the governmental level, it was clear that the Croatian legal framework could not be kept as it was. A new legislation would need to come into place to enable recovery and reconstruction works. Nevertheless, even though the legislator had a clear vision of the regulatory framework that needed to be defined, the disaster recovery and reconstruction regulatory framework that was initially prescribed needed to be adapted in accordance with the needs identified during the practical use of the legislation.

On 21 March 2020, the Croatian Government introduced a "stay at home" order for the whole country due to the COVID-19 pandemic, and the very next day, a magnitude 5.5 earthquake shook the capital city of Zagreb [42]. The regulatory framework for disaster recovery was structured in a series of different measures: the suspension of COVID-19 restricting measures in the affected areas, financial relief and support, disaster emergency housing, emergency repair support in terms of financial and workforce organization, and finally, the framework supporting the recovery and repair of damaged infrastructure and the built environment.

The main goal of the regulatory framework, after the earthquake series, was to assist the owners or co-owners of damaged and destroyed real estate to setup their estates quickly and with less effort in comparison to the previously available legal framework. The first recovery and reconstruction law was created to aid the affected areas of the first earthquake: *A Law on the reconstruction of buildings damaged by earthquakes in the city of Zagreb, Krapina-Zagorje County and Zagreb County* [43]. The main goals of the law were to reduce and simplify the documentation needed for the approval of the reconstruction, and:

- To establish the "Reconstruction fund"—the main governmental executive body for the organization, implementation, and monitoring of the implementation of reconstruction activities of earthquake-damaged buildings [44].
- To define the process of building reconstruction in case the building was only damaged, and the construction of replacement housing in case a house was destroyed or damaged in a way that repair would not possible or would be financially inefficient.
- To prescribe financial support for temporary repair works, building reconstruction and repair works.

In addition to the law, in October 2020, the first program of measures for the reconstruction of earthquake damaged buildings in the city of Zagreb, Krapina-Zagorje County and Zagreb County, was prescribed. This program of measures defines the levels and scopes of repair and/or reconstruction that can be financed from the Reconstruction fund. Furthermore, it defines the organizational structure of the governmental bodies responsible for activities in the reconstruction, the criteria for the project parties' selection, reconstruction priorities, etc., [45]. As the title of the law shows, the law regulates the recovery measures only in the affected areas and cannot be implemented outside of the mentioned counties.

By October 2020, 7 months after the earthquake passed, the emergency repair works were mainly done; besides these, only a few reconstruction projects had started, among which the city of Zagreb was the main investor. By that time, even though there is no official data, the number of reconstruction activities in the affected region was at the minimum.

With the occurrence of the second earthquake series in the area of Petrinja (Sisak-Moslavina county), an amendment of the already existing law on reconstruction was made with the law amendments from February 2021 [46] (just two months after the December earthquake series). As the new situation required a new approach, the amendment of the law was not only used to broaden the area of use to the new affected areas, but also to accommodate new needs. Except for the historic city centers in the affected areas of Sisak-Moslavina County and the other affected areas, these areas are more rural types, with occasional historic buildings and the occasional industrial facilities, which have now sustained major damages, as opposed to the earthquakes of Zagreb where most damages were sustained in the historical buildings which were not designed to withstand seismic activities of any kind.

By the time of the law amendment publication, the Reconstruction fund began to function as intended, resulting in the first 231 finished reconstruction investments with an investment sum of about 1.1 mil EUR [47]. As the earthquake from December 2021 had more serious consequences than the one from Zagreb County (March 2021) the main changes in legislation were oriented towards creating the emergency housing capacities for people whose homes were destroyed or severely damaged. Therefore, a part of the responsibilities and powers which were mainly activities of the Reconstruction fund were transferred to the Central State Office for Reconstruction and Housing to divide the intensity and the activity scope of the Reconstruction fund [46].

During the reconstruction process, several main issues were encountered that were slowing down the reconstruction process:

- The owners (potential investors) were not allowed to start reconstruction on their own as, to be entitled for the governmental funding, the reconstruction process had to start via the governmental administration [48], for which the process was rather sluggish.
- Co-financing measures were limited to 80% of the cost of the structural renovation of a building which, in the whole process of reconstruction, would cover no more than 30% of the whole reconstruction investment, causing many potential investors to give up on the potential reconstruction investment [49]
- There was a problem of unresolved ownership relations for which the process of renewal was entirely disabled, even for cases when real ownership was not in question, but it was not legally implemented, or the legal trace of ownership was difficult to prove (a problem expressed in rural parts of Croatia) [50]
- Construction works' prices rose uncontrollably on the global market, which was more pronounced in Croatia due to a sped-up increase in the demand in construction and reconstruction works and the COVID-19 sanitary crisis. Hereby, the owners' ability to invest was severely diminished [51]
- The affected area was widely marked by cultural heritage buildings, which also made up a significant share of the damaged buildings. The necessary activities of the relevant administration for cultural heritage are poorly defined even by basic laws, which is even more evident in crisis situations [52]
- The reconstruction process indicated some administrative deficiencies in the process [48,52], among which is that, for instance, the demolition of heavily damaged buildings that potentially threaten the environment requires a series of administrative approvals.

Still, even with the flaws of the law, the rate of investments in reconstruction rose to 792 reconstruction investments in total and approximately 5.6 mil EUR [47]. In relation to this, investments rose from 33 cases per month and approximately 160,000 EUR/month to 99 cases/month and 700,000 EUR/month. These numbers cannot be taken as the absolute measure of the success of the laws, but still, they can be taken as an indicator that the reconstruction measures are giving positive results.

These mentioned issues were to be resolved by the latest amendment of the law on reconstruction [53] with the next measures:

- The main and most important change is the reorganization and improved definition of the tasks of governmental bodies included in the process of reconstruction. The improvements also include the definition of the maximum allowed time for decision making in the process of project approval or the definition of requested conditions that must be obeyed (e.g., preservation measures for cultural heritage buildings).
- The governmental financial support for reconstruction increased from 80 to 100% of the construction and reconstruction cost, with the possibility to receive the governmental subsidies in advance (only in cases where the buildings had a legal and official representative). This reduces the initial cost of reconstruction and repairs at the start of the investment process.
- For the cases where family house owners are willing to invest into the recovery of their real estate, they are now allowed to finance the works by themselves with the possibility to request a full refund for the applicable reconstruction costs (only for the construction/reconstruction).
- To improve the implementation rate of the law, the state can buy off the ownership of a building or a part of the ownership to improve the implementation of the law on reconstruction.
- The demolition of heavily damaged buildings is financed completely by the government, and in the case where a building is endangering the surroundings or persons, the building can be demolished through a shortened administrative procedure (with a duration of up to 5 days), where the owners of a demolished real estate have the possibility to receive a financial reimbursement for their real estate or they can request a replacement house (only for real estate where owners were living in at the time of the earthquake).

Hereafter, until the day of writing this paper (28 December 2021) a further 157 reconstruction investments and approximately 1.1 mil EUR [47] were approved. However, the results achieved by the newest addendum to the laws cannot be identified yet as the process of intervention planning, from the decision to the intervention execution, takes at least 2–3 months, as per the experience of the authors. Still, it is important to notice that the regulatory framework needs to accommodate the real case issues, mainly focusing on removing the main barriers for the successful implementation of the disaster relief regulatory framework which is, as evidenced, the main goal of the law on the reconstruction of buildings damaged by earthquakes.

4. Short Overview of Seismic Disaster Risk Management Regulatory Framework of Selected Countries

The results shown here are just shortlisted main conclusions of a wider scope of research conducted in [5].

To identify the possible coverage levels of disaster risk management, different principles and approaches were analyzed. The data collection involved different regions, ranging from earthquake prone regions undertaking almost no preventive measures, to highly developed disaster risk reduction strategies with a high level of systematic integration into everyday use:

- After the disastrous earthquakes in 1999, Turkey introduced an earthquake-resistant design of new buildings with a more stringent design control, on site inspections and as-built revisions, which was the first step forward in seismic risk reduction.

The recovery from the earthquake proved to be a significant burden on Turkey's fiscal policy. At the insistence of the World Bank, a "Turkish Disaster Insurance Company" was established as a preventive measure to take care of the existing buildings in order to transfer the financial responsibility for recovery from the government to the building owner. As an incentive, the Turkish government provided a $17,000 deposit for each insurance policy, and the implementation of these measures has been ensured by the introduction of the Disaster Protection Act. Under this law, the insurance of all private and public buildings is mandatory [54,55].

- Chile has, for a long time, had a seismic resistant design of new buildings (thoughout history), whereby no additional seismic DRR policies are needed, as Dr. Matias Hube from the Civil Engineer Catholic University of Chile has mentioned through personal contact.

- Japan uses a set of different laws and norms to regulate the construction of new and the protection of existing buildings, all of which are accompanied by policies regulating their execution. However, what sets out the Japanese legal framework are the next several elements which are a part of a hazard management plan: the prioritization of buildings and areas before and after the earthquake; the protection and involvement of vulnerable parties; the involvement of all interested parties in risk reduction programs; understanding the possibilities and limits of earthquake risk management; in this, prevention and preparedness are equally important [56]. In addition to these main features of the Japanese disaster risk reduction measures, the Japanese government leaves the final definition and prioritization measures of the disaster risk management approaches to the regional government, leaving the regional governments the ability to improve the regulatory framework and their approach to DRR in accordance with local needs and possibilities [57].

- Romania has proactively protected existing buildings since 1994, when a law on the seismic evaluation of existing buildings was put into effect. By this law, all regions in Romania are obligated to categorize their buildings and create a priority list accordingly. To enable execution of this order, the government has ensured the complete financing of the intervention for tenants with a lower-than-average income. For tenants with an above-average income, government low-interest loans are available [58].

- The Canadian PWGSC, as the owner and manager of all governmental buildings, identified the loss of resources' cost (destruction of a building) caused by an earthquake as a significant problem. This was based on a study of the costs and benefits of seismic building retrofitting which concluded that these interventions do not exceed the total cost of up-keeping works on existing buildings by more than 3–5% [59]. Therefore, the PWGSC has developed a set of handbooks which are mandatory guides for the screening, seismic safety evaluation and seismic upgrade of government owned or leased buildings [60–62]. This model can also be used for privately owned buildings, but several surveys showed that owners are usually not willing to conduct the seismic screening of their buildings [63].

- In the year 2000, the Swiss government recognized the dangers of earthquakes and empowered a decree by which all governmental buildings had to be evaluated and, if needed, strengthened; therefore, they released the SIA-2018 norm [64]. By the governmental decree, the seismic assessment of government owned buildings is obligatory. The governmental decree which defines the seismic risk reduction process was introduced through 4 steps: the introduction of the regulatory framework, the definition of the assessment process, a disaster management plan definition for the case of an earthquake, and post-disaster recovery planning. The main intention of the developed procedure was to cost-effectively assess larger numbers of buildings, and was delegated to the Federal Office for Water and Geology, which developed a three-step building assessment process composing of: a quick seismic vulnerability and loss evaluation; detailed analysis; and a seismic strengthening feasibility assessment [65].

- New Zealand's Society of Civil Engineers has developed a handbook for the Assessment and Improvement of the Structural Performance of Buildings in Earthquakes which uses a three-step assessment process for the evaluation of the seismic resistance of existing buildings. As an additional feature, a list of improvement techniques is given [66,67]. The document was drafted in accordance with the New Zealand Building Act of 2004, which requires all existing buildings to comply with the current New Zealand building code. Additionally to these seismic hazard mitigation-supporting guidelines, New Zealand's government subsidizes insurance policies for existing buildings [68,69].
- The US Government offers a whole scope of programs and measures supporting the improvement of the seismic resilience of existing buildings. Besides these, the US Government promotes seismic safety improvement by setting a good example. Namely, the US Government has been using a specially designed and obligatory procedure for ensuring the seismic safety of federal buildings [70]. Besides these measures, the USA has a whole set of compulsory and non-compulsory guidelines and standards developed by the FEMA and ASCE which had a noticeable impact on the development of the Canadian, Swiss and New Zealand's seismic disaster risk mitigation models. The latest edition of the 3 step seismic assessment and retrofit guidelines is presented in the ASCE/SEI 41-13 standard [71].

5. Discussion

For a functioning disaster risk management system, the legal framework needs to accept and promote the seismic hazard assessment and mitigation activities. However, as the legal framework is different in every county, the creation of a DRR regulatory framework should be done having the existing local legal framework in mind, rather than adopting existing ones from other countries. Nevertheless, the law-making institutions can and should learn from positive examples used in other countries. This is clearly showcased with the post-disaster recovery process of Croatia, where the custom developed recovery framework included laws, bylaws, and execution programs to encompass all the regional specificities. Here, every change in the law had to be followed up with the change of the bylaws and the implementation programs.

All the presented case countries have gone through the seismic disaster risk reduction process, where the whole process can be summarized with the Sendai framework priorities.

5.1. Sendai Framework Priority: Understand Disaster Risk

Regardless of the triggers leading the governmental will to reduce seismic disaster risk, it can safely be concluded that the most crucial element for disaster risk planning is disaster risk awareness. Obviously, the most showcased countries (Turkey, New Zealand, the USA, Japan) have started planning to reduce disaster risk only after a major disaster happened, resulting with material losses and loss of life. Only in rare cases was the disaster risk reduction triggered with a firm political will (Switzerland) and a scientific background (Canada).

The Croatian platform for disaster risk reduction has just recently (within the last decade) started raising risk awareness with various actions [32]. However, the wanted effect was only triggered by the latest series of disasters, and still the complete aftermath of the freshly introduced measures seems to have had a rather temporary effect, with the only goal being to build back better, with no preventive measures for the undamaged, but still vulnerable, buildings of Croatian cities that were not affected by the recent earthquake series. It is, therefore, understandable to suggest further efforts to facilitate disaster risk understanding, where the roles of the government, the profession and researchers are essential.

5.2. Sendai Framework Priority: Strengthening Disaster Risk Governance to Manage Disaster Risk

As the improvement of disaster risk response capacities is the main goal of the presented legal framework, disaster risk governance is well covered within the existing and presented framework. Regarding the 2nd Sendai framework priority, the existing legal framework is well structured and is a good platform for further development in accordance with the Sendai framework's recommendations. The Homeland Security System Act [27] clearly defines the roles and responsibilities of all stakeholders and ensures their involvement. It also establishes an institutional framework at national and local levels by assigning them their role in disaster risk reduction and planning. However, as stated by the national vulnerability assessment [35], the legal framework lacks the institutional coordination of activities and the means to improve the implementation or the control of the implementation of positive disaster risk-reducing measures. Based on the Civil Protection System Act, all public authorities on the local and national level should have disaster recovery capacities, plans and strategies. However, as the Civil Protection System Act causes a fragmentation of efforts and knowledge between local authorities and subordinates, which usually lack the resources for conducting even the simplest tasks, and, taking into account that 55% of Croatian municipalities cannot function without subsidies from the state budget, some of them are not allocated financial resources for the needs of the development or operation of civil protection at all, it is difficult to expect that these plans are sufficient or, in some cases, implemented at all [35].

In the cases of Switzerland, Canada, the USA, or Romania, the strive to strengthen the disaster risk governance requires political will and dedication with an understanding of what is needed. In the case of Croatia, capacity-building interventions focused disproportionately on preparedness, with little attention given to building capacities for prevention and mitigation work, and even less to building capacities for disaster recovery. This is mostly evident in the very long period of 7 months after the earthquake in Zagreb which was needed for the government to release the law on the reconstruction of earthquake damaged buildings [43], which needed 2 further adaptations to fit the regional and local specificities. Here it is important to stress that the recovery process, in the case of Croatia, had not started until the release of the law enabling the reconstruction process.

Hereafter, disaster risk governance cannot be structured only after a disaster happens. It should be ready and prepared before such an event can even occur. The elements of disaster risk should also [8]:

- foster disaster risk ownership, such as, for instance, the responsibility transfer (Turkey, New Zealand) or encouraging building strengthening (reducing risk);
- consider sustainability in disaster management programs to improve disaster management; for instance, creating a regulatory framework for enabling disaster risk recovery as soon as the disaster occurs, unlike the law on the reconstruction of earthquake damaged buildings being released 7 months after the disaster;
- allow longer timescales to accommodate the regulatory framework adaptation process which would, in the case of Croatia, improve the implementation rate and thus shorten the time of the post-disaster recovery.

When observing the governance perspective, it is important to develop the understanding that governance must be continuously improved and adapted, to enable a more effective organization that can achieve its goals.

5.3. Sendai Framework Priority: Investing in Disaster Risk Reduction

According to the interview with the National Sendai framework focal point, Croatia is constantly investing in disaster risk reduction (3rd Sendai framework priority), however, these investments are structured on such very rare occasions as, for instance, the investment in a national fire early warning system. However, the Croatian Sendai Focal point stresses that the Croatian national institutions, such as Croatian forests, Croatian waters, as well as different Ministries, have disaster preparedness and prevention strategies which involve investment for increasing disaster risk resilience. Still, the disaster risk management initiative must start at the very top of the country as the essential need for successful disaster risk management is creating an enabling environment for DRM. Here, various governments can have different approaches, and still the governmental will is usually not enough, but the will needs to be supplemented by the financial support. Nevertheless, even the preparation of the regulatory framework for the worst-case scenario would help avoid the situation where no recovery activities can start while the regulatory framework is expected, as was the case in Croatia immediately after the Zagreb earthquake. The process of creating the regulatory framework in the case of Zagreb was additionally slowed down due to major political changes happening, however, this makes an even stronger case for the necessity to have the regulatory framework ready before a disaster happens.

A positive political will is nicely showcased in Switzerland and Canada where the government decided to set a good example by improving the resilience of the critical infrastructure. Here are generally government owned assets that can, when hazards are considered, include buildings that are occupied or used by larger numbers of people, or buildings which, if not functional after the disaster, can cause more damage than was caused by the hazard itself (such as hospitals, police stations, fire departments, etc.). The private owners were not forced to do the same, although, if the will existed, private owners would, after the example of the government, have an easier adaptation to the reconstruction process. Still, some examples of preventive measures for privately owned buildings and houses are also promoted and supported. Here, good examples are set in Japan, Romania, Italy and New Zealand. Although investment in disaster risk reduction requires significant attention and funding, it is also strongly related to the financial capacity of the community. Still, short-term policies and post-disaster recovery actions are the most expensive scenarios. Therefore, the ground for a disaster risk reduction strategy should be long-term planning, including continuous investments in disaster risk reduction over a long period of time.

5.4. Sendai Framework Priority: Enhancing Disaster Preparedness for Effective Response and to "Build Back Better" in Recovery, Rehabilitation, and Reconstruction

For the 4th Sendai framework priority, it can safely be concluded that Croatia is continuously building on the existing preparedness structure, however, lacks the "Build Back Better" element completely. In terms of disaster preparedness, the Croatian platform for disaster risk reduction focused its work mainly on immediate post-disaster relief and rescue, whereas the preventive measures were rather a part of isolated pilot projects raising the disaster risk awareness. Only the earthquake in Zagreb, and the so far existing legal framework which would not allow the promotion of positive activities in disaster risk reduction, triggered the development of the regulatory framework focusing more on long-term recovery and "Build Back Better".

The law on the reconstruction of buildings damaged during earthquakes, with all its addendums and programs, was necessary to start the recovery process, but also to regulate and stimulate the reconstruction process in accordance with the "Build Back Better" principle. For instance, within the law on reconstruction, the reconstruction of damaged buildings is envisioned with the aim that all damaged buildings can also be upgraded in terms of energy efficiency. Still, these measures are more intended for larger apartment buildings and publicly owned buildings. The downside is that the regulatory framework only enables the energy efficiency works, but they are not stimulated, which leaves it to the investor and their own financial capability to decide if the structural upgrade of the building would be followed by the improvement in the energy efficiency.

Psychologically and socially, disasters are rather quickly forgotten, and politically, not a wanted topic. Still, it is the responsibility of governments and professionals to communicate the issue, and to permanently work on enhancing the disaster preparedness in every aspect.

6. Conclusions

The Croatian case study emphasizes the importance of the disaster risk governance, showcasing the adaptation process for the post-disaster recovery process to start. Here, the process could have evidently been shortened had the post-disaster recovery regulatory framework been ready and waiting in case of an emergency. That the disaster risk recovery governance was weakly developed was already identified by the national disaster risk assessment. This emphasizes the importance of the second Sendai framework priority, which also highlights the importance of the necessary political will and the positive and enabling surroundings for effective disaster risk reduction measures. Without either the political will or the enabling surroundings, disaster risk management is next to impossible.

The national risk assessment clearly states that the government had been strongly and intensively investing in preparedness, and these activities played an important role in the short-term post-disaster process. It can be safely assumed that the disaster risk management disabling surroundings and the nonexistent political will made it tough and demotivating to invest into preventive disaster risk reducing measures, at least when it came to retrofitting the built environment to resist the expected seismic events. Hereby, the amount of investments aimed at reducing the risk of damage to the built environment was severely reduced, making another strong statement that the national governance makes a strong impact on enabling the disaster risk management. One can argue that both issues can be attributed to a weak understanding of the risk at hand, however, it is unclear which awareness raising processes could have achieved the wanted result.

Analysis shows that the disaster risk reduction measures need time to be adopted in a culture, and the Croatian risk raising campaigns started only a decade ago. Still, it is unclear if a longer or more aggressive risk raising campaign would have had a wanted impact and might have enabled a creation of the so-much-needed disaster risk reduction governance.

Whether known or unknown, disaster risk sources are numerous, and their direct impacts are very well known and ever increasing. However, as currently we are living in a globalized world, real unwanted impacts of a particular disaster can only be discovered once the disaster happens. These can have a much more spread out impact than obvious at the first sight. At the time of writing this article, the COVID-19 pandemic has made this global risk landscape more evident than ever. Due to the current global crisis, states must undertake immediate action at community, national, and international levels to reduce the risks. It is all too evident that the four Priority Areas of the Sendai Framework for Disaster Risk Reduction need to be fully implemented: (1) understanding risk in all its multiple dimensions; (2) strengthening disaster risk governance; (3) investing in DRR for resilience; and (4) enhancing preparedness and build back better.

Author Contributions: Conceptualization, Z.S., M.R. and J.A.; methodology, Z.S.; formal analysis, Z.S.; resources, Z.S., M.R. and J.A.; writing—original draft preparation, Z.S.; writing—review and editing, M.R. and J.A.; visualization, Z.S.; funding acquisition, Z.S. All authors have read and agreed to the published version of the manuscript.

Funding: This research received no external funding.

Conflicts of Interest: The authors declare no conflict of interest.

References

1. Statista Research Department Cost of Natural Disaster Losses Worldwide from 2000 to 2020, by Type of Loss. Available online: https://www.statista.com/statistics/612561/natural-disaster-losses-cost-worldwide-by-type-of-loss/ (accessed on 17 October 2021).
2. European Science & Technology Advisory Group. *Socioeconomic and Data Challenges Risk Reduction in Europe*; European Science & Technology Advisory Group: Brussels, Belgium, 2019.
3. Gall, M.; Cutter, S.L. *Governance in Disaster Risk Management (IRDR AIRDR Publication No. 3)*; Integrated Research on Disaster Risk: Beijing, China, 2014. [CrossRef]
4. Hagelsteen, M.; Becker, P. Challenging Disparities in Capacity Development for Disaster Risk Reduction. *Int. J. Disaster Risk Reduct.* **2013**, *3*, 4–13. [CrossRef]

5. Sigmund, Z.; Radujković, P.M.; Lazarević, P.D. Public Buildings Seismic Vulnerability Risk Mitigation Management Model. Ph.D. Thesis, Faculty of Civil Engineering University of Zagreb, Zagreb, Croatia, 2014; p. 349.
6. Bryant, T. Mapping Vulnerability: Disasters, Development and People. *Geogr. Res.* **2006**, *44*, 328–329. [CrossRef]
7. Wisner, C.; Nivaran, D. *At Risk: Natural Hazards, People's Vulnerability and Disasters*; Routledge: London, UK, 2003.
8. Few, R.; Scott, Z.; Wooster, K.; Avila, M.F.; Tarazona, M.; Thomson, A. *Strategic Research into National and Local Capacity Building for DRM Synthesis Report*; Geneva Press: Geneva, Switzerland, 2015.
9. UNDRR Terminology—Online Glossary. Available online: https://www.undrr.org/terminology (accessed on 17 October 2021).
10. Markušić, S. Seismicity of Croatia. In *Earthquake Monitoring and Seismic Hazard Mitigation in Balkan Countries*; NATO Science Series: IV: Earth and Environmental Sciences; Husebye, E.S., Ed.; Springer: Dordrecht, The Netherlands, 2008; Volume 81. [CrossRef]
11. Herak, D.; Herak, M.; Tomljenović, B. Seismicity and Earthquake Focal Mechanisms in North-Western Croatia. *Tectonophysics* **2009**, *465*, 212–220. [CrossRef]
12. Markušić, S.; Stanko, D.; Penava, D.; Ivančić, I.; Oršulić, O.B.; Korbar, T.; Sarhosis, V. Destructive M6.2 Petrinja Earthquake (Croatia) in 2020—Preliminary Multidisciplinary Research. *Remote Sens.* **2021**, *13*, 1095. [CrossRef]
13. Markušić, S.; Stanko, D.; Korbar, T.; Belić, N.; Penava, D.; Kordić, B. The Zagreb (Croatia) M5.5 Earthquake on 22 March. *Geosci. Switz.* **2020**, *10*, 252. [CrossRef]
14. Analiza Naknadnih Potresa. Available online: https://www.pmf.unizg.hr/geof/seizmoloska_sluzba/o_zagrebackom_potresu_2020/pola_godine_od_potresa/analiza_naknadnih_potresa (accessed on 13 October 2021).
15. HCPI Rezultati Procjena Oštećenja Građevina Nakon Potresa u Zagrebu 2020. Available online: https://www.hcpi.hr/rezultati-procjena-ostecenja-gradevina-nakon-potresa-31 (accessed on 13 October 2021).
16. Government of Croatia; World Bank. *Croatia Earthquake Rapid Damage and Needs Assessment*; The World Bank: Washington, DC, USA, 2020.
17. Herak, M.; Ivančić, I.; Kuk, V.; Marić, K.; Markušić, S.; Sović, I. *Seismic Hazard Map*; Republic of Croatia: Zagreb, Croatia, 2011.
18. USGS Earthquake near Petrinja, Croatia. Available online: https://earthquake.usgs.gov/earthquakes/eventpage/us6000d3zh/map (accessed on 17 October 2021).
19. EMSC M 6.4—CROATIA—2020-12-29 11:19:54 UTC. Available online: https://www.emsc.eu/Earthquake/earthquake.php?id=933701#providers (accessed on 17 October 2021).
20. HCPI Hrvatski Centar Za Potresno Inženjerstvo. Available online: https://www.hcpi.hr/ (accessed on 28 December 2021).
21. Government of Croatia; World Bank. *Croatia December 2020 Earthquake—Rapid Damage and Needs Assessment*; The World Bank: Washington, DC, USA, 2021.
22. Stepinac, M.; Lourenço, P.B.; Atalić, J.; Kišiček, T.; Uroš, M.; Baniček, M.; Šavor Novak, M. Damage Classification of Residential Buildings in Historical Downtown after the ML5.5 Earthquake in Zagreb, Croatia in 2020. *Int. J. Disaster Risk Reduct.* **2021**, *56*, 102140. [CrossRef]
23. Altshuler, A.; Amaratunga, D.; Arefyeva, E.; Dolce, M.; Sjastad Hagen, J.; Komac, B.; Migliorini, M.; Mihaljević, J.; Mysiak, J.; Fra Paleo, U.; et al. *Socioeconomic and Data Challenges: Disaster Risk Reduction in Europe*; UNDRR: Brussles, Belgium, 2019.
24. FEMA. *Seismic Retrofit Incentive Programs—A Handbook for Local Governments*; FEMA: Anchorage, AK, USA, 1994.
25. OECD. *OECD Reviews of Risk Management Policies: Boosting Resilience through Innovative Risk Governance*; OECD: Paris, France, 2014.
26. Šimić, M.; Sigmund, Z. The Role of Military Forces in Crisis—Example of Zagreb Earthquake. In Proceedings of the 1st Croatian Conference on Earthquake Engineering—1CroCEE, Zagreb, Croatia, 22–24 March 2021; Lakušić, S., Atalić, J., Eds.; Faculty of Civil Engineering, University of Zagreb: Zagreb, Croatia, 2021.
27. Parliament of Croatia. *Homeland Security System Act*; Parliament of Croatia: Zagreb, Croatia, 2017.
28. Government of the Reoublic of Croatia. *Procjena Rizika od Katastrofa za Republiku Hrvatsku*; Government of the Reoublic of Croatia: Zagreb, Croatia, 2018.
29. Ministry of the Interior of the Republic of Croatia Hrvatska Platforma Za Smanjenje Rizika Od Katastrofa. Available online: https://civilna-zastita.gov.hr/hrvatska-platforma-za-smanjenje-rizika-od-katastrofa/80?impaired=0 (accessed on 13 October 2021).
30. Parliament of Croatia. *Civil Protection System Act (NN82/15, 118/18, 31/20, 20/21)*; Official Gzette: Zagreb, Croatia, 2021.
31. Parliament of Croatia. *Defense Act*; Official Gazette: Zagreb, Croatia, 2019.
32. Civil Protection Direcotrate Civil Protection Direcotrate—Projets. Available online: https://civilna-zastita.gov.hr/projekti-2635/2635 (accessed on 18 October 2021).
33. Ministry of Defense. *Ministry of Defence Strategic Plan 2020–2022*; Ministry of Defense: Zagreb, Croatia, 2019.
34. Ministry of Defense. *The Republic Of Croatia National Security Strategy*; Ministry of Defense: Velika Gorica, Croatia, 2017.
35. State Administration for Protection and Rescue. *Assessment of the Vulnerability of the Republic of Croatia to Natural and Technical Technological Disasters and Major Accidents*; Ministry of the Interior of the Republic of Croatia: Zagreb, Croatia, 2009.
36. Pairlament of Croatia. *Construction Law*; Official Gazette: Zagreb, Croatia, 2013.
37. Pairlament of Croatia. *Law on Protection and Preservation of Cultural Heritage*; Official Gazette: Zagreb, Croatia, 1999.
38. European Committee for Standardisation. *EN 1998-3: Eurocode 8: Design of Structures for Earthquake Resistance—Part 3: Assessment and Retrofitting of Buildings*; European Committee for Standardisation: Brussels, Belgium, 2011.
39. Croatian Standards Institute. *Eurokod 8-Projektiranje Potresne Otpornosti Konstrukcija-Dio: Ocjenjivanje i Obnova Zgrada-Nacionalni Dodatak*; Croatian Standardization Institute: Zagreb, Croatia, 2011.

40. Pairlament of Croatia. *Law on Mitigation and Elimination of the Consequences of Natural Disasters*; Official Gazette: Zagreb, Croatia, 2019.
41. Pairlament of Croatia. *NN 77/14—Law on Remediating the Aftermath of the Disaster in Vukovar-Srijem County*; Pairlament of Croatia: Zagreb, Croatia, 2014.
42. Sgmund, Z.; Uroš, M.; Atalić, J. *The Earthquake in Zagreb Amid the COVID-19 Pandemic: Opinion*; UNDRR: Brussles, Belgium, 2020.
43. Pairlament of Croatia. *NN 102/20—Law on Reconstruction of Earthquake Buildings on the Territory of the City of Zagreb, Krapina-Zagorje County and Zagreb County*; Official Gazette: Zagreb Croatia, 2020.
44. Ministry of Physical Planning, C. and S.P. Fond za Obnovu. Available online: https://mpgi.gov.hr/o-ministarstvu/djelokrug/graditeljstvo-98/obnova-zgrada-ostecenih-potresom-na-podrucju-grada-zagreba-i-krapinsko-zagorske-zupanije/fond-za-obnovu/11220 (accessed on 6 November 2021).
45. Pairlament of Croatia. *First Program of Measures for Reconstruction of Earthquake Damaged Buildings in the City of Zagreb, Krapina-Zagorje County And Zagreb County*; Official Gazette: Zagreb, Croatia, 2020.
46. Pairlament of Croatia. *Amendments to the Law on Reconstruction of Equity Damaged Buildings in the City of Zagreb, Krapina-Zagorje County and Zagreb County*; Official Gazette: Zagreb, Croatia, 2021.
47. Reconstruction fund Reconstruction Fund—Overview of Payments and Costs. Available online: https://www.arcgis.com/apps/dashboards/fd7f27fcda014e97a8ef1238729f837e (accessed on 23 November 2021).
48. Ministry of Physical Planing Construction and State Assets Expert Reconstruction Advice Meeting—03.02. Available online: https://mpgi.gov.hr/vijesti/izmjenama-programa-mjera-dodatno-ce-se-pojednostaviti-procedure-u-obnovi-i-smanjiti-potrebna-dokumentacija/11497 (accessed on 23 November 2021).
49. Latinović, A. GDJE JE ZAPELO? Obnova Nakon Potresa u Italiji Najbolji Je Putokaz Za Hrvatsku, Trebamo Međunarodnu Pomoć, Ali i Onu Iz Dijaspore 2021. Direktno, Zagreb, Croatia. Available online: https://direktno.hr/direkt/gdje-je-zapelo-obnova-nakon-potresa-u-italiji-najbolji-je-putokaz-za-hrvatsku-trebamo-medunarodnu-pomoc-ali-i-onu-iz-dijaspore-247986/ (accessed on 24 March 2022).
50. Pušić, M. Prizori s Banije Prije i Poslije: Novi Krovovi Niču, Ali Postoji Jedan Veliki Problem 2021. Jutarnji List, Zagreb, Croatia. Available online: https://www.jutarnji.hr/vijesti/hrvatska/prizori-s-banije-prije-i-poslije-novi-krovovi-nicu-ali-postoji-jedan-veliki-problem-15107101 (accessed on 24 March 2022).
51. Felić, E. *Građevni Materijal—Cijene Poludjele, a Rast će i Dalje*; Lider: Zagreb, Croatia, 2021. Available online: https://lidermedia.hr/poslovna-scena/hrvatska/gradevni-materijal-cijene-poludjele-a-rast-ce-i-dalje-135844 (accessed on 24 March 2022).
52. Ministry of Physical Planing Construction and State Assets Expert Reconstruction Advice Meeting—27.10. Available online: https://mpgi.gov.hr/vijesti-8/pocetak-obnova-manjih-zgrada-trebao-bi-krenuti-do-proljeca/11155 (accessed on 23 November 2021).
53. Pairlament of Croatia. *NN 117/21—Amendments to the Law on Reconstruction of Earthquake Damaged Buildings on the Territory of the City of Zagreb, Krapina-Zagorje County and Zagreb County*; Official Gazette: Zagreb, Croatia, 2021.
54. Comfort, L.K.; Sungu, Y. *Organizational Learning from Seismic Risk: The 1999 Marmara and Duzce, Turkey Earthquakes*; Graduate School of Public and International Affairs, University of Pittsburgh: Pittsburgh, PA, USA, 2001.
55. Government of Mexico; World Bank Group. *Improving the Assessment of Disaster Risks to Strengthen Financial Resilience*; The World Bank: Washington, DC, USA, 2012.
56. Otani, S. Disaster Mitigation Engineering—The Kobe Earthquake Disaster. In Proceedings of the JSPS Seminar on Engineering in Japan, Royal Society, London, UK, 27 September 1999; Volume 27.
57. Ikeuchi, K.; Isago, N. Earthquake Disaster Mitigation Policy in Japan. In Proceedings of the 39th Joint Meeting, Panel on Wind and Seismic Effects, Tsukuba, Japan, 14–16 May 2007.
58. Lungu, D.; Arion, C. *PROHITECH—Chapter 5 Intervention Strategies*; Structural Safety for Natural Hazard Research Centre, Technical University of Civil Engineering Bucharest: Bucharest, Romania, 2006.
59. Sundararaj, P.R.; Foo, S.; Balazic, J. PWGSC Policy on Seismic Resistance of Existing Buildings. In Proceedings of the 13th WCEE, Vancouver, UK, 1–6 August 2004.
60. IRC; NRCC. *Guideline for Seismic Upgrading of Building Structures*; National Research Council Canada: Vancouver, BC, Canada, 1995.
61. IRC; NRCC. *Guidelines for Seismic Evaluation of Existing Buildings*; National Research Council Canada: Vancouver, BC, Canada, 1993.
62. IRC; NRCC. *Manual for Screening of Buildings for Seismic Investigation*; National Research Council Canada: Vancouver, BC, Canada, 1993.
63. Davy, G.H.; Granadino, J. Seismic Retrofit of Federal Buildings in Canada. In Proceedings of the 12th World Conference on Earthquake Engineering, Auckland, New Zealand, 30 January–4 February 2000.
64. SIA. *SIA-2018 Überprüfung Bestehender Gebäude Bezüglich Erdbeben*; Schweizerischer Ingenieur-Und Architectenverein: Zurich, Switzerland, 2004.
65. Bundesamt für Wasserund Geologie; Bundesamt für Umwelt. *Beurteilung der Erdbebensicherheit Bestehender Gebäude*; Bundesamt für Wasserund Geologie: Bern, Switzerland, 2005.
66. NZSEE. *Assessment and Improvement of the Structural Performance of Buildings in Earthquakes*; New Zealand Society for Earthquake Engineering: Wellington, New Zealand, 2006; p. 343.
67. NZSEE. *Assessment and Improvementof the Structural Performanceof Buildings in Earthquakes—Section 3 Revision—Initial Seismic Assessment*; New Zealand Society for Earthquake Engineering: Wellington, New Zealand, 2013; p. 67.

68. EERI. *The M 6.3 Christchurch, New Zealand, Earthquake of February 22, 2011*; Earthquake Engineering Research Institute: Oakland, CA, USA, 2011.
69. EERI. *The Mw 7.1 Darfield (Canterbury), New Zealand Earthquake of September 4, 2010*; Earthquake Engineering Research Institute: Oakland, CA, USA, 2010.
70. *NISTIR 676,27*; Standards of Seisimc Safety for Federally Owned and Leased Buidlings. NIST: Gaithersburg, ML, USA, 2002.
71. ASCE/SEI. *Seismic Evaluation and Retrofit of Existing Buildings*; American Society of Civil Engineers: Reston, VA, USA, 2014.

 *buildings*

Article

Post-Earthquake Rapid Damage Assessment of Road Bridges in Glina County

Anđelko Vlašić *, Mladen Srbić, Dominik Skokandić and Ana Mandić Ivanković

Faculty of Civil Engineering, University of Zagreb, 10000 Zagreb, Croatia; msrbic@grad.hr (M.S.); dskokandic@grad.hr (D.S.); ana.mandic.ivankovic@grad.unizg.hr (A.M.I.)

* Correspondence: vlasic@grad.hr

Abstract: In December 2020, a strong earthquake occurred in Northwestern Croatia with a magnitude of $M_L = 6.3$. The epicenter of this earthquake was located in the town of Petrinja, about 50 km from Zagreb, and caused severe structural damage throughout Sisak-Moslavina county. One of the biggest problems after this earthquake was the structural condition of the bridges, especially since most of them had to be used immediately for demolition, rescue, and the transport of mobile housing units in the affected areas. Teams of civil engineers were quickly formed to assess the damage and structural viability of these bridges and take necessary actions to make them operational again. This paper presents the results of the rapid post-earthquake assessment for a total of eight bridges, all located in or around the city of Glina. For the assessment, a visual inspection was performed according to a previously established methodology. Although most of the inspected bridges were found to be deteriorated due to old age and lack of maintenance, very few of them showed serious damage from the earthquake, with only one bridge requiring immediate strengthening measures and use restrictions. These measurements are also presented in this paper.

Keywords: earthquake; bridge; damage assessment; strengthening; rehabilitation

check for **updates**

Citation: Vlašić, A.; Srbić, M.; Skokandić, D.; Mandić Ivanković, A. Post-Earthquake Rapid Damage Assessment of Road Bridges in Glina County. *Buildings* **2022**, *12*, 42. https://doi.org/10.3390/buildings12010042

Academic Editors: Chiara Bedon, Mislav Stepinac, Marco Francesco Funari, Tomislav Kišiček and Ufuk Hancilar

Received: 2 December 2021
Accepted: 29 December 2021
Published: 4 January 2022

Publisher's Note: MDPI stays neutral with regard to jurisdictional claims in published maps and institutional affiliations.

1. Introduction

Last year, Northwestern Croatia was shaken by two major earthquakes. The first occurred in March 2020, with an epicenter 10 km north of the capital Zagreb and a magnitude of $M_L = 5.5$. The second occurred in December 2020 with an epicenter near the town of Petrinja 50 km southeast of Zagreb and a magnitude of $M_L = 6.2$. Both earthquakes caused devastating damage to buildings and other structures. The World Bank estimated the total damage to be around 16.5 billion euros for 73,000 affected buildings [1,2]. In the first earthquake in Zagreb, most of the damage was due to the old age and poor maintenance of the buildings over the last 100 years of their existence. These buildings date back to the beginning of the 20th century and were mainly constructed with masonry and timber structural elements [3]. Furthermore, they were built without any seismic design requirements, had undergone many unauthorized reconstructions and adaptations during their lifetime, and had not been properly maintained [4]. The second earthquake in Petrinja was much stronger ($M_L = 6.2$, VIII-IX EMS-98 [5]), with a specific fault mechanism and shallow focal depth. Surface failures that occurred showed damage to linear infrastructure along a 30 km long section of the NE–SW strike [5]. New fault planes occurred along the NW–SE Dinaric strike, activating the 20 km long section of the Pokupsko fault [5,6]. A complex fault system was activated at the intersection of the two main longitudinal and transverse faults (Petrinja and Pokupsko faults) [5]. The PGA (peak ground acceleration) values for the bedrock foundation ranged between 0.29 and 0.44 g, but due to the high nonlinearity of the soil that was composed of clays with medium-to-high plasticity (evident from surface deposits and significant ground fractures [7]), it was estimated that locally amplified PGA values were likely in the range of 0.4–0.6 g [5]. This was also consistent with the observed damage to the buildings, of which approximately 15% sustained severe damage or complete collapse (DG4 and DG5) (Figure 1), 20% sustained significant damage (DG3), and 65% sustained light-to-moderate damage (DG1 and DG2) [5]. These damage grades were assigned according to EMS98 [8,9]. The main earthquake (M_L = 6.2) was preceded by two foreshocks (M_L = 4.7 and 5.2) and a series of aftershocks (up to M_L = 4.9). A strong foreshock helped save lives, as many critical buildings had been evacuated before the main earthquake struck and caused them to collapse. Many significant aftershocks that occurred over the next month (though noticeable earthquakes of up to $M_L \approx 3$ are still recorded weekly even now, a year after the main earthquake) caused subsequent damage to already damaged structures, making it difficult to classify the extent of damage and demanding the reassessment of already examined structures. In contrast to the Zagreb earthquake (March 2020), the Petrinja earthquake (December 2020) showed significant damage to linear infrastructure and underlying soil (such as landslides, liquefaction, suffusion, and sinkholes) that occurred during or as a result of the earthquake over the next few months. Significant damage occurred to structures crossing the activated faults of the fault system, evident on roads, bridges, pipelines, and river embankments. Examinations of the road damage revealed dextral co-seismic strike-slip cracks and displacements [6]. The clearest evidence of an active Petrinja fault was the presence of cracks on the Brest Bridge on the Kupa River, which was under tension along the fault line [6]. Galdovo Bridge over the Sava River in Sisak showed a 10 cm abutment bearing displacement as a consequence of a still-unknown N–S fault line [6]. There was also damage to the pipelines and river embankments of the Kupa and Sava rivers. More than 90 sinkholes occurred within a radius of about 10 km without prior warning of ground deformation. Many of them had a radius of 25 m and a depth of 12 m and endangered the surrounding buildings and infrastructure [7]. Some of them are still active and make any reconstruction work impossible. After the Petrinja earthquake, structural engineers from all over the country were called upon by the Civil Protection Agency to assess the extent of the structural damage, safety, and restrictions on use, as well as to work in collaboration with emergency rescue and demolition services to mitigate the consequences of the disaster. Assessments were prioritized based on the extent of damage and the importance of the structures. Thus, health and infrastructure structures were assessed first. The assessment of bridges was particularly important due to the need for their immediate use by rescue teams working with heavy machinery that needed to be quickly and safely moved. Moreover, due to cold and snowy winter weather, a humanitarian crisis occurred as people evacuated their destroyed or damaged homes and had to be temporarily housed.

Figure 1. DG4 and DG5 damage from the December 2020 earthquake in Sisak-Moslavina county.

Mobile housing units had to be transported on a large scale to the affected areas, which required the use of special heavy vehicles and the crossing of many bridges. The entire area is located at the intersection of four rivers (Sava, Kupa, Glina, and Maja) and many of their tributaries, so many bridges had to be immediately checked for safety. Most of these bridges are more than 50 years old and therefore not designed for the seismic safety required by modern standards. In addition, many of them were already in poor state due to material deterioration, they had not been properly maintained, and they had been subjected to overweight heavy traffic for which they were not designed. There were also very little to no data of their prior examination and assessment, nor any documentation from their design. Many specialized teams of civil engineers with experience in bridge engineering were assigned to this task and sent to different regions of the affected area. This paper provides an overview of the post-earthquake assessment of bridges in the Glina region, which included eight overall bridges.

In the forensic investigation of bridge failures, a sequence of events was identified using an interdisciplinary information gathering approach to identify the main causes of failure [10]. This approach can also be used to determine the current condition of the structure following an event that was not anticipated when the bridge was designed [10]. The first and most important part of any forensic investigation is the visual inspection of a bridge, followed by the collection of existing information about the structure, followed by non-destructive testing (NDT), and finally the analysis of all data using numerical and/or analytical models to determine cause–effect relationships.

It is important to emphasize that the task of this assessment was not to provide a detailed account of the existing load-carrying capacity of the bridge or its past deficiencies, as this could not be done without calculations and NDT for which there was no time in a crisis situation. The task was to identify critical damage as a result of the earthquake, assess the possibility of continued use of the bridge, and establish restrictions and guidelines for the use of the bridge. This work relied heavily on the experience of the commissioned engineers and their good judgment.

2. Theoretical and Practical Background in Bridge Assessment

2.1. Bridges Visual Inspection—Practices Overview

Visual inspection is a fundamental tool for bridge assessment and decision making (Figure 2). The visual inspection of a bridge greatly differs from that of any other structure due to bridges' generally longer life span, exposure to very aggressive environmental conditions, and structural elements made of different materials with different deterioration processes and rates. Improper and untimely maintenance leads to rapid changes in the slope of the time-related deterioration curve that determines the remaining service life of individual bridge elements [11,12], and the failure of any non-structural element (such as waterproofing, drainage, and expansion joints) is critical to the duration of the remaining service life.

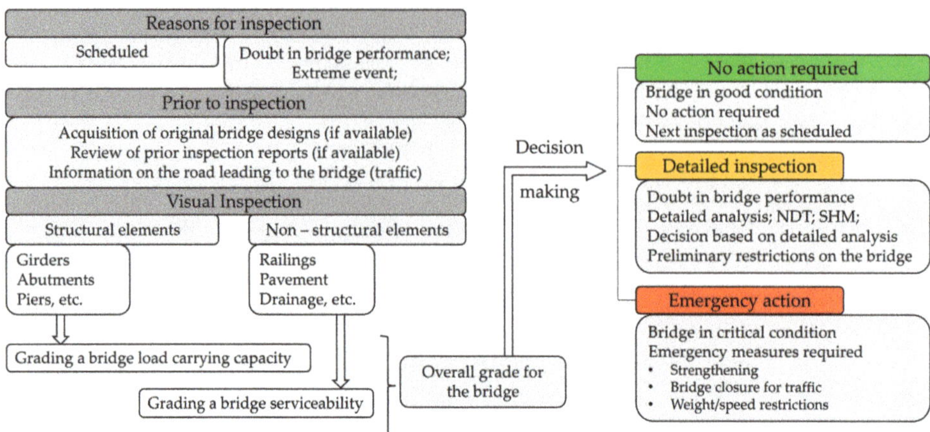

Figure 2. Flowchart for visual inspection protocol and decision making.

Most bridge visual inspections are based on a rating system in which a bridge is divided into elements and each element is assigned a numerical condition state describing its degree of damage. Condition states ranging from "no defects" to "critical defects" often comprise 5 or 6 rating points [13–16], but there are also examples of up to 10 rating points [16]. In a more detailed visual inspection, rating points are assigned not only to bridge elements but also to a location within an element, thus creating a geometrical mesh of damage distribution [13]. Since the damage location determines the failure mechanism, such an approach is beneficial for structural reliability analysis. It is imperative that the point ranking system for each visual inspection procedure provides a detailed description for each point of the condition assessment and that this description of damage is done separately for each bridge element and material. The collected data can be sorted and analyzed using mathematical statistical methods. Then, using probability-based models, one prioritizes the extent and timeframe of maintenance work [14]. A 2002 study [17] examining the reliability of visual inspection of highway bridges concluded that there was a significant spread in ranking points between inspectors, with only 68% of inspectors differing by up to one rating point within a 10-point rating system. This spread was attributed to the inspectors' individual formal training, the thoroughness of the inspection, and (in part) their subjective perception of the significance of the damage. Other limitations and shortcomings of any visual inspection can be summarized in three categories [18]: 1. timing (recognizing the damage at the moment it occurs and detecting the damage propagation rate in time); 2. interpretability (subjective evaluation by different inspectors depending on their training and given guidelines); and 3. accessibility (ability to access all elements and the interior of the structure to detect damage). To mitigate some of these shortcomings, a combination with non-destructive testing (NDT) [19] and structural health monitoring (SHM) [18] is recommended.

In order to obtain useful information for planning appropriate maintenance work, a visual inspection must be standardized within a certain management system and certain documented guidelines for an inspection procedure and frequency must be provided [18]. A 2010 study conducted by the Croatian Road Administration to assess bridge condition based on visual inspections introduced a six-category rule ranging from 0 (undamaged) to 5 (extensive damage) [14]. For each bridge, twelve bridge elements were evaluated, divided into three element groups (substructure, superstructure, and equipment). An example of visual inspection as a tool for evaluating bridge performance and prioritizing bridge repair in the transportation network can be found in [20]. The defined method was applied to six different (in terms of length and structural system) bridges in Croatia.

Although visual inspection is the imperative in any bridge assessment methodology, a detail account of bridge performance can only be obtained by collecting additional data of the bridge structure. These data must include stiffness distribution parameters, material properties, real traffic loads, and modeling analysis [20,21]. For example, a very effective procedure of collecting additional performance indicators is bridge weigh-in-motion (B-WIM), a method that uses real traffic data to determine the effects of maximum load on a particular bridge and later decision making based on value of information (VoI) analysis [16,22]. Additional information can also be obtained through non-destructive testing to assess the damage in the reinforcement of RC bridges and subsequently predict their service life using numerical models [19].

The visual inspection conducted during this rapid assessment of the bridges in Glina county after the earthquake followed the methodology shown in Figure 2 with some modifications. These modifications were made due to the lack of information about the bridges that is typically collected prior to the visual inspection and the need to act very quickly and make decisions. The focus of the inspection was placed on the structural elements (including bearings) that are critical to evaluating the load-carrying capacity of a bridge. The serviceability rating of the bridge was not important in the decision-making process. Nevertheless, the deterioration of non-structural elements was recorded for future reference and is also presented. Since this quick visual inspection immediately after the earthquake only served to answer the question of whether the bridge should continue to be used after the earthquake, no scoring system was used and the ratings were given as "continue to use", "close the bridge", or "issue use restrictions" (Figure 3).

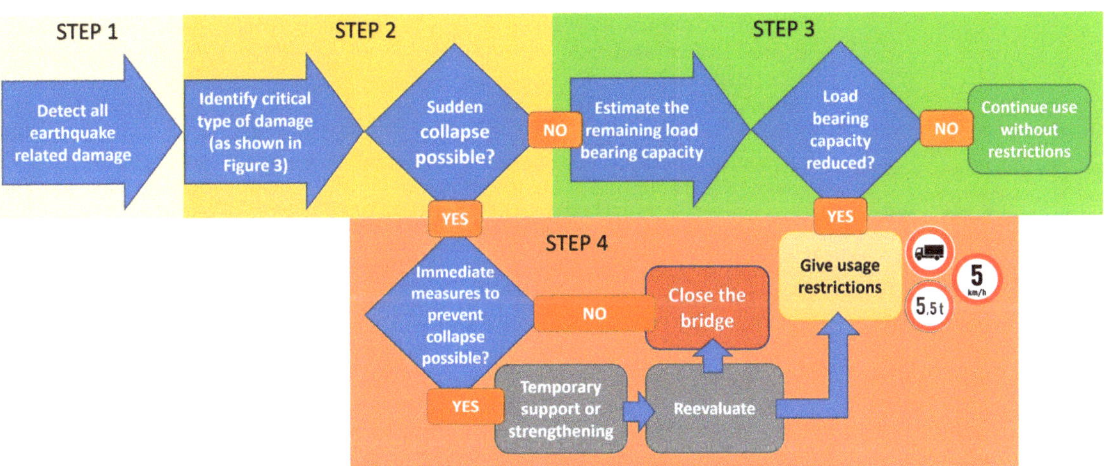

Figure 3. Post-earthquake bridge rapid assessment methodology.

2.2. Bridge Seismic Assessment Methods

There are a number of methods for the seismic assessment of existing road bridges, depending on the degree of complexity and practicality. There is no universal opinion on which is the optimal method, as this depends on a number of parameters mainly related to the characteristics of the bridge (structure, span, material used, etc.).

In general, it can be said that nonlinear analysis is more suitable for existing bridges because it considers the plastic behavior of the elements. Most often, a performance-based assessment such as nonlinear pushover static analysis or nonlinear time history dynamic analysis is used, which utilizes the ductility and energy dissipation characteristics of the real structural behavior. The pushover method measures structural capacity through inelastic displacement, which is then compared to the demands of a particular earthquake ground motion from the response spectrum. The accuracy of such an assessment largely depends on the accurate characterization of the material and dynamic properties of the bridge, which can be experimentally determined using destructive and non-destructive testing [23]. As pointed out in [23,24], this is particularly advantageous for traditional masonry bridges, whose material and dynamic properties are difficult to predict. For reinforced concrete arch bridges, a two-level seismic assessment procedure is possible [25], with evaluation checks at each level. The first, a more conservative level of evaluation uses a linear multimodal analysis, while the second level utilizes nonlinear pushover analysis for a less conservative, easier to meet safety requirement. The failure probability was investigated in a parametric study that included several variables such as geometry, material properties, earthquake records, and intensity levels [26,27]. The probability of failure was expressed by a safety indicator, which shows the difference between the seismic capacity and demand, obtained from a nonlinear dynamic analysis.

Based on the literature review and experience in the practical design and the assessment of existing reinforced concrete road bridges, a list of the most common deficiencies is further presented in this paper. The most critical element of bridges in seismic assessment is often found to be columns. The main causes of column structural deficiencies are a low percentage of longitudinal and transverse reinforcement, poor concrete, inadequate seismic detailing, and a lack of confinement reinforcement [28]. The ductility of columns is important for seismic energy dissipation, but for older bridges that do not comply with modern seismic design standards, it is very difficult to estimate the level of ductility. Research [29] has shown that the use of smooth rebar reinforcement, which is common in bridges older than 30 years, helps to improve the ductility of atypical cross-sections without modern seismic design. The failure mode of columns in earthquake situations is often found to be shear critical brittle failure due to the very limited shear capacity of short piers or flexural failure of tall columns [28]. A comparison of the prediction of shear strength capacity according to various codes models and experimental results for hollow circular piers was shown in [28]. A significant influence on the seismic response of a bridge is the relationship between the soil and the foundation. A rigid foundation-soil model may overestimate the seismic capacity of a bridge. This is especially important for masonry arch bridges. Research [30] has shown that that a 50% increase in safety confidence level can be observed when ignoring soil foundation flexibility effects.

For the earthquake assessment of multiple bridges on a larger scale, research has been carried out to develop the necessary tools to quickly determine the vibration periods for the structural seismic demand and fragility of structures, when only basic geometric variables are known [31]. To this end, a large database of reinforced concrete bridges was created and statistically processed to identify relationships between seismic response parameters and geometric and material input variables. Such relationships can assist in rapid assessment actions.

A holistic probabilistic framework [32] based on visual inspections and fragility curves has been proposed for the assessment of bridges in the network after an earthquake. Fragility curves have shown the relationships between the parameters of a seismic event (PGA, spectral acceleration, and measure of shaking intensity) and the probability of structural damage when the given performance level of a bridge is exceeded. This methodology uses a six step process to determine the interventions needed after a catastrophic event: 1. gathering information (visual inspections); 2. deriving fragility curves; 3. deciding whether a non-destructive evaluation is needed after an earthquake; 4. updating fragility curves for the damaged bridge while considering the uncertainties of visual inspection; 5. deciding whether to allow traffic to cross over a damaged bridge; and 5. deciding for immediate repairs [32]. Fragility curves can be used for the basic evaluation of multiple bridges on a section of a transportation network, but detailed analysis should be based on a more site-specific approach.

The decision-making process based on assessment results is often implemented in bridge management systems for priority rankings. In the case of seismic evaluation, the decisions are primarily based on comparisons between the fragility curves of a bridge with and without seismic retrofit measures. In [32], partial restriction of traffic on a damaged bridge after an earthquake was never considered due to the uncertainties related to the loss of load-carrying capacity. The decision to close a bridge is based on the ratio between the new updated risk of failure of a damaged bridge and the risk of failure before the earthquake event [31]. If this ratio is greater than 1, the bridge must be closed.

3. Petrinja Post-Earthquake Rapid Assessment Actions

3.1. Damage Assessment Management

After the earthquake in Petrinja, groups of volunteer civil engineers for crisis management were quickly formed and started operating throughout Sisak-Moslavina county. The formation and coordination of these groups was led by experts from the Faculty of Civil Engineering in Zagreb, whose previous experience from the Zagreb earthquake was crucial for rapid response and effective management. The methodology previously prepared based on Italian experience [33–35] included six-level classification categories: N1 and N2 as unusable buildings due to external risks or internal damage, respectively; PN1 and PN2 as temporarily unusable buildings due to uncertainties about the extent of damage requiring additional investigations or due to emergency remediation measures, respectively; and U1 and U2 as usable buildings without restrictions or with precautionary advice issued, respectively. Inspection groups were assigned to geographical locations and neighborhoods, and the results of their inspections (category classifications) were recorded via a centrally managed digital database system that was accessible via mobile devices and thus reflected an up-to-date situation in the terrain. However, this usability classification methodology and database did not consider or allow for other non-residential structures (such as infrastructure structures, special engineering structures, or bridges) to be included. Most of the inspection teams were educated by simple guidelines and given examples to use the grading categories for buildings only, leaving the more complex tasks of evaluating infrastructure structures to fewer groups of experts in their respective fields.

As noted earlier, the bridges needed to be assessed quickly because their availability was critical to many emergency services throughout the region. Previous experience in visual inspection, damage detection, and classification [19,20] was an important prerequisite for the bridge assessment team, as was experience with the seismic behavior of bridges [29]. Therefore, the teams with this practical knowledge were called in and conducted their assessments with the help of the road and transportation authorities.

3.2. Bridge Post-Earthquake Rapid Assessment Methodology

The methodology for rapid post-earthquake bridge assessment (Figure 3) in the case of the Petrinja earthquake was established on an emergency basis since there was no time to prepare, distribute, and discuss documented and detailed guidelines. It was imperative to keep bridges in service as long and as much as possible, closing them only when critical damage was detected. For bridges where moderate damage was found, it was recommended that operating restrictions (such as vehicle weight and traffic speed) be placed on their continued use. Where it was possible to provide emergency strengthening to a bridge to keep it operational in any capacity or prevent its complete collapse, services and resources were placed at a priority disposal for this work to be quickly carried out without the need for any design documentation.

The first step in the assessment was to identify all earthquake-related damage. Since none of the eight bridges assessed by this team had information on previous conditions or damage, it was important to identify the damage caused by the earthquake itself and distinguish it from any earlier damage. For example, fresh cracks in asphalt or concrete can be recognized when there is no water sediment or discoloration in or around the crack, fresh bearing displacements can be recognized by uncorroded scratch marks on bearing plates or blocks, abutment movements or rotations can be detected by cracks in the embankment soil or its erosion, and column movements or rotations can be detected visually.

The second step of the assessment was to identify the most critical type of damage that could lead to the collapse of the entire structure without any warning. The most obvious type of such collapse during or after an earthquake would be the slippage of the superstructure from the bearings, shear failure of the column or main girder, loss of stability of the substructure (overturning failure or sliding failure), or massive landslide erosion of the ground soil near the abutments (Figure 4). These types of earthquake bridge failures have been recorded in earthquakes in Japan and Chile [36–39], and all correspond to the bridge types found in this particular post-earthquake seismic assessment. Therefore, they were recognized as most likely to occur in these circumstances.

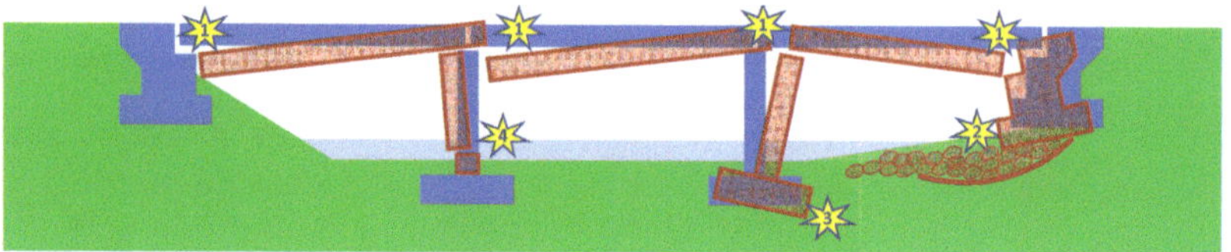

Figure 4. Most common earthquake types of bridge sudden collapses (any or all can occur): 1—bearing slippage; 2—abutment foundation soil landslide; 3—column turnover; 4—column shear failure.

The abovementioned types of collapse may occur independently, or they may be interconnected to form a progressive zipper-type collapse [10] in which the failure of the first element causes the failure of the second element, then that of the third, and so on, thus creating a cascading overall collapse of the bridge. Of course, the possibility of any or all of these collapse scenarios here depended on a bridge's structural system, location, and foundation type. These collapse scenarios could even occur in the coming days, weeks, or months after the main earthquake during the numerous small or moderate magnitude aftershocks that frequently occurred after the main Petrinja earthquake. They could even occur as a result of heavy traffic (axle loads or velocity-induced vibrations) on a structure or foundation soil that had reached a critically unstable equilibrium that is easily unbalanced over the tipping point. If such a possibility was detected, the bridge was to be immediately closed for traffic.

The third step was to evaluate the contribution of all cumulative damage to the reduction in the remaining load-carrying capacity of the bridge. This is undoubtedly the most difficult type of assessment as it had to be conducted without any testing or calculations. It could therefore only be given as an estimate, which had to be conservative enough for safety reasons but not too conservative as to unnecessarily hinder a much-needed use of the bridge in a crisis.

The fourth step of the evaluation was only required if the evaluation from the third step indicated a reduction in load-carrying capacity. In such a case, restrictions on bridge use, namely limits on vehicle axle loads, total vehicle weights, and maximum vehicle movement speeds, were required. In the event of a risk of further damage to the bridge or its sudden collapse, the final step of the assessment was to prescribe immediate measures to prevent this if possible at a given time with the available resources. It is obvious that this method of evaluation lacked the aspect of testing (destructive or non-destructive), static or dynamic analysis, and reliability calculations—all of which are required for any long-term seismic evaluation or seismic retrofit. This was, of course, due to the extreme circumstances of the crisis situation and the mitigation of consequences that would result from protracted decision-making or uncertainties in the required use of the bridge. Despite its shortcomings, this rapid assessment proved quick and effective, and it was undoubtedly a critical part of the post-earthquake emergency life-saving actions.

4. Glina Bridges Assessment and Damage Detection

After the Petrinja earthquake, different teams were deployed to examine bridges in the affected area within a radius of about 50 km. This paper presents the results of this examinations for bridges in and around the town of Glina, which covers a radius of about 10 km. This area is located 10–15 km from the epicenter of the earthquake and was therefore strongly affected by it. Figure 5 shows the map of this area and the total eight bridges that were examined (the numbering on the map follows the order of examination and is kept in the next subsections). All examinations were carried out in one day, with follow-up examinations for the most critical bridge (Section 4.2) in the next two days. Most of these bridges (Figure 6) were built in the previous century and are now 50 or more years old. There are only noted three exception bridges (numbered 1, 7, and 8 in Figure 4) that were built in the last 30 years. All bridges were found to have simply supported or continuous girders and have spans between 7 and 20 m. The superstructures were found to be either concrete slabs or composite steel-concrete ribbed section. The only exception was Glina Bridge (No. 1), which is a steel girder bridge with a span of 40 m. The main problem of all bridges, which became evident during the examinations, was the lack of maintenance, which, in combination with poor waterproofing, led to progressive material deterioration and subsequent damage to the structural and non-structural parts of the bridges. The most common types of this damage were concrete spalling, the corrosion of reinforcement, the corrosion of steel girders, the corrosion of railings, bearing degradation, the clogging of expansion joints, the cracking of asphalt, and the erosion of abutment slopes [40]. Most of these problems could have been avoided if timely maintenance had been performed to prevent further deterioration due to water intrusion and corrosion of the reinforcement. Although all these problems were evident and noted during the examination, the purpose of the examination was to record and evaluate any damage caused by the earthquake that would pose a risk to the continued use of these bridges and endanger the safety of the users. Therefore, it was necessary to accurately identify the nature of the damage according to its cause and significance to the overall load-bearing capacity and/or stability of each bridge. The visual inspection protocol and decision-making process, as shown in the flowchart in Figure 2, were followed as closely as possible. Obviously, a prior review of bridge documentation was lacking because it was not available for most bridges and/or in the critical timeframe. Grading system was only binary, i.e., the bridge was still to be used or it was to be closed and/or prescribed immediate action (as shown in Figure 4). Expert judgement was used for the safety assessment of the continued use of each bridge. A detailed inspection was one of the possible recommended measures, but it was not to be a prerequisite for further bridge use.

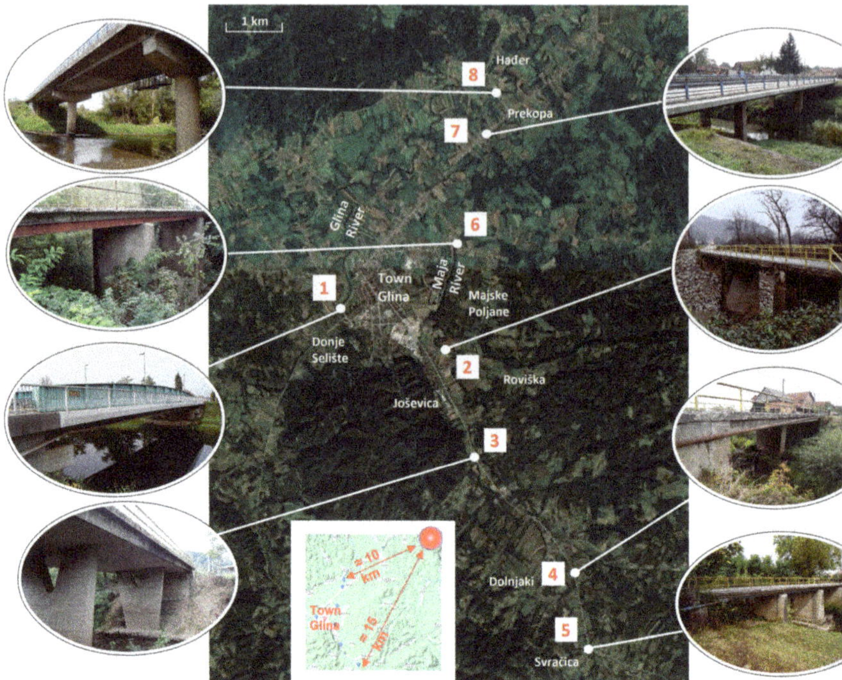

Figure 5. Geographical overview of Glina county bridges for post-earthquake rapid assessment.

4.1. Glina Bridge

Glina road bridge is the main city bridge over the river Glina, located in the south-west part of the town (Figure 5). It is the newest of all the examined bridges, built in 2003. It is a simply supported girder, crossing the river in one 42 m long span (Figure 6). The superstructure comprises two steel girders with variable height from 2.35 to 3.4 m and at 9.2 m apart (Figure 7a). The concrete road deck is supported by cross girders tapered in-between the main girders, and the footways are supported by consoles on the outside of the main girders. The deck slab is 20 cm thick on the carriageway and 12 cm thick on the footways. The overall width of the bridge is 12.6 m. Abutments are massive, 8.7 m high, reinforced concrete structures. Fixed bearing is positioned on the west abutment, and a movable bearing is on the east abutment. The bridge was found to be in an overall good condition, showing signs of medium structural steel and bearing corrosion, but no loss in the section area due to corrosion was detected. The concrete deck was found to be in almost perfect condition, with no reinforcement corrosion detected. Abutment concrete is also without any damage, there were only small parts of stone cladding detached. The partial erosion of the embankment slope around the abutment wings under the footways consoles was present, but it was not critical or caused by the earthquake.

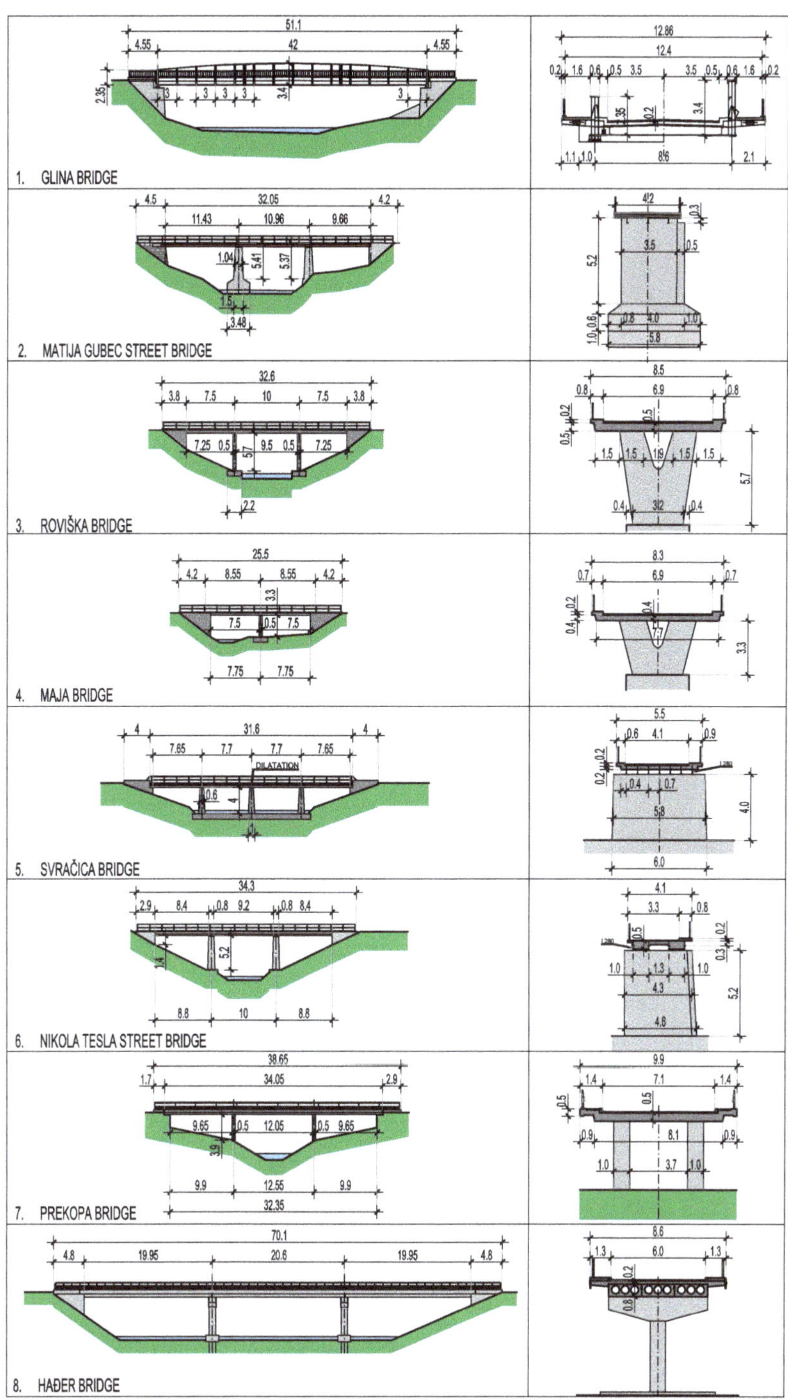

Figure 6. Glina bridges layouts and cross-sections (units in m).

Figure 7. Glina bridge: (**a**) main girder and carriageway; (**b**) fixed bearing; (**c**) movable bearing.

Examination showed no damage due to the earthquake. There were no signs of abutment movements; the fixed bearing successfully transferred the horizontal force on the abutment wall without any damage to the bearing or the wall (Figure 7b,c). Abutment embankments showed no landslide signs, and foundations showed no rotations or settlements. There was no visible damage to the expansion joints other than existing cracking in the asphalt layer due to dynamic loads from traffic. The expansion joints were found to be clogged with dirt and gravel and should be maintained in the future. No further actions were required, and the bridge was maintained for traffic use without any restrictions.

4.2. Matija Gubec Street Bridge

This bridge is located in the south part of the town, crossing the river Maja and leading to Majske Poljane village (Figure 5). The bridge superstructure is a series of simply supported girders over three spans: 11.43 + 10.96 + 9.66 m (Figure 6). The cross-section comprises three steel girders of 355 mm in height placed at 160 cm apart, as well as an 18 cm thick concrete deck (considering the age of the bridge, the level of section's composite behavior was unknown). The width of the superstructure is 4.15 m. The superstructure is directly supported by columns and abutments without any bearings. Abutments are massive structures, about 4 m high and 4.5 m long. It was evident that the bridge had undergone reconstruction in the past since one of the abutments was found to be a reinforced concrete structure and the other was found to be masonry structure from stone blocks. Its columns are massive, reinforced concrete structures that are 8 and 6.4 m high and 4 m wide. The east column was found to have visible scour signs, with parts of the foundation soil missing. The west column was shown to have a much wider and longer foundation (5.8 × 3.5 m) than the east column (4 × 2 m), which suggests that the west column foundation underwent a rehabilitation in the past. This was probably due to scour developing under the west column sooner due to its position in the middle of the riverbed (Figure 8a). It is evident that the bridge was in a poor structural state even before the earthquake: main girders were shown to be heavily corroded, the concrete deck slab was spalling, the reinforcement bars were visible and corroded, the edges of the footways and the cornice were eroded and largely missing, a permanent deflection in the superstructure was evident due to overweight traffic load, and the railings were not anchored in the footways. Maja river has a highly variable water level and flow speed that caused the erosion of the west riverbank and scour developing under the west and east column foundations (Figure 8a,d).

Figure 8. Matija Gubec street bridge: (**a**) columns—large foundation for the column in the middle of the riverbed and scour visible on the east column; (**b**) abutment stone wall damage; (**c**) abutment sliding signs; (**d**) evidence of subsequent abutment sliding—stone wall integrity compromised.

The examination revealed serious deficiencies in the west abutment. Stone wall joints showed cracks and openings up to few centimeters, with mortar missing and stone block movements (Figure 8b). The whole abutment showed signs of translation and rotation towards span opening due to ground movements and soil erosion (Figure 8c). The soil around the abutment wings showed signs of land sliding (Figure 8d). The best course of action at this time was to close the bridge, but this action would have severed the connection to the nearby settlement that was the most affected by the earthquake and needed supplies and help at this time. It was reluctantly decided that the bridge could stay open with restrictions of only 5 ton vehicles at 5 km/h traveling speed. Only one vehicle was permitted on the bridge at the same time. Furthermore, emergency actions were ordered to strengthen the abutment and prevent its further damage (see Section 5). The bridge was placed under continuous monitoring due to aftershocks that were frequent in the coming days. Subsequent inspection the following day showed further degradation of the abutment in which the falling of the stone blocks occurred and the abutment integrity was compromised (Figure 8d). At this time, it was decided that the bridge safety could no longer be assumed, and the bridge was completely closed for traffic. Stabilization measures were undertaken at this time.

4.3. Roviška Bridge

This bridge is located in the south access road to the town Glina (Figure 5). It is a reinforced concrete slab bridge with over three spans of 7.5 + 10 + 7.5 m (Figure 6). The superstructure comprises a 50 cm thick reinforced concrete slab that is directly supported by wide columns branching at the top. The width of the bridge is 8.5 m, columns are of variable cross-section between 3.2 × 0.5 m at the bottom and two branches (arms) at the top, each 1.5 × 0.5 m. The height of the columns is 5.7 m. The abutments are massive reinforced concrete structures with about 3–4 m high walls and 3.8 m long wings. Column foundations are 4.0 × 2.2 m slabs. There are no bearings present on the bridge and bridge is without any drainage elements. There was visible damage on the abutment walls, where the corner part of the side walls was found to be missing on both abutments, and reinforcement bars were protruding out of concrete (Figure 9a).

Figure 9. Roviška bridge: (**a**) damage to the abutment wall and wing, as well as drainage problems; (**b**) degraded footway (left) and new footway (right); (**c**) substructure without signs of earthquake damage.

Since the damaged parts of the wall showed heavy discolorations, traces of long-term water leakage, and algae sedimentations, it was evident that this damage was not caused by this earthquake. The quality of concrete in these fallen off parts of the abutment corners could be described as poor, exhibiting the local segregation of large fractions of aggregate and very low quantity of reinforcement. It is possible that this damage was a consequence of faults during erection since it resembled typical damage observed when improper concreating without vibration is performed. Since this damage was very localized and not in the main load transfer path, it was not considered serious at this stage of post-earthquake bridge evaluation. It did not compromise the load-bearing capacity of the abutment wall. One of the bridge footways and cornices were heavily degraded due to lack of waterproofing and reinforcement corrosion. The other footway had been rehabilitated in the past with a new concrete layer (Figure 9b).

Overall, the bridge was found to be in structurally good condition and showed no signs of damage due to the earthquake. There was no damage to the asphalt joint between the abutment and superstructure, and no displacements were recorded at the superstructure supports. Columns and abutments were without any major cracks, rotations, or settlements (Figure 9c). After the earthquake, the bridge was continued to be used for traffic without any restrictions. Further inspection and rehabilitation were recommended due to prior abutment wall damage, visible reinforcement corrosion, and possible scour on foundation piers.

4.4. Maja Bridge

Maja bridge is located just south of previous Roviška bridge (Figure 5), and it is of similar type but smaller length. The bridge is also a continuous girder slab bridge with over two spans of 7.5 + 7.5 m (Figure 6). The slab girder is 35 cm thick and 7.7 m wide, with short consoles on both sides. The superstructure is skewed in regard to abutments, columns, and the riverbed (Figure 10a). Height of the columns and abutments is about 3.5 m. There were no bearings or expansion joints causing dilatation cracks in the asphalt. There was also a visible vertical dilatation crack between the abutment wall and wing, suggesting that they were not fixed together and moved separately. The superstructure concrete was seen to be in relatively good condition, without signs of progressive reinforcement corrosion or concrete spalling but with visible signs of discoloration due to water leakage. Although drainage was present on the bridge, gutters were clogged and caused water to seep through the concrete slab. The slab consoles (footways) were observed to be heavily degraded due to a lack of waterproofing and poor concrete quality. Abutment walls were exposed to water draining from above (Figure 10b). There was no visible damage due to the earthquake. No new cracks or movements of the substructure or superstructure elements were detected. No restrictions regarding traffic were given. Further inspection and maintenance were recommended due to the noticeable water leakage due to failed waterproofing and drainage.

Figure 10. Maja bridge: (**a**) side view of the bridge; (**b**) abutment wall.

4.5. Svračica Bridge

Svračica bridge is the furthest south bridge from town Glina that was examined (Figure 5). Its superstructure is a series of two continuous composite girders with over four even spans of 7.7 m (Figure 6). Dilatation is in the middle of the bridge. The superstructure comprises a multi-girder composite cross-section with 7 I280 steel girders placed 0.7 m apart, and a 20 cm thick reinforced concrete deck slab. The superstructure is 5.45 m wide in total, with 4.05 m wide roadway and asymmetric footways of 0.55 and 0.85 m partly supported by deck consoles (Figure 6). Columns are massive, 4 m high reinforced concrete structures with a variable cross-section ranging from 6 × 1 m at the bottom to 5.8 × 0.6 m at the top. All columns have a joint foundation slab that covers the whole riverbed (two spans), approximately 18 m long and 10 m wide. Abutments are minimal structures, about 2 m high and 4 m long with variable width and skewed abutment wings due to a road junction located immediately at their end. The superstructure is directly supported on the substructure elements without bearings. The bridge was in fairly good structural condition, showing moderate signs of steel girder corrosion, mostly in the vicinity of their supports (abutments) due to the longitudinal displacements at the ends of the bridge not being properly managed (no bearings) and water leakage from behind the abutment wall (Figure 11a). Since expansions joints are not present on the bridge, cracks were visible in the asphalt at the end of the superstructure. The deck slab concrete was in good condition. The concrete columns showed signs of reinforcement corrosion, with protective layer only locally spalling. The main reason for this problem is heavy water leakage due to non-existent drainage and waterproofing. Since there was a dilatation in the superstructure above the central column, the water is draining through the asphalt directly onto the column. The spalling of concrete showed a different type of concrete underneath and occurred at the contact of these different materials, so it very likely that columns had undergone rehabilitation work in the past. There was no visible damage from the earthquake, either in permanent deformations or movements of the superstructure or substructure elements. Therefore, the bridge was maintained for operation without restrictions. Further inspection and maintenance were recommended due to steel girder corrosion and signs of column reinforcement corrosion.

Figure 11. Svračica bridge main girder support: (**a**) abutment; (**b**) column.

4.6. Nikola Tesla Street Bridge

Nikola Tesla street bridge is located in the northeast part of town (Figure 5). It is a three span continuous girder bridge of 9.2 + 10 + 9.2 m (Figures 6 and 12a). Its superstructure comprises four I280 steel girders and two 52 cm high concrete ribs that are concreted in between two pairs of steel girders (Figure 6). Steel girders are thus partly concreted inside these ribs, also serving as a side formwork for concrete. The width of the concrete ribs is 1.0 m, and the inside distance between the ribs is 1.3 m. The width of the superstructure is 4.1 m, with an asymmetrical traffic area and footway only on one side. The superstructure is directly supported by columns and abutments. Columns are 5.2 m high, 4.3–4.6 m wide in the transverse, and 0.8–1 m wide in the longitudinal direction.

Figure 12. Nikola Tesla street bridge: (**a**) view of the bridge; (**b**) main girders to abutment support; (**c**) superstructure under view.

Abutment walls are 1.4 m high, and abutment wings are 2.9 m long. The bridge superstructure was found to be in a moderate-to-poor condition, with problems regarding structural steel corrosion and water leakage due to non-existent waterproofing and drainage. The asphalt layer was heavily worn out and almost completely missing in the footway area. The edges of superstructure consoles were missing a cornice, and reinforcement bars were protruding out of the concrete. Columns and abutments were found to be in fairly good condition, with no visible cracks or concrete spalling. Discoloration was visible due to water leakage from the superstructure onto columns, thus causing long-term damage to the column concrete and possibly reinforcement corrosion (Figure 12c). The bridge showed no signs of serious damage caused by the earthquake on the superstructure, substructure, or embankment slopes around abutments. Visible cracks were detected in the area of connection between the superstructure and abutment, where the superstructure is supported on the abutments, between the abutment wall and the cross girder (Figure 12b). Since this connection was not assumed as fixed in the statical system, it was expected that the opening of this crack occurred and was of no importance regarding bearing capacity. After inspection, the bridge was maintained for operation without restrictions. Further inspection and maintenance were recommended due to the poor state of the traffic surface (asphalt layer and footways), column reinforcement corrosion, failed waterproofing, and possible scour developing on column foundations.

4.7. Prekopa Bridge

Prekopa bridge crosses the river Maja at the north access road to Glina (Figure 5). The bridge has a continuous girder slab superstructure with over three spans of 10.15 + 12.55 + 10.15 m (Figure 6). Its reinforced concrete slab is 50 cm thick and 8.1 m wide, with 0.9 m consoles on each side. Its superstructure is supported by twin 3.9 m high columns at each side of the riverbed (the cross-section of each column is 1.0×0.5 m) and 2 m high and 1.7 m long abutments. There are no bearings on the bridge, and supports are realized as concrete hinged sections at the top of the substructure elements (Figure 13a). In comparison to the other examined bridges, this bridge was found to be fairly new, erected in 1999. An open drainage system and waterproofing were observed, so no serious long-term water damage was found.

Figure 13. Prekopa bridge: (**a**) side view of the bridge; (**b**) expansion joint.

Only traces of water leakage were visible on the abutment wall, probably due to the failed waterproofing of expansion joints at the ends of the bridge. There were hints of reinforcement corrosion on the abutment wall due to this water leakage. Expansion joints were also found to be clogged with dirt and gravel, with visible cracks in the asphalt layer around them (Figure 13b). The bridge was reported to be in very good condition; the superstructure and substructure concrete showed no signs of degradation or reinforcement corrosion. Except for expansion joints needing maintenance, no other notable problems were found. Due to it dating from a newer generation of bridges, it was certainly designed with seismic loads and seismic detailing, so no earthquake damage was expected nor found. This bridge performed exceptionally in this seismic event.

4.8. Hađer Bridge

Hađer bridge is just north of town Glina, leading to a nearby settlements west of the river Glina (Figure 5). It crosses river Glina just at the mouth of river Maja. It is the longest of all the bridges, with relatively large spans and tallest columns, so the seismic action was certainly the highest here. Being built in 1987, it was presumably designed with a certain degree of seismic behavior in mind. The bridge comprises a series of one continuous slab girder with over two spans of 19.95 + 20.6 m and one simply supported girder spanning 19.95 m (Figure 6). Between these girders, there is a visible dilatation above one pier (Figure 14d). The superstructure comprises three hollow girders (each girder with a 210 × 75 cm cross-section) connected by longitudinal in situ concrete joints and in situ 20 cm thick concrete deck plate above them. The deck plate continues to consoles at each side of the cross-section to support footways. This type of cross-section can be regarded as a slab cross-section. The overall width of the superstructure is 8.6 m. The substructure comprises two single circular cross-section columns with a wide consoled head cross girder to accommodate the supports of each girder. The abutments are massive reinforced concrete structures. The bridge has no drainage system, no bearings, and no expansion joints, all of which resulted in durability problems and limited displacement capacity (as is elaborated later). The bridge showed signs of heavy water leakage from the superstructure on abutment walls, causing concrete degradation and spalling, and reinforcement corrosion on both the superstructure and abutment (Figure 14c). The same problem was present at the dilatation above the central column, where the water is leaking through the dilatation and causing damage to the ends of the girder slabs. The head of the column, as well as the abutment wall and wing console (Figure 14c,d), already showed progressive reinforcement corrosion, with parts of the concrete protective layer missing due to delamination. There were wide visible cracks in the column head girder and visible reinforcement bars due to corrosion (Figure 14d). It was also noticeable that the girders were not symmetrically supported on the column head, i.e., one girder was found to have a longer support length than the other, which was not correctly executed in the erection process. This poses a potential danger in an event of even stronger earthquake since inadequate support length could cause the girder to slip from the column head. No horizontal seismic limiting element was found, so only support length insured this from happening.

Earthquake-related damage was recorded on several elements of the bridge. It was evident that the bridge superstructure moved both longitudinally and transversely at a notable rate. The first proof of such movements could be seen on the asphalt layer at the ends of the bridge superstructure, which was heavily cracked, waved, and delaminated upwards (Figure 14a). This movement also caused the fracture of bridge cornice, which was executed without any dilatation between the abutment and superstructure (Figure 14c,e). It is surprising that a bridge this long does not have expansion joints at the ends that would accommodate for such movements without causing any damage. Secondly, transverse movements caused heavy damage to abutment side walls, thus cracking and completely fracturing parts of them (Figure 14b). It is also incorrect that the abutment side wall is so close to the superstructure with no tolerances for transverse movements of any degree (the abutment was found to be too narrow for this width of the superstructure). The structure of this bridge is very flexible due to tall single piers of circular cross-section, positioned centrally along bridge axis. Therefore, it is not surprising that the bridge exhibited large movements during the earthquake; it is more surprising that these movements were not accounted for in the design of both the expansion joints and spacing tolerances between the substructure and superstructure elements.

Figure 14. Hađer bridge: (**a**) damage caused by earthquake in the asphalt; (**b**) fractured abutment side wall due to the earthquake; (**c**) water leakage from superstructure end, causing damage to abutment; (**d**) uneven support length for the girders; (**e**) cracked cornice due to earthquake movement.

It is evident that all the damage due to the earthquake could have been easily avoided if proper design rules had been followed. Besides the described damage, which can be regarded as non-structural, the bridge performed well in this earthquake event. Further inspection and rehabilitation were recommended due to aforementioned earthquake damage and heavy reinforcement corrosion of column heads and abutment walls.

5. Immediate Strengthening Measures

Only one of the assessed bridges needed closure and immediate strengthening measures. As discussed in Section 4.2, Matija Gubec Street bridge leading to Majske Poljane showed signs of west abutment slippage and rotation, and its stability and integrity were compromised. Following the closure of the bridge for traffic, emergency measures were prescribed for the immediate strengthening of the abutment to prevent its total collapse. The immediate danger of collapse was even more emphasized due to heavy rains that followed in the days after the earthquake, which caused the water level to rise and further erode the west riverbank that had already shown scour signs under the abutment foundation. Additionally, multiple aftershocks threatened the unstable balance of the abutment wall. The decision was made to use large stone block material and to fill the slope of the riverbank up to the top of the abutment, around its wings, and as far as the middle columns in the riverbed (Figure 15a). The stone material needed to be of very large fractions (from 50 to 100 cm) to prevent it from being washed away by the river flow. The abutment foundation, wall, and wings were enveloped by this stone infill, thus preventing the further erosion of the soil and acting as a support for the abutment (Figure 15b). The whole work was performed over just two days in hard working conditions due to soaked soil from continuous rains. The measure was proven to be effective since it stopped the further movement of the abutment. Nevertheless, the bridge was closed for traffic as a precautionary measure until it was to be evaluated further and permanent solutions were found.

Figure 15. Immediate measures to secure the abutment of Matija Gubec street bridge: (**a**) beginning of the work; (**b**) completed infill.

6. Recommendations for Further Rehabilitation Work and Current Progress

Following the detailed evaluation and assessment of the previously strengthened bridge, it was concluded that extensive rehabilitation work was needed. The bridge stone wall abutment was deemed unsalvageable and to be replaced with new reinforced concrete abutment. The new reinforced concrete abutment will have a deeper and wider foundations, with the stone and shotcrete cladding of the embankment slope and river bank to prevent scour. One of the columns that showed scour signs and had an insufficient foundation-bearing capacity is also to be replaced. The new reinforced concrete column will have an 80 × 400 cm cross-section and a 300 × 550 cm, 100 cm thick foundation slab. This foundation will also be protected by stone and shotcrete cladding. One column and east abutment are to be salvaged and jacketed with a new 10 cm thick layer of reinforced concrete. Since the superstructure was heavily degraded, with progressive structural steel and reinforcement corrosion, missing parts of the footways, and low remaining load-bearing capacity, it was decided that it also needs to be replaced. The new 50 cm thick reinforced concrete slab superstructure will also be wider (600 cm), accommodating more traffic width for vehicles and pedestrians. The superstructure and substructure elements will be integrally connected without any bearings to achieve better durability. The bridge will be equipped with waterproofing and closed drainage system. Figure 16 shows the current progress of this rehabilitation.

Figure 16. Progress of rehabilitation work of Matija Gubec street bridge.

Regarding the recommendations for the other examined bridges, the Hađer Bridge most critically needs detailed inspection and rehabilitation following earthquake damage. Since this damage was caused by improper movement management, it is recommended that expansion joints allowing for seismic movements are added at the bridge ends and above the central dilatation column. The inelastic movements should also be checked by nonlinear calculations to determine if the safe tolerances against slippage of the girders are met.

An extensive detailed inspection and NDT due to durability issues were recommended for Roviška bridge, Maja bridge, Svračica bridge, Nikola Tesla Street bridge, and Hađer bridge, as previously stated in the corresponding sections.

Thus far, only the rehabilitation of Matija Gubec Street bridge has been undertaken.

7. Conclusions

On 29 December 2020, a devastating $M_L = 6.2$ earthquake hit the Sisak-Moslavina county of Croatia. Immediately after the earthquake, structural engineers' teams were dispatched to conduct rapid damage assessment and evaluate the usability of structures. Eight evaluated bridges located in Glina county have been discussed in this paper as case studies. Only one bridge with major damage was closed for traffic, and others were opened for continued use without restrictions. Most of the bridges performed well in the earthquake (Table 1), with major damage attributed to Matija Gubec Street bridge and minor damage attributed to Hađer bridge. Seismic retrofitting is recommended for both bridges. For the former, this retrofitting has already been undertaken, and half of the substructure and the whole superstructure will be replaced. For the second bridge with minor damage, it has been recommended to add retrofitting measures to allow for seismic superstructure horizontal movements and prevent excess movement that could result in catastrophic failure.

Table 1. Overview of seismic damage, degradation, and design flaws of examined bridges.

Bridge	Seismic Damage	Degradation	Design Flaws
Glina bridge	No damage	Expansion joints clogged, steel corrosion	Not evident
Matija Gubec Street bridge	Abutment sliding and block wall damage	Heavy steel corrosion, concrete spalling and reinforcement corrosion, scour, footways and cornice partly missing	Insufficient foundation for one column, one abutment from stone blocks and one from reinforced concrete, no waterproofing and drainage
Roviška bridge	No damage	Abutment wall reinforcement corrosion, parts of side wall missing, heavily degraded footway and cornice, possible column foundation scour	Abutment side wall with insufficient reinforcement and poor concrete, no waterproofing and drainage
Maja bridge	No damage	Clogged drainage, heavily degraded footway, abutments exposed to water damage	Poor waterproofing, dilatation between abutment wall and wings
Svračica bridge	No damage	Steel girder corrosion, columns reinforcement corrosion	No waterproofing or drainage
Nikola Tesla Street bridge	No damage	Steel girder corrosion, columns reinforcement corrosion, heavily degraded footway and cornice, heavy asphalt damage	No waterproofing or drainage
Prekopa bridge	No damage	Expansion joint clogged and with failed waterproofing, abutment wall water leakage	Not evident
Hađer bridge	Excessive movements, abutment wall damage, asphalt damage, cornice damage	Heavy water leakage on abutment walls and column head, concrete spalling and reinforcement corrosion, cracked column head	No expansion joints, no bearings, no drainage, girders not symmetrically supported

As a benefit of these examinations, many other durability-related problems and design flaws were also discovered (Table 1), demonstrating that the progressive deterioration of materials and elements had already started. All the examined bridges were lacking in regular maintenance or even periodical inspection to such a degree that rehabilitation work has been recommended for some. The common deficits observed for all the bridges are as follows: a lack of superstructure waterproofing, non-existent or failed drainage, the corrosion of reinforcement and/or structural steel, footways and consoles with missing parts of concrete and cornice, the cloggage of expansion joints (when they are present), and damage to the asphalt layer.

However, despite long service lives and insufficient maintenance, most of the bridges performed well during this earthquake event and continued to be used after the earthquake for rescue and evacuation purposes.

Author Contributions: Conceptualization, A.V., M.S., D.S. and A.M.I.; methodology, A.V. and M.S.; software, A.V.; validation, A.V., M.S., D.S. and A.M.I.; investigation, A.V. and M.S.; resources, A.V. and M.S.; data curation, A.V., M.S. and D.S.; writing—original draft preparation, A.V.; writing—review and editing, A.V., M.S., D.S. and A.M.I.; visualization, A.V., M.S. and D.S.; supervision, A.V., M.S., D.S. and A.M.I.; project administration, A.V. and A.M.I. All authors have read and agreed to the published version of the manuscript.

Funding: This research was founded in the framework of the project "Key Performance Indicators for Existing Road Bridges". The project team consists of project leader prof. Ana Mandić Ivanković and team members which include all remaining authors.

Institutional Review Board Statement: Not applicable.

Conflicts of Interest: The authors declare no conflict of interest.

References

1. Government of Croatia World Bank Report. *Croatia Earthquake Rapid Damage and Needs Assessment*; Government of Croatia World Bank Report: Zagreb, Croatia, 2020.
2. Perković, N.; Stepinac, M.; Rajčić, V.; Barbalić, J. Assessment of Timber Roof Structures before and after Earthquakes. *Buildings* **2021**, *11*, 528. [CrossRef]
3. Stepinac, M.; Kisicek, T.; Renić, T.; Hafner, I.; Bedon, C. Methods for the Assessment of Critical Properties in Existing Masonry Structures under Seismic Loads-the ARES Project. *Appl. Sci.* **2020**, *10*, 1576. [CrossRef]
4. Novak, M.Š.; Uroš, M.; Atalić, J.; Herak, M.; Demšić, M.; Baniček, M.; Lazarević, D.; Bijelić, N.; Crnogorac, M.; Todorić, M. Zagreb Earthquake of 22 March 2020—Preliminary Report on Seismologic Aspects and Damage to Buildings. *Gradjevinar* **2020**, *72*, 843–867. [CrossRef]
5. Markušić, S.; Stanko, D.; Penava, D.; Ivančić, I.; Oršulić, O.B.; Korbar, T.; Sarhosis, V. Destructive M6.2 Petrinja Earthquake (Croatia) in 2020—Preliminary Multidisciplinary Research. *Remote Sens.* **2021**, *13*, 1095. [CrossRef]
6. Korbar, T.; Markušić, S.; Stanko, D.; Penava, D. Petrinja M6.2 earthquake in 2020 damaged also solid linear: Are there similar active faults in Croatia? In Proceedings of the 1st Croatian Conference on Earthquake Engineering 1CroCEE, Zagreb, Croatia, 22 March 2021; Faculty of Civil Engineering, University of Zagreb: Zagreb, Croatia, 2021; pp. 353–362.
7. Bacic, M.; Sasa Kovacevic, M.; Librić, L.; Žužul, P. Sinkholes induced by the Petrinja M6.2 earthquake and guidelines for their remediation. In Proceedings of the 1st Croatian Conference on Earthquake Engineering 1CroCEE, Zagreb, Croatia, 22 March 2021; Faculty of Civil Engineering, University of Zagreb: Zagreb, Croatia, 2021; pp. 341–352.
8. Grünthal, G.; European Seismological Commission; Working Group "Macroseismic Scales". *European Macroseismic Scale 1998: EMS-98*; European Seismological Commission, Subcommission on Engineering Seismology, Working Group Macroseismic scales: Luxembourg, 1998.
9. Spence, R.; Foulser-Piggott, R. The international macroseismic scale-extending EMS-98 for global application. In Proceedings of the Second European Conference on Earthquake Engineering and Seismology, Istambul, Turkey, 25 August 2014.
10. Milic, I.; Ivankovic, A.M.; Syrkov, A.; Skokandic, D. Bridge Failures, Forensic Structural Engineering and Recommendations for Design of Robust Structures. *Gradjevinar* **2021**, *73*, 717–736.
11. Yokoyama, K.; Inaba, N.; Honma, A.; Ogata, N. Development of Bridge Management System for Expressway Bridges in Japan. *Tech. Memo. Public Work. Res. Inst.* **2006**, *4009*, 99–104.
12. Goran Puž, A.; Radić, J.; Puž, G.; Tenžera, D. Bridge Condition Forecasting for Maintenance Optimisation. *Građevinar* **2013**, *65*, 1079–1088. [CrossRef]
13. Estes, A.C.; Asce, M.; Frangopol, D.M.; Asce, F. Updating Bridge Reliability Based on Bridge Management Systems Visual Inspection Results. *J. Bridge Eng.* **2003**, *8*, 374–382. [CrossRef]
14. Tenžera, D.; Puž, G.; Radić, J. Visual Inspection in evaluation of Bridge Condition. *Gradjevinar* **2012**, *64*, 717–726.
15. Gattulli, V.; Chiaramonte, L. Condition Assesment by Visual Inspection for a Bridge Management System. *Comput.-Aided Civ. Infrastruct. Eng.* **2005**, *20*, 95–107. [CrossRef]
16. Quirk, L.; Matos, J.; Murphy, J.; Pakrashi, V. Visual Inspection and Bridge Management. *Struct. Infrastruct. Eng.* **2018**, *14*, 320–332. [CrossRef]
17. Graybeal, B.A.; Phares, B.M.; Rolander, D.D.; Moore, M.; Washer, G. Visual Inspection of Highway Bridges. *J. Nondestruct. Eval.* **2002**, *21*, 67–83. [CrossRef]
18. Agdas, D.; Rice, J.A.; Martinez, J.R.; Lasa, I.R. Comparison of Visual Inspection and Structural-Health Monitoring as Bridge Condition Assessment Methods. *J. Perform. Constr. Facil.* **2016**, *30*, 04015049. [CrossRef]
19. Marić, M.K.; Ivanković, A.M.; Vlašić, A.; Bleiziffer, J.; Srbić, M.; Skokandić, D. Assessment of Reinforcement Corrosion and Concrete Damage on Bridges Using Non-Destructive Testing. *Gradjevinar* **2019**, *71*, 843–862.
20. Ivanković, A.M.; Skokandić, D.; Marić, M.K.; Srbić, M. Performance-Based Ranking of Existing Road Bridges. *Appl. Sci.* **2021**, *11*, 4398. [CrossRef]
21. Mandić Ivanković, A.; Skokandić, D.; Žnidarič, A.; Kreslin, M. Bridge Performance Indicators Based on Traffic Load Monitoring. *Struct. Infrastruct. Eng.* **2019**, *15*, 899–911. [CrossRef]
22. Skokandić, D.; Mandić Ivanković, A. Value of Additional Traffic Data in the Context of Bridge Service-Life Management. *Struct. Infrastruct. Eng.* **2020**, 1–20. [CrossRef]
23. Pelà, L.; Aprile, A.; Benedetti, A. Seismic Assessment of Masonry Arch Bridges. *Eng. Struct.* **2009**, *31*, 1777–1788. [CrossRef]
24. Pelà, L.; Aprile, A.; Benedetti, A. Comparison of Seismic Assessment Procedures for Masonry Arch Bridges. *Constr. Build. Mater.* **2013**, *38*, 381–394. [CrossRef]
25. Franetović, M.; Ivanković, A.M.; Radić, J. Seismic Assessment of Existing Reinforced-Concrete Arch Bridges. *Gradjevinar* **2014**, *66*, 691–703. [CrossRef]
26. Monteiro, R.; Delgado, R.; Pinho, R. Probabilistic Seismic Assessment of RC Bridges: Part I—Uncertainty Models. *Structures* **2016**, *5*, 258–273. [CrossRef]
27. Monteiro, R. Sampling Based Numerical Seismic Assessment of Continuous Span RC Bridges. *Eng. Struct.* **2016**, *118*, 407–420. [CrossRef]
28. Cassese, P.; de Risi, M.T.; Verderame, G.M. Seismic Assessment of Existing Hollow Circular Reinforced Concrete Bridge Piers. *J. Earthq. Eng.* **2020**, *24*, 1566–1601. [CrossRef]

29. Srbić, M.; Ivanković, A.M.; Vlašić, A.; Kovačević, G.H. Plastic Joints in Bridge Columns of Atypical Cross-sections with Smooth Reinforcement without Seismic Details. *Appl. Sci.* **2021**, *11*, 2658. [CrossRef]
30. Homaei, F.; Yazdani, M. The Probabilistic Seismic Assessment of Aged Concrete Arch Bridges: The Role of Soil-Structure Interaction. *Structures* **2020**, *28*, 894–904. [CrossRef]
31. Zelaschi, C.; Monteiro, R.; Pinho, R. Parametric Characterization of RC Bridges for Seismic Assessment Purposes. *Structures* **2016**, *7*, 14–24. [CrossRef]
32. Morbin, R.; Zanini, M.A.; Pellegrino, C.; Zhang, H.; Modena, C. A Probabilistic Strategy for Seismic Assessment and FRP Retrofitting of Existing Bridges. *Bull. Earthq. Eng.* **2015**, *13*, 2411–2428. [CrossRef]
33. Stepinac, M.; Lourenço, P.B.; Atalić, J.; Kišiček, T.; Uroš, M.; Baniček, M.; Šavor Novak, M. Damage Classification of Residential Buildings in Historical Downtown after the ML5.5 Earthquake in Zagreb, Croatia in 2020. *Int. J. Disaster Risk Reduct.* **2021**, *56*, 102140. [CrossRef]
34. Lulić, L.; Ožić, K.; Kišiček, T.; Hafner, I.; Stepinac, M. Post-Earthquake Damage Assessment-Case Study of the Educational Building after the Zagreb Earthquake. *Sustainability* **2021**, *13*, 6353. [CrossRef]
35. Baggio, C.; Bernardini, A.; Colozza, R.; Pinto, A.; Taucer, F. *Field Manual for Post-Earthquake Damage and Safety Assessment and Short Term Countermeasures (AeDES)*; Translation from Italian: Maria ROTA and Agostino GORETTI; European Commission: Luxembourg, 2007.
36. Setunge, S.; Li, C.Q.; Mcevoy, D.; Zhang, K.; Mullett, J.; Mohseni, H.; Mendis, P.; Ngo, T.; Herath, N.; Karunasena, K.; et al. *Failure Mechanisms of Bridge Structures under Natural Hazards*; RMIT: Melbourne, Australia, 2018.
37. Tanimura, S.; Sato, T.; Umeda, T.; Mimura, K.; Yoshikawa, O. A Note on Dynamic Fracture of the Bridge Bearing Due to the Great Hanshin-Awaji Earthquake. *Int. J. Impact Eng.* **2002**, *27*, 153–160. [CrossRef]
38. Mitchell, D.; Huffman, S.; Tremblay, R.; Saatcioglu, M.; Palermo, D.; Tinawi, R.; Lau, D. Damage to Bridges Due to the 27 February 2010 Chile Earthquake. *Can. J. Civ. Eng.* **2013**, *40*, 675–692. [CrossRef]
39. Kawashima, K.; Buckle, I. Structural Performance of Bridges in the Tohoku-Oki Earthquake. *Earthq. Spectra* **2013**, *29*, 315–338. [CrossRef]
40. Hajdin, R.; Kušar, M.; Mašović, S.; Linneberg, P.; Amado, J.; Tanasić, N. *COST Action TU1406: Quality Specifications for Roadway Bridges, Standardization at a European Level*; WG3 Technical Report: Establishment of a Quality Control Plan; COST: Brussels, Belgium, 2018.

 *buildings*

Article

Seismic Vulnerability Assessment for Masonry Churches: An Overview on Existing Methodologies

Mattia Zizi *, Jafar Rouhi, Corrado Chisari, Daniela Cacace and Gianfranco De Matteis

Department of Architecture and Industrial Design, University of Campania "Luigi Vanvitelli", CE 81031 Aversa, Italy; jafar.rouhi@unicampania.it (J.R.); corrado.chisari@unicampania.it (C.C.); daniela.cacace@unicampania.it (D.C.); gianfranco.dematteis@unicampania.it (G.D.M.)
* Correspondence: mattia.zizi@unicampania.it

Abstract: The present manuscript deals with the seismic vulnerability assessment of existing masonry churches, which is a fundamental process for risk and consequent prioritization analyses, as well as application of effective retrofitting strategies. In the past, different approaches with various levels of accuracy and application ranges have been developed to assess the vulnerability to damage of such structures in case of seismic events. Based on the classification provided in the Italian Guidelines for the Cultural Heritage, in this paper a review of seismic vulnerability assessment methodologies for existing masonry churches is presented. The main goal of the current study is to provide a critical comparative overview about these procedures, highlighting the main issues related to the application of each detail level. Moreover, particular attention is focused on the applications present in literature, allowing for the definition of a potential systematic procedure for smart management policy aimed at preserving cultural, architectural and historical heritage.

Keywords: masonry churches; earthquakes; seismic vulnerability assessment; typological methods; macro-element approach; numerical analysis

Citation: Zizi, M.; Rouhi, J.; Chisari, C.; Cacace, D.; De Matteis, G. Seismic Vulnerability Assessment for Masonry Churches: An Overview on Existing Methodologies. *Buildings* **2021**, *11*, 588. https://doi.org/10.3390/buildings11120588

Academic Editors: Chiara Bedon, Tomislav Kišiček, Ufuk Hancilar, Marco Francesco Funari and Mislav Stepinac

Received: 8 October 2021
Accepted: 16 November 2021
Published: 26 November 2021

Publisher's Note: MDPI stays neutral with regard to jurisdictional claims in published maps and institutional affiliations.

1. Introduction

Nowadays, interest in the preservation of cultural built heritage, and in particular of existing masonry churches, is globally increasing [1]. Among different sources of risk threatening historical structures, seismic motion represents one of the main causes of damage and overall destruction. The mitigation of seismic risk is complex, involving hazard, exposure and vulnerability. If, on the one hand, it is not possible to manage hazard and exposure due to the intrinsic characteristics affecting sites and uses of buildings, efforts are being made by civil engineers to reduce the vulnerability of existing built heritage. Unfortunately, large areas of Europe, including Italy, are characterised by a high level of seismic hazard, and the vulnerability of ancient masonry structures is often relevant [2]. This has been widely demonstrated in survey campaigns carried out after disastrous seismic events occurred in Italy, and all over the world, during the last half century, which dramatically highlighted notable damage experienced by masonry churches. Therefore, there is a strong necessity to increase theoretical and technical knowledge aimed at improving existing methodologies for preserving valuable architectural and cultural heritage by means of risk mitigation.

The first step for seismic risk mitigation of masonry churches is represented by a seismic vulnerability assessment process, which, to date, is still difficult [3]. The complexity of studying churches and, in general, historical masonry structures mainly depends on their peculiar material characteristics and structural features, which cause significant structural deficiencies under seismic loads. In addition, it is worth noting that these heritage buildings were mainly built by skilled manufacturers based on empirical rules addressed to resist gravity loads only, and thus, in most cases, masonry churches cannot resist horizontal forces arising from seismic shaking.

In the literature, a large number of approaches are available, corresponding to different levels of accuracy of the seismic assessment process and strongly dependent on the aim of the study and the field of application. An exhaustive classification of these methods is provided by the Italian Guidelines for the Cultural Heritage (hereinafter stated as Italian Guidelines), in which three different Evaluation Levels (EL) are distinguished [4]:

- EL1: qualitative analysis and assessment by means of simplified mechanical and statistical models.
- EL2: assessment of single macro-elements (local collapse mechanisms).
- EL3: global assessment of the seismic response of the structure.

In recent times, the adoption of even more simplified methods for masonry churches has been promoted. In this sense, it is possible to consider a lower level of accuracy, namely EL0, usually based on very limited typological information and adopted mainly for applications on a larger scale [5,6]. EL0 methods provide a risk scale rather than quantifying the seismic safety of the structures considered, which is possible with the higher-level methods. Conversely, the other Italian Guidelines' levels of accuracy allow for the determination of a seismic safety index, defined as the demand-to-capacity ratio in terms of both accelerations and return period referred to each relevant limit state. Thus, their use has been conceived at the building scale, although with different detail levels.

The first level, EL1, is based on simplified mechanical formulations, based on the estimation of a vulnerability index. In case of both churches and historical buildings, a macro-element approach is suggested within this lower-accuracy level.

The second level of accuracy, EL2, is based on the assessment of collapse mechanisms of single parts of the building, which is mostly performed by resorting to a limit analysis approach.

Finally, the third level, EL3, assesses the seismic behaviour of the entire structure by means of linear or nonlinear numerical analyses (e.g., Finite Element Method, Discrete Element Method, among others). As a consequence, it is characterised by larger accuracy than previous analyses, and a wide range of information is required for its application, including detailed geometrical, structural and mechanical characteristics of the structure.

The reliability of such methods is strictly linked to the available data and the scale of application, as well as the scope of the analyses. The scheme in Figure 1 summarizes the components of seismic vulnerability assessment related to the different levels of accuracy.

In general, both EL0 and EL1 are recognised as suitable for large-scale approaches, and thus their uses are addressed to define prioritization in the decision-making processes. On the other hand, EL2 and EL3 are considered when local or global interventions, respectively, must be realised. Nevertheless, this difference may be smoothed and, in particular, EL1 and EL2 can be profitably adopted for either territorial or building scale analyses.

In the present study, an overview of the most adopted methods for assessing the seismic vulnerability of masonry churches at a territorial scale (i.e., EL0 and EL1) and their applications is provided. More detailed methods (i.e., EL2 and EL3) are also critically discussed. The study allows for identifying the application ranges of the different methods, and finally a potential management strategy for preserving existing masonry churches is suggested.

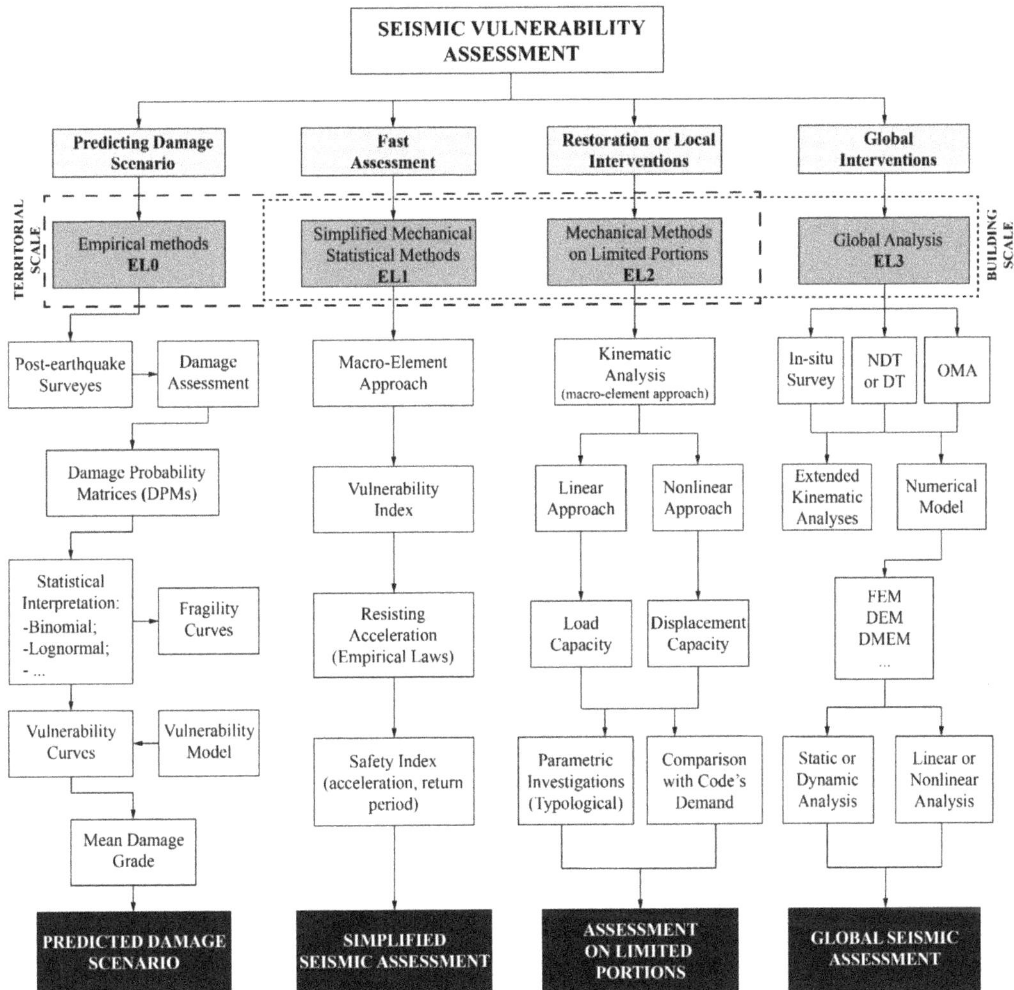

Figure 1. Components of seismic vulnerability assessment at different scales.

2. Seismic Response of Existing Masonry Churches

Many studies are present in the literature in which it has been noted that seismic events occurring in the last 25 years in Italy caused non-usability or access restrictions to more than 80% of investigated masonry churches in Umbria and Marche 1997 [7], Molise 2002 [8], L'Aquila 2009 [9–12], Emilia 2012 [13], Central Italy 2016–2017 [14–17], and Ischia 2017 [18]. This was confirmed after notable seismic events occurred all over the world, e.g., Perù 2007 [19], Chile 2010 [20,21], New Zealand 2010–2011 [22,23] and Mexico 2017 [24].

As noted in some of the mentioned studies, seismic vulnerability of historical masonry churches is higher than in other types of structures, including monumental buildings [2]. In particular, historical masonry churches are usually characterised by recurrent geometrical features generally favouring the occurrence of local mechanisms rather than an overall response [25], such as large wall height-to-thickness and length-to-thickness ratios, large roofing systems and openings, complex shapes and the absence of box-like behaviour, as well as insufficient connection between structural elements.

The fact that local mechanisms are strongly favoured by such features is confirmed by post-earthquake inspections. In Figure 2, pictures of damaged churches that experienced the Central Italy earthquake of 2016–2017 are shown as examples.

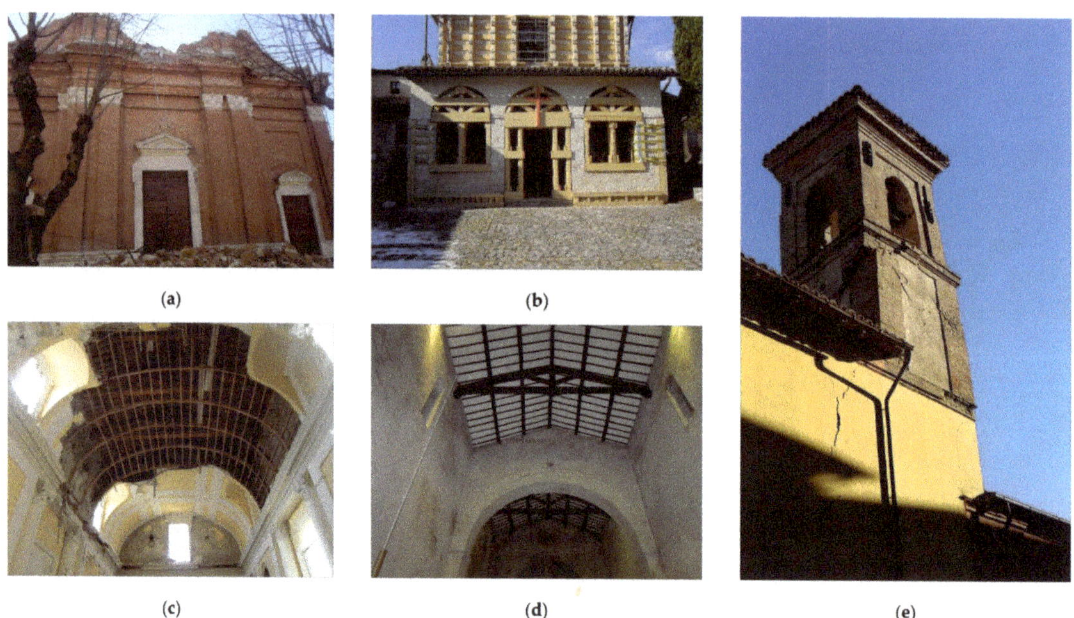

Figure 2. Examples of activation or collapse related to damage mechanisms during 2016-2017 Central Italy earthquake: (**a**) overturning of top façade in San Filippo church (Camerino, MC), (**b**) damage in the prothyrum in Santa Maria di Costantinopoli church (Cerreto di Spoleto, PG), (**c**) vault collapse in Santa Chiara Monastery (Camerino, MC), (**d**) triumphal arch damage in SS Felice and Mauro Abbey (Sant'Anatolia di Narco, PG), and (**e**) severe damage in the bell tower of San Carlo church (Camerino, MC).

This corroborates the approach proposed in the Italian Guidelines, where the study of masonry churches and, in particular, damage detection, is traced back to 28 possible failure mechanisms grouped in nine macro-elements. In Table 1, the identified macro-elements, and the corresponding damage mechanisms typical of Italian masonry churches, are listed.

Table 1. Damage mechanisms and macro-element of masonry churches according to Italian Guidelines.

Macro-Element	Damage Mechanism
Façade	M1. Façade overturning M2. Overturning of the top façade M3. In-plane mechanism of façade
Naves	M4. Narthex M5. Transversal response M6. Shear mechanisms in the lateral walls M7. Longitudinal response M8. Central nave vaults M9. Aisles vaults
Transept	M10. Overturning of the transept façade M11. Shear mechanisms in the transept wall M12. Transept vaults

Table 1. *Cont.*

Macro-Element	Damage Mechanism
Triumphal arch	M13. Triumphal arch
Dome	M14. Dome M15. Lantern
Apse	M16. Apse overturning M17. Shear mechanisms in the apse M18. Apse vaults
Roof	M19. Mechanisms in nave roof M20. Mechanisms in transept roof M21. Mechanisms in apse roof
Chapels and annexed bodies	M22. Chapel overturning M23. Shear mechanisms in chapels M24. Chapel vaults M25. Plain-height irregularities
Bell tower	M26. Decorations M27. Bell tower M28. Belfry

3. Territorial Scale Approaches: EL0 and EL1

3.1. Empirical and Statistical Methods: EL0

3.1.1. From Observed Damage to Predictive Models: Fragility and Vulnerability Functions

The lowest level of the seismic vulnerability assessment, EL0, also called the macroseismic approach, is based on empirical methods aimed at predicting the expected mean damage grade, due to a seismic event of a certain intensity, on a homogeneous population of buildings having similar geometrical and constructive features. The methods belonging to this level of accuracy are calibrated based on the real damage experienced by existing buildings after an earthquake and have been particularly suitable and adopted in the recent past for masonry churches. In particular, extensive surveys allow for assigning a given church to a specific damage level D_k, which is estimated according to the EMS-98 scale ranging between 0–5 [26] (see Table 2).

Table 2. Macroseismic Intensity Scale. Classifications used in the European Macroseismic Scale (EMS) for masonry structures (reproduced with permission from G. Grünthal [26]).

Classification of Damage to Masonry Buildings	
	Grade 1: Negligible to slight damage (no structural damage, slight non-structural damage) Hair-line cracks in very few walls. Fall of small pieces of plaster only. Fall of loose stones from upper parts of buildings in very few cases.
	Grade 2: Moderate damage (slight structural damage, moderate non-structural damage). Cracks in many walls. Fall of fairly large pieces of plaster. Partial collapse of chimneys
	Grade 3: Substantial to heavy damage (moderate structural damage, heavy non-structural damage) Large and extensive cracks in most walls. Roof tiles detach. Chimneys fracture at the roof line; failure of individual non-structural elements (partitions, gable walls).

Table 2. *Cont.*

Classification of Damage to Masonry Buildings

Grade 4: Very heavy damage (heavy structural damage, very heavy non-structural damage)

Serious failure of walls.

Partial structural failures of roofs and floors.

Grade 5: Destruction (very heavy structural damage)

Total or near total collapse.

The global damage level D_k is assessed as a function of a global damage index i_d (ranging between 0–1) as suggested in Table 3.

Table 3. Definition of structural damage levels for the churches, based on the damage score.

D_k	i_d	Description
0	$i_d \leq 0.05$	No damage: light damage only in one or two mechanisms
1	$0.05 < i_d \leq 0.25$	Negligible to slight damage: light damage in some mechanisms
2	$0.25 < i_d \leq 0.4$	Moderate damage: light damage in many mechanisms, with one or two mechanisms activated at medium level
3	$0.4 < i_d \leq 0.6$	Substantial to heavy damage: many mechanisms have been activated at medium level, with severe damage in some mechanisms
4	$0.6 < i_d \leq 0.8$	Very heavy damage: severe damage in many mechanisms, with the collapse of some macroelements of the church
5	$i_d > 0.8$	Collapse: at least 2/3 of the mechanisms exhibit severe damage

In turn, the global damage index i_d is calculated for each church based on the experienced damage according to Equation (1), which is proposed in the Italian Guidelines:

$$i_d = \frac{1}{5} \cdot \frac{\sum_{i=0}^{28} \rho_{k,i} \cdot d_{k,i}}{\sum_{i=0}^{28} \rho_{k,i}} \tag{1}$$

where: $d_{k,i}$ ($0 \div 5$) is the damage level observed for the i-th damage mechanism of Table 1 and $\rho_{k,i}$ ($0 \div 1$) is the corresponding score factor indicating the importance that each damage mechanism has on the global safety of the church.

The score factors $\rho_{k,i}$ were firstly defined in [8], and those values were assumed as referenced in the Italian Guidelines. Nonetheless, in some works such values are modified according to the expert judgment of the authors and thus are not fully consistent with the Guidelines' provisions (e.g., [17,27]), while in other works the damage mechanisms are all assumed with the same importance, with $\rho_{k,i}$ values constant and equal to 1 (e.g., [14,15,18]). In Table 4, the values of $\rho_{k,i}$ adopted in some reference works are shown in comparison with the ranges provided by the Italian Guidelines.

The damage assessment allows for defining fragility curves, which relate the probability of damage being larger than a specified level to the intensity of the earthquake (measured with a macroseismic or peak ground acceleration scale) and are formulated based on an observational vulnerability model. One of the most adopted function for describing the probability of damage exceedance is the binomial probability function (BDPF) shown in Equation (2).

$$p_k = \frac{5!}{k! \cdot (5-k)!} \cdot \left(\frac{\mu_D}{5}\right)^k \cdot \left(1 - \frac{\mu_D}{5}\right)^{(5-k)} \tag{2}$$

Table 4. Values for the importance score factor $\rho_{k,i}$ assumed in reference works.

Damage Mechanism	$\rho_{k,i}$			
	[8]	[4]	[17]	[27]
M1. Façade overturning	1	1	1	1
M2. Overturning of the top façade	1	1	1	1
M3. In-plane mechanism of façade	1	1	0.5	1
M4. Narthex	0.5 ÷ 1	0.5	0.25	0.9
M5. Transversal response	1	1	1	0.9
M6. Shear mechanisms in the lateral walls	1	1	1	0.9
M7. Longitudinal response	1	1	1	1
M8. Central nave vaults	1	1	0.75	1
M9. Aisles vaults	1	1	0.75	0.5
M10. Overturning of the transept façade	0.5 ÷ 1	0.5 ÷ 1	0.75	1
M11. Shear mechanisms in the transept wall	0.5 ÷ 1	0.5 ÷ 1	0.5	1
M12. Transept vaults	0.5 ÷ 1	0.5 ÷ 1	1	0.9
M13. Triumphal arch	1	1	1	1
M14. Dome	1	1	0.75	0.9
M15. Lantern	0.5	0.5	0.25	1
M16. Apse overturning	1	1	0.75	0.9
M17. Shear mechanisms in the apse	1	1	0.5	0.9
M18. Apse vaults	0.5 ÷ 1	0.5 ÷ 1	0.75	0.9
M19. Mechanisms in nave roof	1	1	0.5	0.9
M20. Mechanisms in transept roof	0.5 ÷ 1	0.5 ÷ 1	0.5	0.8
M21. Mechanisms in apse roof	1	1	0.5	1
M22. Chapel overturning	0.5 ÷ 1	0.5 ÷ 1	0.25	1
M23. Shear mechanisms in chapels	0.5 ÷ 1	0.5 ÷ 1	0.25	1
M24. Chapel vaults	0.5 ÷ 1	0.5 ÷ 1	0.5	1
M25. Plain-height irregularities	0.5 ÷ 1	0.5 ÷ 1	1	1
M26. Decorations	0.8	0.5 ÷ 1	0.25	0.9
M27. Bell tower	1	1	1	0.9
M28. Belfry	1	1	1	0.9

In the above equation p_k (with $k = 0 \div 5$) is the probability of reaching a specific damage level D_k, and μ_D is the observed mean damage grade in the population calculated as in Equation (3), where n represents the number of considered buildings:

$$\mu_D = \frac{\sum\limits_{k=1}^{n} D_k}{n} \tag{3}$$

The binomial distribution was firstly introduced for damage statistical analyses by Braga, Dolce and Liberatore [28], who based their study on the damage distribution matrixes (DPMs) obtained for ordinary buildings after the Irpinia earthquake (1980). By means of this methodology, the fragility curve is assumed as the cumulative probability to reach a specific damage grade, as shown in Equation (4).

$$P(D \geq D_k) = \sum\limits_{k=k}^{5} p_k \tag{4}$$

Although, to date, the binomial distribution of the damage of Equation (2) represents the most adopted fragility function [7,8,12,29–32], it has the disadvantage of not allowing for an independent definition of the scatter of the expected damage, which is dependent on the only free parameter of the distribution, the mean damage μ_D [33]. Hence, continuous beta or lognormal distributions can be successfully adopted in order to statistically interpret the observed damage. For masonry churches, recent studies carried out after the Central-Italy 2016–2017 earthquake showed that the observed damage as a function of seismic intensity was well interpreted by a lognormal distribution.

In Equation (5) the function adopted in [14,15] is shown as an example.

$$P(D \geq D_k \mid PGA = x) = \phi \left(\frac{\ln(x/\mu|\theta)}{\beta} \right) \tag{5}$$

In particular, in the above equations, $P(D \geq D_k \mid PGA = x)$ is the probability of exceeding a specific damage grade D_k as function of a certain seismic intensity $PGA = x$, ϕ is the normal cumulative distribution, μ and θ are the mean and median values, respectively, (which can be alternatively adopted) and β is the standard deviation. Hence, in these cases, the shape of lognormal fragility curves strongly depends on the parameter β, indicating the dispersion of the observed data. Thus, although unlike binomial functions they can provide several distributions of the damage, it should be noted that these formulations strongly depend on the parameter β.

Moving from an observational to a predictive approach, fragility curves can be adopted to estimate the probability of damage occurrence if the mean damage grade μ_D is determined a-priori. Hence, this method is suitable in combination with vulnerability curves, which can return the mean damage grade according to the seismic intensity and the vulnerability of the buildings. Some of the first authors who proposed vulnerability curves by means of a hyperbolic function were Sandi and Floricel [34]. In their study, they proposed a vulnerability function for ordinary buildings, as indicated in Equation (6):

$$\mu_D = 2.5 \cdot \left[1 + \tanh \left(\frac{I + 6.25 \cdot V_i - 13.1}{Q} \right) \right] \tag{6}$$

where the expected damage μ_D is evaluated as a function of the seismic intensity I in macroseismic scale (I_{MCS}), the vulnerability index V_i ranges between 0–1 according to the vulnerability classification in EMS-98 [24] and the ductility factor, Q, for ordinary buildings can be assumed as 2.3.

Then, based on the damage that occurred after the 1997 Umbria and Marche seismic event, Lagomarsino and Podestà [29] calibrated vulnerability curves for existing masonry churches on the basis of a wide post-earthquake survey activity including about 2000 churches. The law proposed in this study is shown in Equation (7):

$$\mu_D = 2.5 \cdot \left[1 + \tanh \left(\frac{I + 3.4375 \cdot i_v - 8.9125}{Q} \right) \right] \tag{7}$$

In particular, the vulnerability law proposed in this study involves the use of a new vulnerability measure i_v, again ranging from 0 to 1. Hence, in place of the vulnerability index V_i, Lagomarsino and Podestà [29] proposed adoption of a value higher than those estimated for ordinary buildings V_i, their correlation being the one indicated in Equation (8):

$$V_i = 0.67 + 0.55 \cdot i_v \tag{8}$$

By inserting Equation (8) in Equation (6), Equation (7) is s obtained, showing the equivalence between the two formulations. In practice, Equation (7) can be considered as the result of Equation (8) introduced in Equation (6).

Moreover, in this case, Q was assumed equal to 3. It is worth mentioning that, according to Equation (7), a higher ductility factor than the one adopted for ordinary buildings (i.e., $Q = 2.3$) means a higher mean damage grade for low-intensity (i.e., $I_{MCS} < $ VII) earthquakes, with less damage for high-intensity seismic event [32].

After this study, in the recent years, and with reference to different seismic events occurring in Italy and all over the world, many research activities have been carried out with the aim of verifying the reliability of these methodologies for predictive purposes by comparing the observed damage with the predicted damage [35,36]. While agreement on the type of vulnerability function is generally present, specific regional and typological features may be better represented by slight modifications in the coefficients. For instance, De Matteis, Brando and Corlito [32] proposed a modification to the vulnerability function provided in Equation (7) for three-nave masonry churches damaged by the L'Aquila 2009 earthquake (Equation (9)):

$$\mu_D = 2.5 \cdot \left[1 + \tanh\left(\frac{I + 6.20 \cdot i_v - 11}{Q} \right) \right] \tag{9}$$

Again, a ductility factor $Q = 3$ was assumed, while the argument numerator of the hyperbolic tangent function was modified to obtain a better correspondence with the observed data.

The use of intensity measures in the vulnerability function implies a conceptual short-circuit since vulnerability depends on intensity, defined based on the effects of the ground motion on the built environment. However, those effects depend in turn on the vulnerability of the stock. In this context, De Matteis and Zizi [17] recently proposed the adoption of vulnerability functions based on a PGA-approach. In their work, the authors studied 68 one-nave churches damaged after the 2016–2017 Central Italy earthquake and highlighted good correspondence between observed and predicted damage obtained if Equation (7) is modified by means of the empirical correlation between I_{MCS} and PGA proposed by Faenza and Michelini [37].

In recent years, research has moved toward the adoption of PGA-based approaches, solving the issue highlighted above and increasing the feasibility of applying this low-detail level of accuracy (i.e., EL0) for predictive purposes. In this sense, one of the most notable examples of application is the national MaRS project promoted by the Department of Civil Protection and the consortium ReLUIS [38]. The main aim of this activity is the realization of national seismic risk maps related to several building typologies, among them churches. Although for these kinds of structures the results are still in an embryonic phase, an interesting method for deriving fragility curves from observational data has been proposed. In particular, the MaRS fragility curves are assumed with a lognormal distribution and are derived from two parameters only: (i) the median value of the PGA_{D2} related to a damage level D2 assigned to the specific vulnerability class, and (ii) the free parameter α (in the range 0.36–0.66 indicating the brittle or ductile behaviour, respectively) for determining the PGA values related to the damage level D_k according to Equation (10).

$$PGA_{Dk} = PGA_{D2} \cdot e^{\alpha(k-2)} k = 1, \dots, 5 \tag{10}$$

Also in this case, the fragility curves assigned to a specific building typology strongly depend on the dispersion β assumed for the lognormal distribution.

Nonetheless, to date, the vulnerability function of Equation (7) still appears the most adopted and robust law for fitting the observed damage, and thus could be adopted for predictive purposes, too. A graphical representation of the vulnerability curves obtained by applying Equation (7), and the consequent generic fragility curves (with a binomial distribution, see Equation (4)), are provided in Figures 3 and 4, respectively.

3.1.2. Vulnerability Assessment: EL0 Methods and Applications

In the past decades, the application of vulnerability and fragility curves has seen wide application not limited to masonry churches. Nowadays, the research world is still moving to corroborate such empirical formulations to adopt them for predictive (and thus preventive) aims within a territorial approach.

It is clear, now, that one of the most complex issues in this field is the definition of a vulnerability parameter, which should be based on few typological characteristics to allow a fast large-scale assessment. In the following, some relevant literature works, not limited to the Italian context and addressing this issue, are reported. It must be pointed out that these methodologies have been adopted in combination with various vulnerability and/or fragility functions. Thus, it must be admitted that the different vulnerability models suggested in each work only make sense within the scope of the specific framework to which they refer.

Generally, within the low-detail level EL0, the vulnerability parameter is estimated according to an initial value (V_0), which is modified accounting for typological and geometrical characteristics as indicated in Equation (11), where the choice of the modifying parameters (V_k) and relative scores are empirically determined on the basis of statistical analyses and expert judgements [1,30].

$$V_i = V_0 + \sum V_k \tag{11}$$

In the literature, many examples of similar EL0 vulnerability models are present, which are based on Equation (11) or its modifications. A notable example of this method has been implemented within the European Risk-UE project [39], which involved seven European cities (Barcelona, Bitola, Bucharest, Catania, Nice, Sofia and Thessaloniki). Therein, a vulnerability model suitable for several ancient masonry construction typologies is provided and, in particular, $V_0 = 0.89$ is defined for churches. The value is then modified according to Equation (11) and by accounting for seven parameters: (i) state of maintenance, (ii) quality of materials, (iii) regularity in plan, (iv) regularity in elevation, (v) position in the urban context, (vi) retrofitting interventions, and (vii) site morphology. As shown in Table 5, these modifiers may have an increasing or decreasing effect on the vulnerability, based on the quality of the feature. Examples of application of this methodology can be found, among others, in [1,30,40].

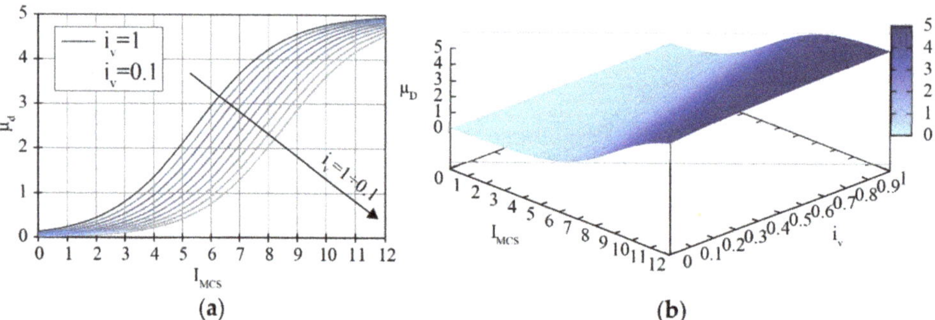

Figure 3. Vulnerability function in (a) I_{MCS}-μ_D plane and (b) I_{MCS}-μ_D-i_v space.

It is worth specifying that the main aim of the European Risk-UE project was to provide unified vulnerability functions regardless of the investigated structural typologies (e.g., towers, bridge, churches) and with this method a vulnerability score in the range 0.63–1.22 can be obtained, which is consistent with Equation (8).

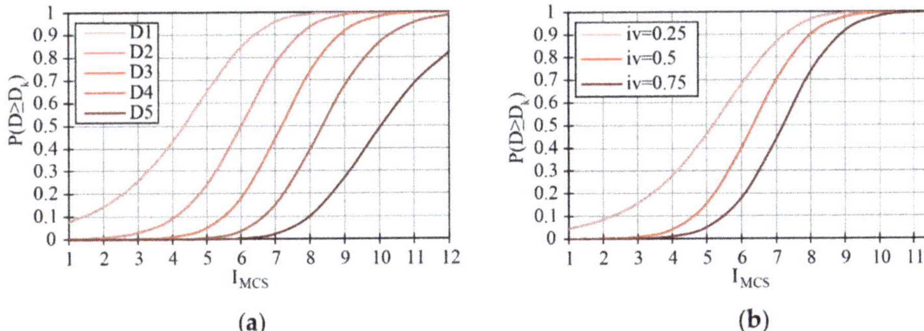

Figure 4. Fragility function: (**a**) curves for $i_v = 0.5$, and (**b**) curves related to $D_k = 3$ for different levels of vulnerability.

Table 5. Reference values of modifiers scores (V_k) (adapted from [4]).

Parameter	V_k			
State of Maintenance	very bad = 0.08	bad = 0.04	Medium = 0	good = −0.04
Quality of Materials	bad = 0.04		medium = 0	good = −0.04
Planimetric Regularity	irregular = 0.04		regular = 0	symmetrical = −0.04
Regularity in elevation	irregular = 0.02			regular = −0.02
Interactions (aggregate)	corner position = 0.04		isolated = 0	included = −0.04
Retrofitting Interventions		effective interventions = −0.08		
Site Morphology	ridge = 0.08		slope = 0.04	flat = 0

A similar approach is proposed in [27]. In this work the authors proposed a simplified seismic risk model assessed by means of hazard and vulnerability scores. Vulnerability is estimated by examining thirteen parameters, ten of which are derived from the Italian GNDT II vulnerability datasheet [41]. In this case, the vulnerability model assumes the form of Equation (12):

$$V_i = \sum_{k=1}^{13} \rho_k \cdot v_{k,i} \tag{12}$$

where $v_{k,i}$ is the score value i of the class selected for the generic k-th parameter, while ρ_k is the weight, representing the importance that each parameter has on the global vulnerability of the church.

Geometrical-based simplified methods, which consider geometrical features to obtain approximate vulnerability indexes, are worth mentioning, too. For example, Lourenço and Roque [42] proposed three simplified safety indexes based on a study concerning 58 Portuguese churches: (i) in-plan area ratio, (ii) area-to-weight ratio, and (iii) base shear ratio. In this work, a vulnerability score is suggested as an indicator for fast screening aimed at prioritizing deeper assessment studies. Similarly, in the work of Salzano et al. [18], vulnerability classes for 27 churches damaged after the Ischia earthquake (2017) are defined according to a fictitious slenderness parameter, namely nave height to square root of the plan area ratio. In this case, the authors examined a wide set of vulnerability functions in order to obtain the best correlation with the observed data.

Another significant study carried out by Palazzi et al. [43] deals with 106 masonry churches that experienced the 2010 Maule earthquake (Central Chile). In this work, the authors tried to find a correlation between the experienced seismic intensity and damage level, and four typological parameters, i.e., (i) masonry type, (ii) architectural layout, (iii) architectural style, and (iv) foot-print area.

3.2. Simplified Mechanical, Statistical and Qualitative Models: EL1

3.2.1. Safety Assessment

Whereas with the previous level of accuracy it is possible to estimate an expected damage, or the probability of attaining a certain damage grade on a homogeneous population of buildings, with the EL1 approach a simplified safety check can be carried out. In particular, this approach is aimed at estimating a safety index I_S to define a suitable risk classification and possibly highlight the need for further studies and planning interventions for the mitigation of seismic risk. Hence, its application is mainly referred to the regional diocesan and municipal scale, since for its application an accurate inspection by expert practitioners is required.

The method is widely described in the Italian Guidelines, and it is based on empirical rules calibrated on the data observed in the aftermath of Marche 1997 and Molise 2002 earthquakes [5,6,27]. A brief outline is provided below.

The safety index $I_{S,LS}$ can be estimated, for each limit state LS, according to Equation (13), where $T_{R,LS}$ is the prescribed return period of the seismic action (demand), and T_{LS} is the actual return period of the seismic action for which LS is attained (capacity):

$$I_{S,LS} = \frac{T_{LS}}{T_{R,LS}} \qquad (13)$$

The Italian Guidelines require masonry churches to consider three limit states:

- The Life Safety Limit State (LSLS), which is considered attained when the building, after a seismic event, experiences collapse of nonstructural elements and relevant damage of the structural components, thus provoking a significant loss of global stiffness with respect to horizontal actions.
- The Damage Limit State (DLS), which is considered attained when the building, after a seismic event, experiences a global damage level (including both structural and non-structural elements) so that the safety of people and the capacity of the structure in bearing vertical and horizontal loads are not endangered.
- The Damage Limit State for Artistic Assets (ALS), which is considered attained when the artistic assets (such as such as frescoes, wall paintings, stone carving, etc.) in the building suffer low damage so that they can be restored without a significant loss of their cultural value.

To define the demand in LSLS and DLS conditions, and thus the corresponding return periods $T_{R,SL}$, the exceedance probability of the seismic action $P_{VR} = 10\%$ and $P_{VR} = 63\%$, respectively, are considered. On the other hand, for assessing the ALS attainment, a $P_{VR} = 63\%$ is, again, considered, while the nominal life is reduced according to a η factor, accounting for the number of checks usually carried out on the specific artistic asset. Since the main aim of this methodology is to perform a fast assessment for prioritization processes on a territorial scale, the same nominal life for the considered churches (e.g., $V_N = 50$ years) is suggested.

In addition, seismic safety in terms of the acceleration factor is also provided by the Italian Guidelines (namely $f_{a,LS}$), which is defined as the ratio between the demand peak ground acceleration (PGA) for the limit state (a_{LS}) and the corresponding capacity ($a_{g,LS}$), as indicated in Equation (14).

$$f_{a,LS} = \frac{a_{LS}}{a_{g,LS}} \qquad (14)$$

Within the application of the EL1 method for assessing the seismic safety of a masonry church, the LSLS and DLS are considered. A direct correlation between a_{LS} and the vulnerability score i_v ranging between 0–1, is proposed. The empirical formulations provided by the Italian Guidelines for masonry churches and the two considered limit states are provided in Equation (15):

$$a_{DLS} \cdot S\,[g] = 0.025 \cdot 1.8^{2.75-3.44\cdot i_v}$$
$$a_{LSLS} \cdot S\,[g] = 0.025 \cdot 1.8^{5.1-3.44\cdot i_v} \tag{15}$$

In the above equations, S is the coefficient accounting for the subsoil category and the topographic conditions defined according to the Italian Technical Code.

Hence, in order to estimate the safety index I_S (Equation (13)), it is necessary to evaluate the return period corresponding to the attainment of the considered limit state by interpolating between two known values related to predefined return periods, peak accelerations on rigid soil and soil condition as in Equation (16).

$$T_{LS} = T_{R1} \cdot 10^{\log(T_{R2}/T_{R1})\cdot\log(a_{LS}S/F_C a_1 S_1)/\log(a_2 S_2/a_1 S_1)} \tag{16}$$

In Equation (16), T_{R1} and T_{R2} are the return periods for which the seismic hazard is provided and that define the range including T_{SL}, a_1 and a_2 are the corresponding peak ground acceleration on rigid soil and F_C is the confidence factor, which for such analyses can be assumed $F_C = 1.35$, while S_1 and S_2 are the soil coefficients, as mentioned above.

The exponential trends obtainable by applying Equations (15) and (16) are shown in Figure 5. In the figure, a flat rigid soil has been considered. Moreover, a generic site has been selected and the return periods related to the different limit states have been normalized with respect to the maximum value referred to the DLS (Figure 5b).

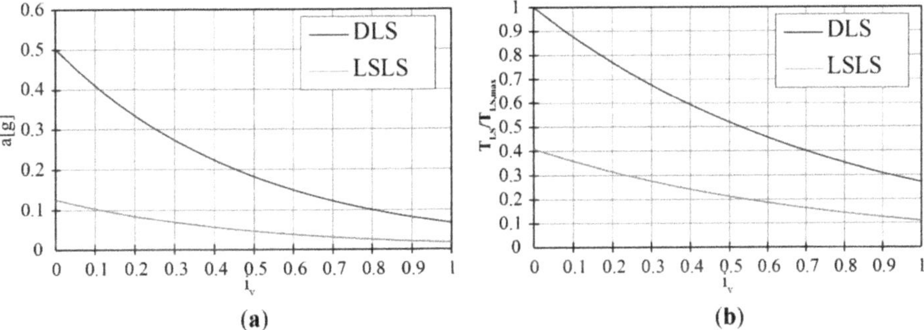

Figure 5. EL1 methodology according to Italian Guidelines related to DLA and LSLS: (**a**) acceleration capacity versus vulnerability index and (**b**) return periods versus vulnerability index.

3.2.2. Vulnerability Assessment: EL1 Methods and Applications

It is evident that, similar to EL0, a parameter describing the vulnerability of the considered structure plays a fundamental role in the EL1 approach. In this case, the Italian Guidelines provide an empirical formulation based on the so-called macro-element approach. In particular, the vulnerability index is evaluated as a weighted average of the scores related to the potential damage mechanisms of Table 1, accounting for the presence of fragility indicators and anti-seismic devices:

$$i_v = \frac{1}{6} \cdot \frac{\sum_{k=1}^{28} \rho_k \cdot \left(v_{k,i} - v_{k,p}\right)}{\sum_{k=1}^{28} \rho_k} + \frac{1}{2} \tag{17}$$

Here, ρ_k is the same importance score factor adopted in the damage assessment (see Table 2) and assigned to each mechanism according to the influence that it has on the global behaviour of the structure (i.e., $\rho_k = 0$ if the mechanism cannot occur), while $v_{k,i}$ and $v_{k,p}$ are the scores related to both the presence (and the severity) of fragility indicators, and the presence (and the efficiency) of antiseismic devices related to the k-th mechanism, respectively. $v_{k,i}$ and $v_{k,p}$ can be estimated according to the number of elements influencing the seismic vulnerability of the damage mechanism and the expert judgment on their effectiveness (ranging between 1–3), as shown in Table 6.

The formulation for assessing the vulnerability index according to the EL1 procedure is only partly consistent with that adopted for EL0. Indeed, starting from a base value (i.e., 0.89 for EL0, 0.5 for EL1), the index is modified according to fragility indicators and the presence of antiseismic devices, in EL1 based on a more accurate survey of the potential damage mechanisms of the macro-elements. In past works, it has been shown that the application of this method to Italian churches generally returns vulnerability indices close to the median 0.5. In this context, among others, the following contributions are worth mentioning: the study of De Matteis, Brando and Corlito [32], which deals with churches that experienced the L'Aquila earthquake, the work of De Matteis and Zizi [17] that focuses on one-nave churches of Central Italy, the analyses performed by Salzano et al. [18] on churches damaged after the Ischia seismic event of 2017, and the study of D'Amato, Laterza and Diaz Fuentes [27] on Matera's churches. In addition, by applying Equation (17) it can be noted that under the assumptions of rigid and flat soil, the resisting ground acceleration in SLSL conditions passing from $i_v = 0.6$ to $i_v = 0.4$ is almost doubled. This indeed means that a small variation of the vulnerability index could lead to significant differences in expected strength.

Table 6. Evaluation of the vulnerability modifiers (adapted from [4]).

Number of Vulnerability Indicators or Antiseismic Devices	Judgment of Effectiveness	v_k
At least 1	3	3
At least 2	2	3
1	2	2
At least 2	1	2
1	1	1
None	0	0

Overall, since the EL1 vulnerability assessment foresees accurate surveys of masonry churches, it can be considered mainly suitable for territorial applications with a lower scale than EL0, such as a provincial or regional scale. Examples of application of this method can be found, among others, in [18,27].

4. Building Scale Approaches: EL2 and EL3

4.1. Assessment Based on Limit Analysis Concept: EL2

4.1.1. Field of Applications and Fundamentals of Limit Analysis

Whereas previous levels of accuracy (EL0 and EL1) are mainly adopted in large-scale applications, an EL2 method is, in principle, suitable for structural-scale applications. EL2 methods are based on the limit analysis by adopting static and kinematic approaches. Hence, the mechanical characteristics of the masonry are disregarded with such approaches, and Heyman's assumptions of the masonry behaviour can be adopted: (i) infinite compressive strength, (ii) zero tensile strength, and (iii) absence of sliding failures [44]. These simplified hypotheses can be assumed within the static and kinematic theorems, which respectively explore statically balanced and kinematically compatible solutions.

Static approaches consist of graphical methods based on the theories developed in France during the 18th and 19th centuries concerning the structural safety of masonry arches founded on the line-of-thrust concept [45,46]. These methods try to find the unique kinematically compatible solution among those admissible that are in equilibrium with external loads. On the other hand, limit analyses with a kinematic approach are widely utilized to study any portion of the entire building that can exhibit recognizable failure mechanisms. This holds true not only for masonry churches, since such an approach is considered by technical codes also for generic masonry structures. The kinematic approach tries to find the unique balanced solution among those corresponding to kinematically compatible mechanisms.

When a linear approach is assumed, the method allows determination of the multiplier of a load distribution by simulating the seismic action for which a failure mechanism is activated in the considered structural element, and thus can be used for a check in Damage Limit State (DLS) conditions. On the contrary, the attainment of the Life Safety Limit State (LSLS) can be verified with a simplified methodology (linear approach with behaviour factor) or by resorting to nonlinear kinematic analysis, which involves the evaluation of the load multiplier at different levels of mechanism amplitude.

4.1.2. Limit Analysis by Means of Kinematic Approach

Kinematic limit analysis consists of subdividing the investigated structural component in a set of rigid blocks linked to one another by means of a minimum number of non-dissipative hinges creating a mechanism (i.e., kinematic chain). Hence, by adopting the principle of virtual works, it is possible to identify the multiplier of a load distribution which brings the system into equilibrium. The mechanism corresponding to the minimum multiplier is identified as the collapse multiplier. Since infinite hinges distributions can be considered, the analysis can be performed by following two approaches: (i) by conveniently minimising the load multiplier by means of automatic procedures, or (ii) by a-priori defining a block subdivision for a predetermined failure mechanism, which is recurrent or compatible with real crack pattern already present in the considered element.

The load multiplier (a_0) for which the activation of a failure mechanism is achieved can be estimated according to Equation (18):

$$a_0 = \frac{\sum\limits_{k=1}^{N} P_k \cdot \delta_{Py,k} - \sum\limits_{k=1}^{m} F_k \cdot \delta_{F,k} + L_i}{\sum\limits_{k=1}^{N} (P_k + Q_k) \cdot \delta_{PQx,k}} \tag{18}$$

where N is the number of blocks constituting the kinematic chain, m is the number of external loads applied on each block, P_k is the resultant weight-force vector applied on the k-th block; Q_k is the resultant weight-force vector not loading the k-th block but whose mass provokes an horizontal seismic action on it, F_k is the external load applied on the k-th block, $\delta_{Py,k}$ is the vertical virtual displacement of the k-th block centroid, $\delta_{F,k}$ is the virtual displacement dual to F_k at its application point, $\delta_{PQx,k}$ is the horizontal virtual displacement of the centroid of forces P_k and Q_k applied on the k-th block, and L_i is the work produced by internal forces.

The so-defined load multiplier has to be converted in spectral terms (a_0^*) by means of Equation (19):

$$a_0^* = \frac{a_0 \cdot g}{FC \cdot e^*} \tag{19}$$

Here, g is the gravity acceleration, FC is the confidence factor ranging between 1–1.35 according to the level of knowledge (KL) of the structure, and e^* is the mass participating in the considered mechanism, which is estimated as indicated in Equation (20).

$$e^* = \frac{\left[\sum\limits_{k=1}^{N} (P_k + Q_k) \cdot \delta_{PQx,k}\right]^2}{\left[\sum\limits_{k=1}^{N} (P_k + Q_k)\right] \cdot \left[\sum\limits_{k=1}^{N} (P_k + Q_k) \cdot \delta_{PQx,k}^2\right]} \tag{20}$$

Hence, safety checks with respect to DLS and LSLS can be carried out by comparing the demand in terms of spectral acceleration and the capacity a_0^*, which in the second case can be increased by means of the behaviour factor q (usually adopted equal to 2 for these analyses).

On the contrary, if a nonlinear approach is adopted, the evolution of the mechanism, described by the displacement of a control point, is followed until the load multiplier becomes zero. In this case, the failure mechanism is described by means of a capacity curve (acceleration versus displacement) assumed as a single degree of freedom system (SDOF). Such an approach allows the estimation of the post-peak response, thus not being particularly suitable for assessing the seismic vulnerability of masonry churches, for which the Collapse Limit State (CLS) is usually not considered.

Many examples of EL2 applications can be found in the literature. In particular, they are referred to several damage mechanisms of entire churches [21,27,47] or single macro-elements [48–50]. Whereas in the first cases the approach can lead to designing local retrofitting interventions, the second approach entails performing parametric studies on macro-elements, e.g., arches and vaults characterised by typical dimensional features. If extended to the complete set of vulnerable macro-elements, this has the potential to become the basis for systematic fast assessment tools based on sole geometrical features.

4.2. Detail Global Seismic Assessment: EL3

4.2.1. General

The most accurate level for assessing the seismic vulnerability of existing masonry churches and historical masonry buildings, is based on the analysis of detailed numerical models. At the scale of the single structure, the Italian Guidelines also suggest the adoption of simplified methods (EL2) applied to each element of the construction. Nevertheless, studying the entire building by means of its numerical representation allows the practitioner to design a global retrofitting intervention rather than localized measures aimed at improving the seismic response of single macro-elements.

4.2.2. Numerical Models and Fields of Application

Numerical modelling of the seismic response assessment of complex masonry structures, such as churches, which are often characterized by inhomogeneous materials with poor connections between orthogonal walls, is not an easy task. The approaches most generally used fall in two main categories: Finite Element Methods (FEM) and Discrete Element Methods (DEM).

Among FEM, differences are related to:

- the representation of masonry, i.e., detailed or simplified micromodelling and macro-modelling [51].
- the geometrical discretisation, i.e., monodimensional [52,53], two-dimensional [54,55] and three-dimensional elements [56,57].
- the material models, i.e., discrete or smeared approaches [58] in combination with plastic, damage or plastic-damage materials [59].

With reference to masonry representation, it can be certainly asserted that the macro-modelling approach is the most adopted one for masonry churches [60–64]. With such an approach, units and joints are not individually modelled, but the material is represented as an equivalent homogeneous continuum, and thus computational efficiency is attained at the expenses of some simplifications. For instance, macro-models are not perfectly able to reproduce sliding failures and out-of-plane mechanisms, which, conversely, can be well interpreted with more detailed modelling approaches. Nonetheless, the latter issue can be profitably solved by introducing contact-like constraints between walls rather than guarantying a perfect node congruence of the mesh [51,65]. More detailed approaches (meso and micro-modelling, in which the masonry bond is entirely represented) entail more significant modelling and computational efforts, which can be not particularly suitable when the seismic responses of entire buildings are assessed and thus, they can be profitably adopted for single portions (e.g., vaults, walls, arches, etc.) [66,67].

As regards the geometrical discretisation, it must be pointed out that some of the simplified hypotheses usually assumed for ordinary buildings are not fully appropriate in case of masonry churches, given the absence of stiff a horizontal diaphragm and the significant slenderness of the structures favouring out-of-plane mechanisms. Hence, the Equivalent Frame Method (EFM), which is the analysis method generally suggested by Eurocode 6 for masonry structures, cannot be considered suitable, and the choice should fall on models having two-dimensional or three-dimensional elements [68]. When masonry elements are characterized by significant thicknesses, which is typical of ancient masonry churches, three-dimensional elements are preferred since flexural, shear and rocking failures can be better interpreted.

As far as material modelling is concerned, generally models developed for simulating the behaviour of concrete and other quasi-brittle material with low tensile strength can be adopted in FE simulations. The choice can fall among material models entailing discrete or smeared approaches, which simulate the cracking process by introducing mesh discontinuities or by nonlinear behaviour of the homogenised material, respectively [58]. In both cases, one of the most notable issue is the definition of the tensile behaviour, which greatly affects the analysis in terms of both global seismic response and stability (convergence problems) [69]. As for all materials showing softening, mesh dependency can be reduced or avoided by using a variety of approaches including nonlocal approaches [70], higher-order continuum models [71], as well as regularization processes [72]. Overall, the latter formulations, which are based on the concept of fracture energy, are preferred since they are relatively easy to implement and little sensitive to mesh, which in case of complex geometries is usually very variable.

With DE models the structure is modelled as an assemblage of distinct blocks (rigid or deformable) interacting with one another according to contact constitutive laws, which, in turn, account for possible failure mechanisms. Such a method, which was first introduced for geotechnical problems by Cundall and Strack [73], has formerly found wide application in civil engineering problems, especially related to granular materials such as masonry [74–76]. Several aspects distinguish different DE models. Among others, should be mentioned: (i) block representation; (ii) contact laws; (iii) solution methods, and (iv) material properties [77]. As mentioned above, the block in DEMs can consist of rigid or deformable elements. In the first case, the deformation is totally accounted for in joints, while in the second case a finite element mesh should be introduced. The interaction between blocks, can be, in turn, schematized by single-point or multi-point contacts characterized by normal and tangential stiffness and strength parameters. In general, the higher the complexity of the interaction, the more accurate results are expected, and the more computational efforts are required. As far as solution methods are concerned, explicit solutions are generally preferred to implicit ones since, with such models, large displacements are almost always expected.

Generally, one of the most notable issues in adopting DEMs for simulating the seismic response of complex structures, such as masonry churches, is the requirement of a significant modelling effort. Nonetheless, recent studies have demonstrated that by adopting block sizes larger than the actual masonry units, acceptable results can be obtained until the discretization is representative enough [78,79]. The recent Discrete Macro-Element Method is an extension of DEM proposed by researchers at the University of Catania [80] to increase computational efficiency. With this strategy, the discrete elements, originally referring to the masonry units, are extended to represent entire masonry components (walls and spandrels) having shear deformability and mechanically interacting with the adjacent elements by means of zero-thickness cohesive interfaces. The strong reduction of degrees of freedom compared to equivalent FEM and the adoption of suitable uniaxial cyclic models allow for computational efficiency and remarkable accuracy.

As far as the fields of application are concerned, the detailed level of accuracy EL3 is mainly referred to single churches [60–65], although in some studies it is extended to more church types in order to identify the typological features most influencing the structural response. A notable example in this sense is the work performed by Valente and Milani [3], who studied the seismic responses of seven masonry churches in Ferrara. A similar approach was followed in [81] but with reference to a regional context (Abruzzi region).

4.2.3. Types of Analysis and Typical Issues

Apart from the modelling features outlined above, EL3 assessment can be based on linear or nonlinear analysis and different load types (static or dynamic). Examples of different approaches can be found, among other, in [63,64,82–84]

The most common strategies in this field are static nonlinear analyses (i.e., pushover-analyses). With pushover analyses, structures are subjected to increasing horizontal loads until failure, considering both a mass-proportional and modal-proportional distribution. One of the most critical aspect in such kinds of analysis is the definition of the ultimate displacement, which is assumed when the base shear-control displacement plot exhibits a strength degradation up to 15–20% of the maximum force. However, since the response is mainly governed by out-of-plane and local failures, this definition of collapse is often very difficult to obtain. This holds even truer if simplified post-linear behaviour is assumed, as in the elastic-perfectly plastic material models allowed by the Italian Guidelines. Hence, standard procedures for ordinary buildings consisting of comparing demand and capacity displacements could be difficult to apply. A critical, and not always straightforward, interpretation of the results is thus always required.

5. Critical Discussion and Proposal of a Smart Management Policy
5.1. Inclusion of Strengthening Solutions

The seismic risk mitigation of existing masonry churches is based on an accurate process of vulnerability assessment that, as discussed above, can be performed at different detail levels. On the other hand, vulnerability can be definitively reduced by realizing strengthening interventions, which may be designed by resorting to the most detailed evaluation levels only (i.e., EL2 and EL3). To preserve the architectural, cultural and historical values of existing masonry churches, the choice should fall on interventions guaranteeing the principles of low invasiveness, reversibility and compatibility with existing surfaces.

Each of the analysed levels can, however, account for the effects of strengthening interventions for the vulnerability assessment. In the case of EL0 they can strongly modify the vulnerability parameter, as it can be noted for example in Table 5, where the score related to the presence of effective interventions has the lowest (and thus most influencing) value (i.e., −0.8).

With reference to the Guidelines' method EL1, the presence of antiseismic devices should be accounted for according to their effectiveness, which should be estimated based on expert judgment (see Table 6). An example of efficiency of anti-seismic devices can be made by considering tie rods, whose effectiveness can be estimated according to the anchor plates typology. In particular, it can be assumed that the larger the masonry-plate contact surface, the higher is the effectiveness of the tie rods. Examples of differently effective tie rods are provided in Figure 6.

On the contrary, with EL2 and EL3 the effects of strengthening interventions can be assessed in a closed form, evaluating the improvements in terms of both strength and ductility. Valuable examples of studies aimed at estimating the effects of retrofitting interventions based on limit analysis (EL2) are proposed, among others, in [85] for towers and in [86] for buttressed arches.

With reference to EL3 methodologies, many examples [87–89], just to mention a few, are present in the literature where masonry churches are studied both in ante and post-operam conditions, since this is the most natural and unique way to design global interventions (see Figure 1).

(a)   (b)   (c)

Figure 6. Different anchorages of tie rods for masonry buildings: low (**a**), medium (**b**) and high (**c**) efficiency.

5.2. Pros and Cons of Different Detail Levels

As regards empirical and statistical methods, it can be stated that reliable EL0 models are more suitable than EL1 models for territorial-assessment approaches, given that vulnerability models for estimating a damage scenario by means of vulnerability and fragility functions are based on few typological parameters, and their application does not require accurate in-situ inspections and expert judgment.

In this sense, it should be highlighted that only recently the research community has been definitively moving toward a PGA-based approach, which can be adopted for predictive purposes. These methods require an accurate calibration based on more precise models or statistical interpretation of the observed damage, and they have shown high reliability when homogeneous populations of buildings are investigated. In this sense, it is evident that, together with appropriate exposure and hazard models, they represent a strong tool for management policies since they can provide a sort of priority scale returning the riskiest population of churches. Nonetheless, inconsistencies have been observed among the different proposals of vulnerability and fragility curves present in the literature. This can be related to several factors, such as the several regional, provincial or diocesan contexts analysed, the subjectivity of the users called to assign both damage and vulnerability levels to each considered church, and the absence of a unified and robust methodology for interpreting damage distribution (e.g., binomial, lognormal, beta continuous, etc.).

Hence, the accuracy of the several formulations analysed in this overview are strongly dependent on the cases used for their calibration. A possible solution to this problem could entail PGA-based laws for predicting the damage scenario, which differ to one another in function of the considered context (e.g., regional or diocesan), although this approach suffers from a generalised lack of data. On the whole, the debate concerning the definition of the most reliable model is still open, since it is the authors' opinion that, in the light of above, it is a very complex challenge.

On the other hand, the EL1 methodology provided by the Italian Guidelines can be considered as one of the few attempts for a unified methodology for large-scale applications. Nonetheless, the limits of such a methodology are worth discussing. As a matter of fact, the application of the EL1 method needs a deep knowledge based on expert judgement of the considered churches and thus the accuracy of the outputs (i.e., an empirical estimation of the resisting acceleration) may not be proportionate to the assessment efforts needed for the application of the method. Another aspect is that the empirical laws provided in Italian Guidelines for EL1 analyses have been calibrated on a set of data not including the last seismic events, and thus the knowledge acquired in recent decades in this field. In this sense, a continuous effort of revision and updating is envisaged to increase its accuracy. Another critical aspect is represented by the fact that, according to this empirical method, the vulnerability values obtained in recent applications were too often close to the median value. It is reasonable to think, therefore, that a hybrid strategy could be developed in which extensive parametric EL2 and/or EL3 analyses on different macro-elements could be used to calibrate mechanics-based EL1 strategies. This could provide a more reliable tool based on a simplified mechanical approach rather than on an empirical one, especially in those cases where observation of real damage due to past earthquakes is not available.

With reference to EL2 methodologies, and in particular kinematic limit analysis, one of the most significant issues in their applications is the definition of block subdivisions when the in-plane behaviour of a structural macro-element is investigated. Whereas in curved structures this could be overcome by an iterative procedure entailing all the possible position of hinges [50], in other cases, such as façades and triumphal arches, this can be more complex since the block subdivisions depend not only on hinge positions but also on both direction and path of subdivision lines. Thus, in this sense the implementation of an automatic tool or software able to solve a wide set of general cases by assuming the hypothesis of the limit analysis is an active direction for research [90]. On the other hand, it must be pointed out that, according to Italian Code, the acceleration returned by a linear limit analysis corresponds to damage mechanism activation (i.e., Damage Limit State—DLS). In general, it should be remarked that, in principle, Heyman's assumptions for masonry behaviour should lead to an underestimation of the effective load capacity, since tensile strength is assumed null. However, unconservative results may also be obtained if the underlying hypotheses regarding negligible sliding and infinite compressive strength are not fulfilled. In this context, enhanced limit analysis formulations accounting for more sophisticated material descriptions are worth mentioning [91,92]. Moreover, since the results achievable with kinematic limit analysis represent an upper bound solution, special care should be taken in ensuring that the mechanism under consideration is effectively the one minimising the corresponding load multiplier.

Nonetheless, EL2 analyses can be hardly extended to all the structural components of masonry churches. In this sense, a combined approach consisting of assessment of the most vulnerable elements by means of EL3 global analyses, and subsequent performance of limit analyses on such elements, can represent a good compromise [93].

As far as EL3 methodologies are concerned, several aspects are worth mentioning. Compared to EL2 assessment strategies, EL3 needs a higher level of knowledge of the structure and entails a significant computational demand. This means that applications to scales larger than the structural one are hardly feasible due to either lack of data or insufficient computational resources. Hybrid strategies for the calibration of EL1 models as the ones envisaged above for EL2 methods [94] could, nonetheless, be potentially developed, and this appears as a promising research direction.

Concerning in depth FE analysis of historical masonry churches, several issues are still open. First, the number of different material models developed over the years and implemented in structural computer codes has raised concerns on the objectivity of the results obtained. As a consequence, the assessment of advanced numerical models has been become a topic of increasing interest in the scientific community [95,96]. Acknowledging the different hypotheses at the base of the models, and the selection of material parameters should be considered model-dependent [97], and appropriate tests should be carried out to estimate them. This also brings additional problems. Since invasive testing is not possible in case of buildings of artistic value, non-destructive testing such as dynamic testing is often the only possibility to achieve deeper knowledge on material characterisation [98–100]. In principle this prevents the designer from obtaining parameters other than elastic. Some recent studies tried to capture other aspects, such as degradation of the material, and calibrate nonlinear parameters [101]. To date, however, the selection of post-elastic behaviour for masonry (historical, in particular) is still an open issue that is not standardised in building codes. This is even more relevant since both Italian and European codes implicitly push toward the adoption of the equivalent frame method, which, as discussed above, is not reliable for masonry churches. Hence, also in this case, standardised rules would be of great help for practitioners.

The above critical discussion about the different detail levels of seismic vulnerability assessment of masonry churches addressed in this paper is summarized in Table 7.

5.3. Proposal of a Smart Management Policy

The vulnerability assessment strategies outlined in this review seem to rationally suggest that suitable policies for historical church management should be based on a multilevel methodology, following a qualitative flowchart as shown in Figure 7. A preliminary low-level screening of the wide church heritage is necessary to adopt preventive, rather than reparative, approaches. In this sense, identifying the riskiest dioceses and municipalities by means of EL0 methods is a fundamental step to move from a territorial to a building scale approach. Then, empirical or simplified-mechanical approaches (i.e., EL1 and EL2, respectively) can be followed to define, among homogeneous populations of churches, those requiring urgent retrofitting interventions, whose design must be based on an accurate study based on EL3 assessment.

Table 7. Pros and cons of different detail level of assessment.

Evaluation Level (EL)	Pros	Cons
EL0	• Based on few typological parameters. • Does not require accurate in-situ inspections and expert judgment.	• Requires an accurate calibration. • Reliable for homogeneous populations of buildings only. • Strongly case-dependent.
EL1	• Closed-form methodology. • Applicable at the building scale.	• Outmoded formulations (Italian Guidelines approach). • Requires accurate in-situ inspections and expert judgment. • Vulnerability values often close to the median value.
EL2	• Applicable at both single and territorial scales. • Does not require mechanical parameters of masonry. • Possibility to design local retrofitting interventions.	• Difficulty to be applied to general geometries. • Computing demand when applied to all damage mechanisms of single buildings. • Lack of accuracy in the cases when Heyman's hypotheses are not fulfilled
EL3	• Possibility to perform accurate assessment. • Possibility to design both local and local retrofitting interventions. • Possibility to calibrate the models based on dynamical characterizations.	• Uncertainties of mechanical parameters. • Difficulty in interpreting out-of-plane response (FEM). • Computational and modelling costs in case of sophisticated modelling.

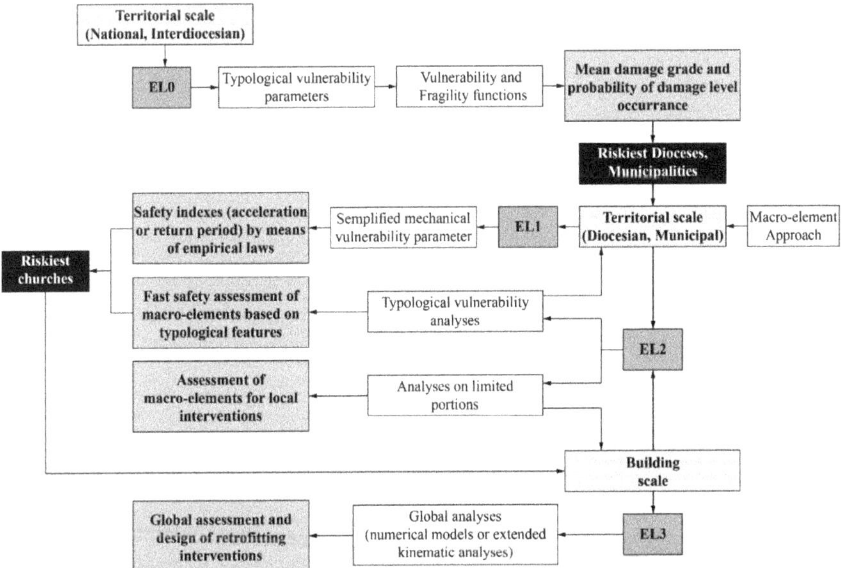

Figure 7. Potential management policy for a smart decision-making process of masonry church assets.

6. Conclusions

In this paper, a wide overview of the methods concerning the seismic assessment of historical masonry churches at different levels of detail has been provided. Of course, each level allows for pursuing specific objectives that range from risk priority scale to design of retrofitting interventions. The study addresses several issues with reference to each detail level of assessment and proposes possible future research directions, generally suggesting a stronger interaction between different methodologies and the definition of hybrid strategies for calibration.

Based on the study and the main outputs, a multilevel procedure for a smart management policy of existing masonry churches has been proposed. Within this, the most simplified assessment procedures EL0 and EL1 are used to rank built assets based on their vulnerability and highlight the need for more accurate investigation and analysis to be performed at EL2 and EL3.

Author Contributions: Conceptualization, J.R. and G.D.M.; methodology, M.Z. and C.C.; investigation, M.Z. and J.R.; data curation, J.R. and M.Z.; writing—original draft preparation, J.R. and M.Z.; writing—review and editing, M.Z., C.C., D.C. and G.D.M.; supervision, C.C. and G.D.M. All authors have read and agreed to the published version of the manuscript.

Funding: Not applicable.

Institutional Review Board Statement: Not applicable.

Data Availability Statement: No new data were created or analyzed in this study. Data sharing is not applicable to this article.

Acknowledgments: This study has been developed within the national project, promoted by the agreement between the Italian consortium ReLUIS and the Department of Civil Protection (DPC), "Seismic Hazard Maps and Damage Scenario (MARS)". C.C. is funded by MUR (Ministry of University and Research) through PON FSE 2014-2020 program (project AIM1879349-2).

Conflicts of Interest: The authors declare no conflict of interest.

References

1. Despotaki, V.; Silva, V.; Lagomarsino, S.; Pavlova, I.; Torres, J. Evaluation of Seismic Risk on UNESCO Cultural Heritage sites in Europe. *Int. J. Archit. Herit.* **2018**, *12*, 1231–1244. [CrossRef]
2. Lagomarsino, S. On the vulnerability assessment of monumental buildings. *Bull. Earthq. Eng.* **2006**, *4*, 445–463. [CrossRef]
3. Valente, M.; Milani, G. Seismic response and damage patterns of masonry churches: Seven case studies in Ferrara, Italy. *Eng. Struct.* **2018**, *177*, 809–835. [CrossRef]
4. dei Ministri, P.D.C. *Valutazione e Riduzione del Rischio Sismico del Patrimonio Culturale con Riferimento alle Norme Tecniche per le Costruzioni di Cui al D.M. 14/01/2008*; G.U. No. 47; Ministero per i Beni e le Attività Culturali: Rome, Italy, 2011. (In Italian)
5. Burton, H.V.; Deierlein, G.; Lallemant, D.; Lin, T. Framework for Incorporating Probabilistic Building Performance in the Assessment of Community Seismic Resilience. *J. Struct. Eng.* **2016**, *142*, C4015007. [CrossRef]
6. Vona, M.; Mastroberti, M.; Mitidieri, L.; Tataranna, S. New resilience model of communities based on numerical evaluation and observed post seismic reconstruction process. *Int. J. Disaster Risk Reduct.* **2018**, *28*, 602–609. [CrossRef]
7. Lagomarsino, S.; Podestà, S. Seismic Vulnerability of Ancient Churches: I. Damage Assessment and Emergency Planning. *Earthq. Spectra* **2004**, *20*, S377–S394. [CrossRef]
8. Lagomarsino, S.; Podestà, S. Damage and Vulnerability Assessment of Churches after the 2002 Molise, Italy, Earthquake. *Earthq. Spectra* **2004**, *20*, S271–S283. [CrossRef]
9. D'Ayala, D.; Paganoni, S. Assessment and analysis of damage in L'Aquila historic city centre after 6th April 2009. *Bull. Earthq. Eng.* **2011**, *9*, 81–104. [CrossRef]
10. Lagomarsino, S. Damage assessment of churches after L'Aquila earthquake. *Bull. Earthq. Eng.* **2012**, *10*, 73–92. [CrossRef]
11. Da Porto, F.; Silva, B.; Costa, C.; Modena, C. Macro-scale analysis of damage to churches after earthquake in Abruzzo (Italy) on April 6, 2009. *J. Earthq. Eng.* **2012**, *16*, 739–758. [CrossRef]
12. De Matteis, G.; Criber, E.; Brando, G. Damage Probability Matrices for Three- Nave Masonry Churches in Abruzzi After the 2009 L'Aquila Earthquake. *Int. J. Archit. Herit.* **2016**, *10*, 120–145.
13. Sorrentino, L.; Liberatore, L.; Decanini, L.D.; Liberatore, D. The performance of churches in the 2012 Emilia earthquakes. *Bull. Earthq. Eng.* **2014**, *12*, 2299–2331. [CrossRef]
14. Cescatti, E.; Salzano, P.; Casapulla, C.; Ceroni, F.; Da Porto, F.; Prota, A. Damages to masonry churches after 2016–2017 Central Italy seismic sequence and definition of fragility curves. *Bull. Earthq. Eng.* **2020**, *18*, 297–329. [CrossRef]

15. Hofer, L.; Zampieri, P.; Zanini, M.A.; Faleschini, F.; Pellegrino, C. Seismic damage survey and empirical fragility curves for churches after the August 24, 2016 Central Italy earthquake. *Soil Dyn. Earthq. Eng.* **2018**, *111*, 98–109. [CrossRef]
16. Penna, A.; Calderini, C.; Sorrentino, L.; Carocci, C.F.; Cescatti, E.; Sisti, R.; Borri, A.; Modena, C.; Prota, A. Damage to churches in the 2016 central Italy earthquakes. *Bull. Earthq. Eng.* **2019**, *17*, 5763–5790. [CrossRef]
17. De Matteis, G.; Zizi, M. Seismic Damage Prediction of Masonry Churches by a PGA-based Approach. *Int. J. Archit. Herit.* **2019**, *13*, 1165–1179. [CrossRef]
18. Salzano, P.; Casapulla, C.; Ceroni, F.; Prota, A. Seismic Vulnerability and Simplified Safety Assessments of Masonry Churches in the Ischia Island (Italy) after the 2017 Earthquake. *Int. J. Archit. Herit.* **2020**, 1–27. [CrossRef]
19. Claudia, N.; Cancino, C. Damage assessment of historic earthen sites after the 2007 earthquake in Peru. *Adv. Mater. Res.* **2010**, *133–134*, 665–670. [CrossRef]
20. D'Ayala, D.; Benzoni, G. Historic and traditional structures during the 2010 Chile Earthquake: Observations, codes, and conservation strategies. *Earthq. Spectra* **2012**, *28*, S425–S451. [CrossRef]
21. Palazzi, N.C.; Favier, P.; Rovero, L.; Sandoval, C.; De La Llera, J.C. Seismic damage and fragility assessment of ancient masonry churches located in central Chile. *Bull. Earthq. Eng.* **2020**, *18*, 3433–3457. [CrossRef]
22. Leite, J.; Lourenco, P.B.; Ingham, J.M. Statistical assessment of damage to churches affected by the 2010–2011 Canterbury (New Zealand) earthquake sequence. *J. Earthq. Eng.* **2013**, *17*, 73–97. [CrossRef]
23. Marotta, A.; Sorrentino, L.; Liberatore, D.; Ingham, J.M. Vulnerability Assessment of Unreinforced Masonry Churches Following the 2010-2011 Canterbury Earthquake Sequence. *J. Earthq. Eng.* **2017**, *21*, 912–934. [CrossRef]
24. Peña, F.; Chávez, M.M. Seismic behavior of Mexican colonial churches. *Int. J. Archit. Herit.* **2016**, *10*, 332–345. [CrossRef]
25. Giuffrè, A. *Restauro e Sicurezza in Zona Sismica. La Cattedrale di S. Angelo dei Lombardi*; Nuova serie 1; Palladio: Rome, Italy, 1998; pp. 95–120. (In Italian)
26. Grünthal, G.; Musson, R.M.W.; Schwarz, J.; Stucchi, M. (Eds.) European Macroseismic Scale 1998. (EMS-98). In *Cahiers du Centre Européen de Géodynamique et de Séismologie*; Centre Européen de Géodynamique et de Séismologie: Luxembourg, 1998; volume 15, p. 99. [CrossRef]
27. D'Amato, M.; Laterza, M.; Diaz Fuentes, D. Simplified Seismic Analyses of Ancient Churches in Matera's Landscape. *Int. J. Archit. Herit.* **2020**, *14*, 119–138. [CrossRef]
28. Braga, F.; Dolce, M.; Liberatore, D. A Statistical Study on Damaged Buildings and an Ensuing Review of the MSK-76 Scale. In Proceedings of the Seventh European Conference on Earthquake Engineering, Athens, Greece, 20–25 September 1982; pp. 431–450.
29. Lagomarsino, S.; Podestà, S. Seismic Vulnerability of Ancient Churches: II. Statistical Analysis of Surveyed Data and Methods for Risk Analysis. *Earthq. Spectra* **2004**, *20*, S395–S412. [CrossRef]
30. Bernardini, A.; Lagomarsino, S. The seismic vulnerability of architectural heritage. *Proc. Inst. Civ. Eng. Struct. Build.* **2008**, *161*, 171–181. [CrossRef]
31. Canuti, C.; Carbonari, S.; Dall'Asta, A.; Dezi, L.; Gara, F.; Leoni, G.; Morici, M.; Petrucci, E.; Prota, A.; Zona, A. Post-Earthquake Damage and Vulnerability Assessment of Churches in the Marche Region Struck by the 2016 Central Italy Seismic Sequence. *Int. J. Archit. Herit.* **2021**, *15*, 1000–1021. [CrossRef]
32. De Matteis, G.; Brando, G.; Corlito, V. Predictive model for seismic vulnerability assessment of churches based on the 2009 L'Aquila earthquake. *Bull. Earthq. Eng.* **2019**, *17*, 4909–4936. [CrossRef]
33. Lagomarsino, S.; Giovinazzi, S. Macroseismic and mechanical models for the vulnerability and damage assessment of current buildings. *Bull. Earthq. Eng.* **2006**, *4*, 415–443. [CrossRef]
34. Sandi, H.; Floricel, I. Analysis of seismic risk affecting the existing building stock. In Proceedings of the 10th European Conference on Earthquake Engineering, Vienna, Austria, 28 August–2 September 1994; A.A. Balkema: Rotterdam, The Netherlands, 1994; Volume 2, pp. 1105–1110.
35. Brandonisio, G.; Lucibello, G.; Mele, E.; De Luca, A. Damage and performance evaluation of masonry churches in the 2009 L'Aquila earthquake. *Eng. Fail. Anal.* **2013**, *34*, 693–714. [CrossRef]
36. Lagomarsino, S.; Cattari, S.; Ottonelli, D.; Giovinazzi, S. Earthquake damage assessment of masonry churches: Proposal for rapid and detailed forms and derivation of empirical vulnerability curves. *Bull. Earthq. Eng.* **2019**, *17*, 3327–3364. [CrossRef]
37. Faenza, L.; Michelini, A. Regression analysis of MCS intensity and ground motion parameters in Italy and its application in ShakeMap. *Geophys. J. Int.* **2010**, *180*, 1138–1152. [CrossRef]
38. Masi, A.; Lagomarsino, S.; Dolce, M.; Manfredi, V.; Ottonelli, D. Towards the updated Italian seismic risk assessment: Exposure and vulnerability modelling. *Bull. Earthq. Eng.* **2021**, *19*, 3253–3286. [CrossRef]
39. Mouroux, P.; Le Brun, B. Presentation of RISK-UE Project. *Bull. Earthq. Eng.* **2006**, *4*, 323–339. [CrossRef]
40. Goded, T.; Lewis, A.; Stirling, M. Seismic vulnerability scenarios of Unreinforced Masonry churches in New Zeland. *Bull. Earthq. Eng.* **2018**, *16*, 3957–3999. [CrossRef]
41. National Group for Earthquakes Defense (GNDT). *First and Second Level Form for Exposure, Vulnerability and Damage Survey (Masonry and Reinforced Concrete)*; GNDT: Rome, Italy, 1994.
42. Lourenço, P.B.; Roque, J.A. Simplified indexes for the seismic vulnerability of ancient masonry buildings. *Constr. Build. Mater.* **2006**, *20*, 200–208. [CrossRef]
43. Palazzi, N.C.; Rovero, L.; De La Llera, J.C.; Sandoval, C. Preliminary assessment on seismic vulnerability of masonry churches in central Chile. *Int. J. Archit. Herit.* **2019**, *14*, 829–848. [CrossRef]

44. Heyman, J. *The Masonry Arch*; Ellis Horwood Ltd.: Chichester, UK, 1982.

45. De La Hire, P. Sur la construction des voûtes dans les édifices. In *Mémoires de l'Académie Royale des Sciences de Paris*; Académie Royale des Sciences: Paris, France, 1712. (In French)

46. Mèry, E. Sur l'equilibre des voutes en berceu. In *Annales de Ponts et Chaussess*; PYC Édition: Paris, France, 1840. (In French)

47. Criber, E.; Brando, G.; De Matteis, G. The effects of L'Aquila earthquake on the St. Gemma church in Goriano Sicoli: Part I: Damage survey and kinematic analysis. *Bull. Earthq. Eng.* **2015**, *13*, 3713–3732. [CrossRef]

48. Cavalagli, N.; Gusella, V.; Severini, L. Lateral loads carrying capacity and minimum thickness of circular and pointed masonry arches. *Int. J. Mech. Sci.* **2016**, *115–116*, 645–656. [CrossRef]

49. Brandonisio, G.; Mele, E.; De Luca, A. Limit analysis of masonry circular buttressed arches under horizontal loads. *Meccanica* **2017**, *52*, 2547–2565. [CrossRef]

50. Zizi, M.; Cacace, D.; Rouhi, J.; Lourenço, P.B.; De Matteis, G. Automatic procedures for the safety assessment of stand-alone masonry arches. *Int. J. Archit. Herit.* **2021**, 1–19. [CrossRef]

51. Lourenço, P.B. Computational Strategies for Masonry Structures. Ph.D. Thesis, University of Porto, Porto, Portugal, 1996.

52. Roca, P.; Molins, C.; Mari, A.R. Strength capacity of masonry wall structures by the equivalent frame method. *J. Struct. Eng.* **2005**, *131*, 1601–1610. [CrossRef]

53. Lagomarsino, S.; Penna, A.; Galasco, A.; Cattari, S. TREMURI program: An equivalent frame model for the nonlinear seismic analysis of masonry buildings. *Eng. Struct.* **2013**, *56*, 1787–1799. [CrossRef]

54. Dejong, M.J.; Beletti, B.; Hendriks, M.A.N.; Rots, J.G. Shell elements for sequentially linear analysis: Lateral failure of masonry structures. *Eng. Struct.* **2009**, *31*, 1382–1392. [CrossRef]

55. Noor-E-Khuda, S.; Dhanasekar, M.; Thambiratnam, D.P. An explicit finite element modelling method for masonry walls under out-of-plane loading. *Eng. Struct.* **2016**, *113*, 103–120. [CrossRef]

56. Casolo, S.; Milani, G. Semplified out-of-plane modelling of three-leaf masonry walls accounting for the material texture. *Constr. Build. Mater.* **2013**, *40*, 330–351. [CrossRef]

57. Betti, M.; Galano, L.; Vignoli, A. Finite Element Modelling for Seismic Assessment of Historic Masonry Buildings. In *Earthquakes and Their Impact on Society*; D'Amico, S., Ed.; Springer: Cham, Switzerland, 2013; pp. 377–415.

58. Borst, R.; Remmers, J.; Needleman, A.; Abellan, M.A. Discrete vs smeared crack models for concrete fracture: Bridging the gap. *Int. J. Numer. Anal. Methods Geomech.* **2004**, *28*, 583–607. [CrossRef]

59. Alfarah, B.; López-Almansa, F.; Oller, S. New methodology for calculating damage variables evolution in Plastic Damage Model for RC structures. *Eng. Struct.* **2017**, *132*, 70–86. [CrossRef]

60. Betti, M.; Vignoli, A. Numerical assessment of the static and seismic behaviour of the basilica of Santa Maria all'Impruneta (Italy). *Constr. Build. Mater.* **2011**, *25*, 4308–4324. [CrossRef]

61. Formisano, A.; Vaiano, G.; Fabbrocino, F.; Milani, G. Seismic vulnerability of Italian masonry churches: The case of the Nativity of Blessed Virgin Mary in Stellata of Bondeno. *J. Build. Eng.* **2018**, *20*, 179–200. [CrossRef]

62. Grazzini, A.; Chiarabrando, F.; Foti, S.; Sammartano, G.; Spanò, A. Multidisciplinary Study on the Seismic Vulnerability of St. Agostino Church in Amatrice following the 2016 Seismic Sequence. *Int. J. Archit. Herit.* **2020**, *14*, 885–902. [CrossRef]

63. De Matteis, G.; Mazzolani, F.M. The Fossanova Church: Seismic Vulnerability Assessment by Numeric and Physical Testing. *Int. J. Archit. Herit.* **2010**, *4*, 222–245. [CrossRef]

64. Kujawa, M.; Lubowiecka, I.; Szymczak, C. Finite element modelling of a historic church structure in the context of a masonry damage analysis. *Eng. Fail. Anal.* **2020**, *107*, 104233. [CrossRef]

65. Karanikoloudis, G.; Lourenço, P.B. Structural assessment and seismic vulnerability of earthen historic structures. Application of sophisticated numerical and simple analytical models. *Eng. Struct.* **2018**, *160*, 488–509. [CrossRef]

66. Alforno, M.; Monaco, A.; Venuti, F.; Calderini, C. Validation of simplified micro-models for the static analysis of masonry arches and vaults. *Int. J. Archit. Her.* **2021**, *15*, 1196–1212. [CrossRef]

67. Gaetani, A.; Bianchini, N.; Lourenço, P.B. Simplified micro-modelling of masonry cross vaults: Stereotomy and interface issues. *Int. J. Mason. Res. Innov.* **2021**, *6*, 97–125. [CrossRef]

68. Milani, G.; Valente, M. Comparative pushover and limit analyses on seven masonry churches damaged by the 2012 Emilia-Romagna (Italy) seismic events: Possibilities of non-linear finite elements compared with pre-assigned failure mechanisms. *Eng. Fail. Anal.* **2015**, *47*, 129–161. [CrossRef]

69. Betti, M.; Vignoli, A. Modelling analysis of a Romanesque church under earthquake loading: Assessment of seismic resistance. *Eng. Struct.* **2008**, *30*, 352–367. [CrossRef]

70. Addessi, D.; Marfia, S.; Sacco, E. A plastic nonlocal damage model. *Comput. Methods Appl. Mech. Eng.* **2002**, *19*, 1291–1310. [CrossRef]

71. Trovalusci, P.; Masiani, R. Non-linear micropolar and classical continua for anisotropic discontinuous materials. *Int. J. Solids Struct.* **2003**, *40*, 1281–1297. [CrossRef]

72. Petracca, M.; Pelà, L.; Rossi, R.; Oller, S.; Camata, G.; Spacone, E. Regularization of first order computational homogeni-zation for multiscale analysis of masonry structures. *Comput. Mech.* **2016**, *57*, 257–276. [CrossRef]

73. Cundall, P.A.; Strack, O.D.L. A discrete numerical model for granular assemblies. *Géotechnique* **1979**, *29*, 47–65. [CrossRef]

74. Pulatsu, B.; Ercdogmus, E.; Lourenço, P.B.; Lemos, V.; Tuncay, K. Simulation of the in-plane structural behavior of unreinforced masonry walls and buildings using DEM. *Structures* **2020**, *27*, 2274–2287. [CrossRef]

75. Foti, D.; Vacca, V.; Facchini, I. DEM modeling and experimental analysis of the static behavior of a dry-joints masonry cross vaults. *Constr. Build. Mater.* **2018**, *170*, 111–120. [CrossRef]
76. Mclnerney, J.; Dejong, M.J. Discrete Element Modeling of Groin Vault Displacement Capacity. *Int. J. Archit. Her.* **2015**, *9*, 1037–1049. [CrossRef]
77. Lemos, J.V. Discrete Element Modeling of the Seismic Behavior of Masonry Construction. *Buildings* **2019**, *9*, 43. [CrossRef]
78. Gonen, S.; Pulatsu, B.; Erdogmus, E.; Karaesmen, E.; Karaesmen, E. Quasi-Static Nonlinear Seismic Assessment of a Fourth Century, A.D. Roman Aqueduct in Istanbul, Turkey. *Heritage* **2021**, *4*, 401–421. [CrossRef]
79. Ferrante, A.; Loverdos, D.; Clementi, F.; Milani, G.; Formisano, A.; Lenci, S.; Sarhosis, V. Discontinuous approaches for nonlinear dynamic analyses of an ancient masonry tower. *Eng. Struct.* **2021**, *230*, 111626. [CrossRef]
80. Pantò, B.; Cannizzaro, F.; Caddemi, S.; Caliò, I. 3D macro-element modelling approach for seismic assessment of historical masonry churches. *Adv. Eng. Softw.* **2016**, *97*, 40–59. [CrossRef]
81. Zizi, M.; Corlito, V.; Lourenço, P.B.; De Matteis, G. Seismic vulnerability of masonry churches in Abruzzi region, Italy. *Structures* **2021**, *32*, 662–680. [CrossRef]
82. Yacila, J.; Camata, G.; Salsavilca, J.; Tarque, N. Pushover analysis of confined masonry walls using a 3D macro-modelling approach. *Eng. Struct.* **2019**, *201*, 109731. [CrossRef]
83. Jain, A.; Acito, M.; Chesi, C. Seismic sequence of 2016–17: Linear and non-linear interpretation models for evolution of damage in San Francesco church, Amatrice. *Eng. Struct.* **2020**, *211*, 110418. [CrossRef]
84. Papa, G.S.; Tateo, V.; Parisi, M.A.; Casolo, S. Seismic response of a masonry church in Central Italy: The role of interventions on the roof. *Bull. Earthq. Eng.* **2021**, *19*, 1151–1179. [CrossRef]
85. Milani, G.; Shehu, R.; Valente, M. A kinematic limit analysis approach for seismic retrofitting of masonry towers through steel tie-rods. *Eng. Struct.* **2018**, *160*, 212–228. [CrossRef]
86. Chisari, C.; Cacace, D.; De Matteis, G. Parametric Investigation on the Effectiveness of FRM-Retrofitting in Masonry Buttressed Arches. *Buildings* **2021**, *11*, 406. [CrossRef]
87. Lourenço, P.B.; Ciocci, M.P.; Greco, F.; Karanikoloudis, G.; Cancino, C.; Torrealva, D.; Wong, K. Traditional techniques for the rehabilitation and protection of historic earthen structures: The seismic retrofitting project. *Int. J. Archit. Her.* **2019**, *13*, 15–32. [CrossRef]
88. Milani, G.; Shehu, R.; Valente, M. Possibilities and limitations of innovative retrofitting for masonry churches: Advanced computations on three case studies. *Constr. Build. Mater.* **2017**, *147*, 239–263. [CrossRef]
89. Angjeliu, G.; Coronelli, D.; Cardani, G.; Boothby, T. Structural assessment of iron tie rods based on numerical modelling and experimental observations in Milan Cathedral. *Eng. Struct.* **2020**, *206*, 109690. [CrossRef]
90. Chiozzi., A.; Grillanda, N.; Milani, G.; Tralli, A. UB-ALMANAC: An adaptive limit analysis NURBS-based program for the automatic assessment of partial failure mechanisms in masonry churches. *Eng. Fail. Anal.* **2018**, *85*, 201–220. [CrossRef]
91. Portioli, F.; Casapulla, C.; Gilbert, M.; Cascini, L. Limit analysis of 3D masonry block structures with non-associative frictional joints using cone programming. *Comput. Struct.* **2014**, *143*, 108–121. [CrossRef]
92. Li, H.X.; Yu, H.S. Kinematic limit analysis of frictional materials using nonlinear programming. *Int. J. Solids Struct.* **2005**, *42*, 4058–4076. [CrossRef]
93. Funari, M.F.; Spadea, S.; Lonetti, P.; Fabbrocino, F.; Luciano, R. Visual programming for structural assessment of out-of-plane mechanisms in historic masonry structures. *J. Build. Eng.* **2020**, *31*, 101425. [CrossRef]
94. Shakya, M.; Varum, H.; Vicente, R.; Costa, A. Seismic vulnerability assessment methodology for slender masonry structures. *Int. J. Archit. Herit.* **2018**, *12*, 1297–1326. [CrossRef]
95. Cattari, S.; Magenes, G. Benchmarking the software packages to model and assess the seismic response of unreinforced masonry existing buildings through nonlinear static analyses. *Bull. Earthq. Eng.* **2021**. [CrossRef]
96. Zizi, M.; Chisari, C.; Rouhi, J.; De Matteis, G. Comparative analysis on macroscale material models for the prediction of masonry in-plane behavior. *Bull. Earth. Eng.* **2021**, 1–34. [CrossRef]
97. Chisari, C.; Macorini, L.; Amadio, C.; Izzuddin, B.A. Identification of mesoscale model parameters for brick-masonry. *Int. J. Solids Struct.* **2018**, *146*, 224–240. [CrossRef]
98. Gentile, C.; Saisi, A.; Cabboi, A. Structural identification of a masonry tower based on operational modal analysis. *Int. J. Archit. Herit.* **2015**, *9*, 98–110. [CrossRef]
99. Ramos, L.F.; Marques, L.; Lourenço, P.B.; De Roeck, G.; Campos-Costa, A.; Roque, J. Monitoring historical masonry structures with operational modal analysis: Two case studies. *Mech. Syst. Signal Process.* **2010**, *24*, 1291–1305. [CrossRef]
100. Aloisio, A.; Di Pasquale, A.; Alaggio, R.; Fragiacomo, M. Assessment of Seismic Retrofitting Interventions of a Masonry Palace Using Operational Modal Analysis. *Int. J. Archit. Herit.* **2020**, 1–13. [CrossRef]
101. Pellegrini, D.; Girardi, M.; Lourenço, P.B.; Masciotta, M.G.; Mendes, N.; Padovani, C.; Ramos, L.F. Modal analysis of historical masonry structures: Linear perturbation and software benchmarking. *Constr. Build. Mater.* **2018**, *189*, 1232–1250. [CrossRef]

 *buildings*

Article

Assessment and Rehabilitation of Culturally Protected Prince Rudolf Infantry Barracks in Zagreb after Major Earthquake

Mija Milić [1], Mislav Stepinac [2,*], Luka Lulić [2], Nataša Ivanišević [3], Ivan Matorić [1], Boja Čačić Šipoš [1] and Yohei Endo [4]

1   Peoples Associates Structural Engineers—Zagreb Office, 10000 Zagreb, Croatia; mija.milic@pase.com (M.M.); ivan.matoric@pase.com (I.M.); boja.cacic.sipos@pase.com (B.Č.Š.)
2   Faculty of Civil Engineering, University of Zagreb, 10000 Zagreb, Croatia; luka.lulic@grad.unizg.hr
3   A&A Architects, 20000 Dubrovnik, Croatia; natasa.ivanisevic@gmail.com
4   Department of Architecture, Shinshu University, 4-17-1 Wakasato, Nagano 380-8553, Japan; endii@shinshu-u.ac.jp
*   Correspondence: mislav.stepinac@grad.unizg.hr

Abstract: Recently, Zagreb was struck by a strong earthquake. Damage throughout the city was tremendous due to numerous aged and vulnerable masonry buildings. Many damaged buildings are under a certain level of cultural heritage protection. Hence, reliable assessment and effective rehabilitation are important to preserve cultural significance and mitigate risk for human life. With that in mind, the procedure of a detailed condition assessment of the building under heritage protection is presented. A detailed historical background of the case study building is shown, and observed damage and conducted in situ tests are discussed. The nonlinear static seismic analysis performed in the 3Muri software is extensively elaborated. Four different levels of reconstruction according to new Croatian law are briefly presented. Additionally, several strengthening scenarios are proposed with various strengthening techniques.

Keywords: earthquake; cultural heritage; nonlinear analysis; existing structures; masonry; flat-jack; strengthening

check for
updates

Citation: Milić, M.; Stepinac, M.; Lulić, L.; Ivanišević, N.; Matorić, I.; Čačić Šipoš, B.; Endo, Y. Assessment and Rehabilitation of Culturally Protected Prince Rudolf Infantry Barracks in Zagreb after Major Earthquake. *Buildings* **2021**, *11*, 508. https://doi.org/10.3390/buildings11110508

Academic Editor: Rita Bento

Received: 21 September 2021
Accepted: 25 October 2021
Published: 27 October 2021

Publisher's Note: MDPI stays neutral with regard to jurisdictional claims in published maps and institutional affiliations.

1. Introduction

In March 2020, in the early morning hours, a strong earthquake occurred in Zagreb with a magnitude of $M_L = 5.5$ and an intensity of VII on the Mercalli scale. The earthquake's epicenter was located 10 km from Zagreb, with a hypocenter at a depth of about 10 km. The quake was felt throughout Croatia and in neighboring countries. In addition to great material damage, the earthquake took one young human life. Shortly after the main quake, a series of aftershocks followed. The quake was unexpected for the population, and the disaster response system was unprepared. Based on citizens' reports, civil engineers inspected the facilities according to a pre-established methodology (EMS-98) and issued recommendations to citizens on the usability of about 26,000 facilities. The World Bank estimates the total financial damage from the Zagreb earthquake as EUR 11.3 billion [1]. Moderate to severe structural damage was sustained by 118 buildings, and heavy structural damage was reported in 41 buildings under heritage protection. The total damage to cultural heritage buildings is about EUR 1.38 billion, most of which was incurred in the city of Zagreb.

Most of the buildings in the center of Zagreb are traditional masonry buildings that are not designed for seismic actions. Such buildings were mostly constructed as interconnected load-bearing masonry walls with wooden floor structures [2]. Damage to such buildings occurs due to uneven stiffness distribution, inappropriate or nonexistent connections between the walls and poor connection to the roof and floor structure. An additional disadvantage is the absence of vertical and horizontal confining elements (e.g., reinforced concrete columns and beams on all corners and wall intersections as it is required today for this type of building and for such high seismic demand according to European seismic regulations), poor load-bearing capacity in its plane and insufficient load-bearing capacity of roof and floor structures [3]. Furthermore, most of the buildings in Zagreb are very old, so the degradation of mechanical properties should be considered. Commonly observed damage was: collapse and damage of chimneys, collapse and damage of attic gable walls, separation of gable walls, damage to the roof, damage to the cantilever elements, damage of the walls (out of and in the plane), damage to lintels and vaults, damage to partition walls, cracks in ceilings and damage to stairs [4]. More information about the earthquake itself, the level of preparedness and immediate actions and, finally, the consequences of the Zagreb earthquake can be found in [5–8].

After a strong earthquake, buildings should go through a well-established assessment process. A key part of this assessment should be the high-precision evaluation of the mechanical properties of masonry. This will reduce the number of unknowns related to the structure's resistance [4]. Technology development facilitates improvements in the field of assessment methods, which then allow a more adequate, economic and safer assessment of existing buildings. A lot of research on this topic has been conducted in different parts of the world. Procedures, methods and norms related to the assessment of existing structures are constantly being improved [9–14]. Special attention is also paid to cultural heritage buildings that represent the identity of historic urban cores [15–21].

New methods such as drone imaging and laser scanning could ease and complement the regular assessment process. Unmanned aerial vehicles (UAVs) can be used for crisis management, crack identification, seismic damage, architectural assessment of cultural heritage and structural assessment of buildings. Laser scanning of structures (light detection and ranging (LiDAR)) is used to scan structures damaged by earthquakes to identify cracks and ways of failure of the element and the entire structure [22]. Additionally, the benefit of these two techniques is particularly visible when inspecting heritage buildings. Digital twins can be produced to preserve the state of the building and for its reconstruction (if needed). Many scientific articles and studies have been published regarding the modeling of the behavior of existing structures and their reconstruction. The reader is referred to the following articles related to the reconstruction of cultural heritage buildings. Case studies like the one described in this manuscript can be found in [2,18,23–27].

This paper presents the procedure of a detailed inspection of a building under cultural heritage protection. The case study building was damaged in the earthquake and needs to be renovated according to new laws in Croatia to ensure the safe and functional future use of the building.

2. Case Study of Rudolf's Barracks

2.1. Historical Background

The case study building is located within the historic complex of buildings in the western part of Zagreb 'Lower Town' called the infantry barracks of Prince Rudolf. The entire complex of Rudolf's barracks is protected as an immovable individual cultural property and is entered in the Register of Cultural Heritage of the Republic of Croatia. The protection of the complex refers to the main building and the entire area of the former pedestrian barracks complex with the existing quality greenery, unbuilt areas and peripheral buildings of high ambient values. The Rudolf's barracks complex is located within the A protection zone of the Historical and Urban Entity of the city of Zagreb, protected as a cultural asset and entered in the Register of Cultural Heritage of the Republic of Croatia—List of Protected Cultural Heritage.

The infantry barracks complex was constructed in the period from 1887 to 1889 according to the project of the Viennese architects Franz Gruber and Carl Völckner. The complex consisted of 13 buildings (Figures 1 and 2), most of which were two-story buildings, and was named after the son of Emperor Francis Joseph I and Empress Sisi, Prince Rudolf [28,29]. The entire complex was built within 15 months of Prince Rudolf laying the foundation stone, and the construction of the complex was triggered by tensions over the Austro-Hungarian occupation of Bosnia and Herzegovina and the need to house the army.

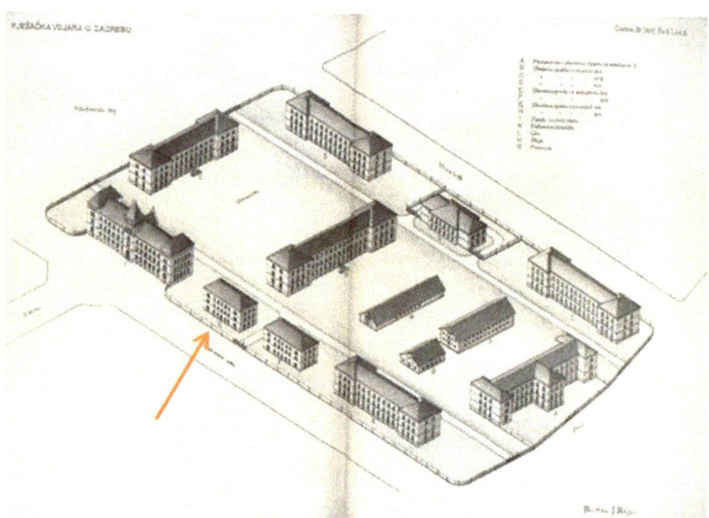

Figure 1. Rudolf's barracks complex (archive sketch) with marked case study building.

(**a**)

(**b**)

(**c**)

Figure 2. (**a**) Postcard from 1898 with a view of Rudolf's barracks [30]; (**b**) Postcard from the beginning of the 20th century with case study building visible on the right [30]; (**c**) Demolition of part of the Rudolf's barracks complex in the late 1970s.

In the late 1970s, a decision was made to demolish Rudolf's barracks (Figure 2c) to make the area a secondary city center, but it was converted into a park without new constructions. Part of the complex was demolished in 1978, and what is left are four buildings, the main representative and three more modest buildings, all built in the neo-romantic style. One of these buildings is the case study of this paper located at Republic Austria Street No. 18 (Figure 3).

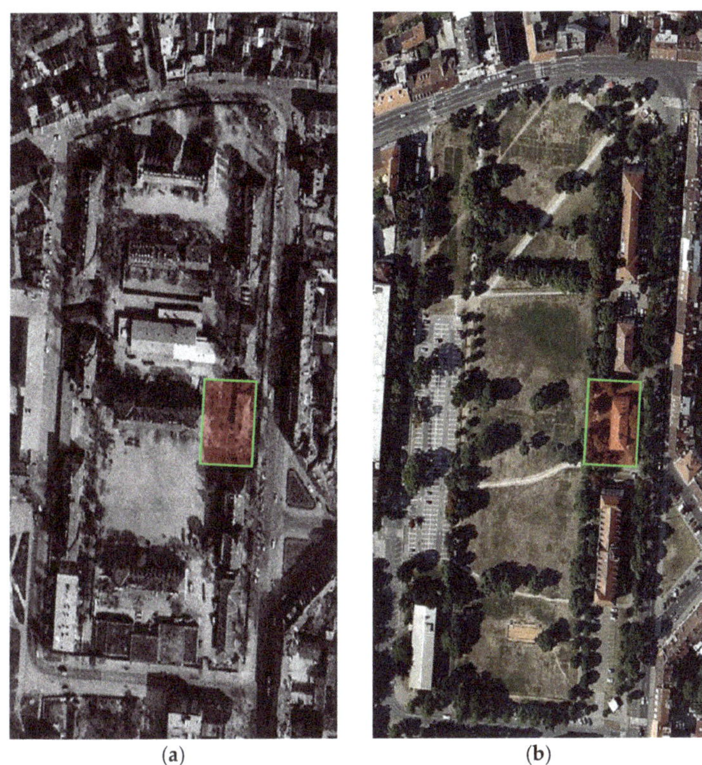

Figure 3. (a) Barracks from the end of the 19th century [30]; (b) Today's situation.

Figure 4 shows the original drawings of the building in question, obtained from the State Archives in Zagreb.

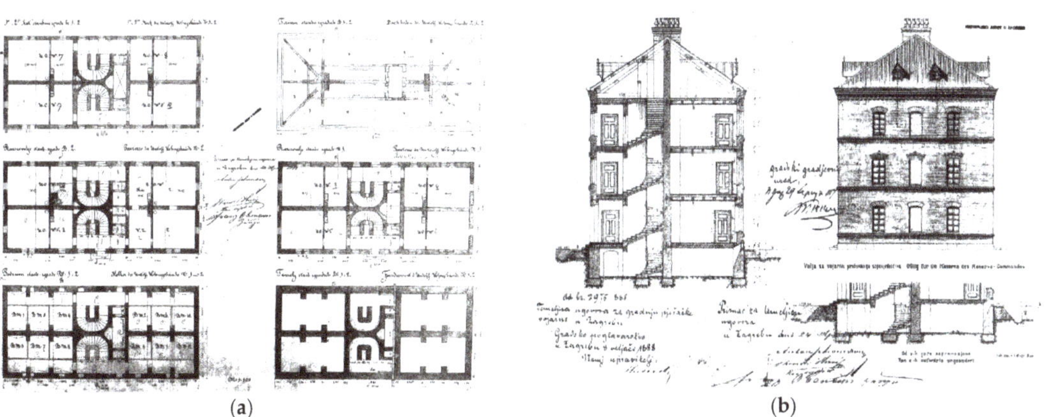

Figure 4. (a) Original floor plans of the buildings from the archive; (b) Original cross-section plans of the buildings from the archive.

2.2. Today's Building

The case study building (Figure 5) is a public-purpose building with a rectangular floor plan of 25.18 m × 11.42 m and a height of approximately 15.50 m. The floor area is approximately 290.00 m^2, and the total gross area is approximately 1450 m^2. The building consists of five floors, all floors of the building are used as office space. The building has undergone minor changes in the original geometry and space over time and has been properly maintained.

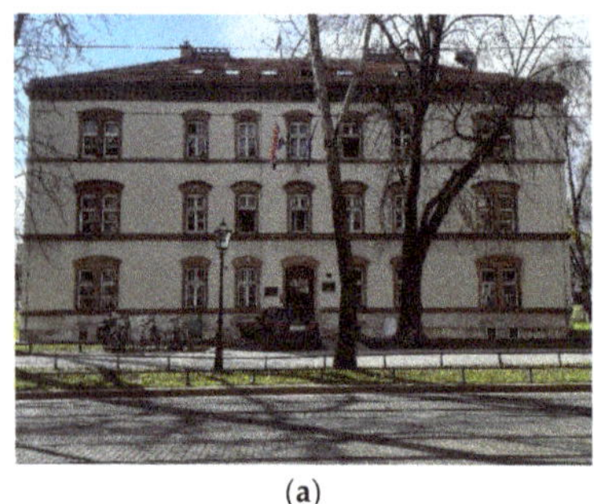

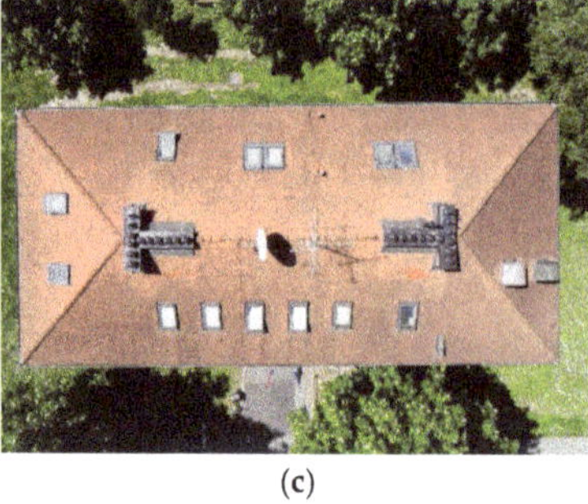

(a)   (b)   (c)

Figure 5. (**a**) View of the east façade; (**b**) View of the north façade; (**c**) View of the roof (photo credit: Mislav Stepinac).

The load-bearing walls are made of solid bricks of the old Austro-Hungarian format 14 × 6.5 × 29 cm. The thicknesses of the load-bearing walls in the basement are 78 cm, 65 cm and 50 cm, at the ground floor 63 cm and 50 cm, and in the other aboveground floors, the thickness is 50 cm. The partition walls are made of solid brick, and the thickness is between 14 cm and 20 cm.

The ceiling structure in the basement is a brick vault supported by brick arches. The structure of the other floors consists of wooden beams and steel beams. The width of the wooden beams is 14 cm, and the height is 20 cm. The steel beam is an "I" profile, 200 mm high.

In the central part of the building, there is a new reinforced concrete cantilever three-legged staircase system.

As the building is under the protection of the Ministry of Culture, a detailed survey of the external dimensions and façade was made to preserve its architectural value. Laser scanning was performed with the Leica BLK360 device and processed in the Cyclone Register 360 software. With the help of laser scanning, a point cloud with a precision of 3 mm was obtained, and the façade with external geometric contours was preserved for the future. In addition, a digital "twin" of the building has been created, which will be used for further restoration works if needed (Figure 6).

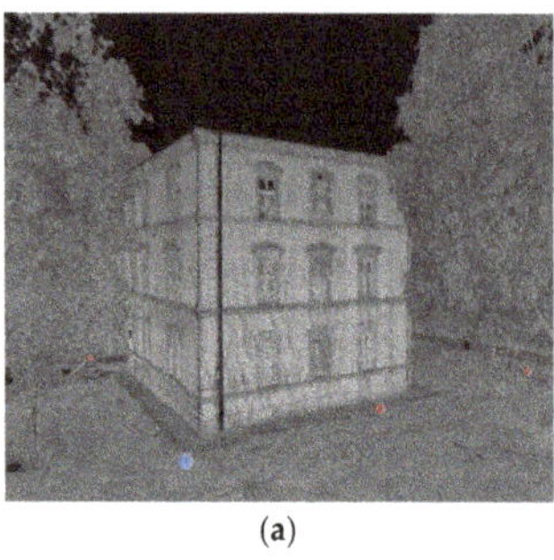

(a)   (b)   (c)

Figure 6. (**a**) Display of laser scanner measuring points Leica BLK 360; (**b**) Presentation of the laser scanning model; (**c**) Cyclone Register model.

After a detailed survey of the external geometry, the interior was recorded and measured. Finally, 2D and 3D models of the building were made (Figure 7).

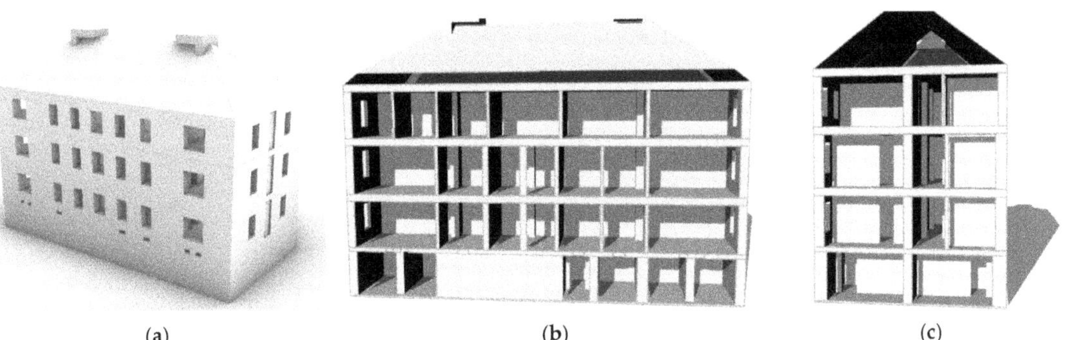

Figure 7. (**a**) Three-dimensional (3D) model; (**b**) 3D model—longitudinal cross-section; (**c**) 3D model—transversal cross-section.

2.3. Damage Detection after Earthquake

The building was inspected after the earthquake on 22 March, 2020. It was assigned the usability mark PN2. The mark PN2 refers to buildings with moderate damage without risk of collapse, but the usability is questionable due to the potential risk of collapse of some elements. The following damage was found:

- Several minor cracks were observed on the façades of the building. Due to their slenderness and low vertical load chimneys failed predominantly by shear sliding and overturning. Additionally, roof displacement and collision with chimneys increased failure occurrences.
- On the ground floor, small cracks were noticed at the places of the lintel and at the connections of the walls and ceiling. Lintels are weakened parts of the masonry walls and are therefore vulnerable since the damage is usually concentrated in them.
- On the first floor, major damage was noticed at the connection of partition and load-bearing walls and at the connection of walls and ceilings. Observed damage is not surprising because at the partition and load-bearing wall connections and wall and ceiling connections, there is a discontinuity of material and contact of different materials that have different behavior, and thus, there are different displacements that cause cracking. Often, such cracks do not pose a significant hazard.
- The original staircase has not been preserved, and the existing staircase has minor damage that does not indicate a threat to mechanical resistance and stability.

Some of the damaged elements are shown in Figure 8.

This inspection established conservation guidelines for the repair of load-bearing and partition walls, staircases and floor structures. The façade and roof design should be preserved along, with the reparation of existing damage after conducting detailed conservation and restoration research. In addition, this paper analyzed retrofitting strategy as one of the methods to preserve the "outer look" of the building.

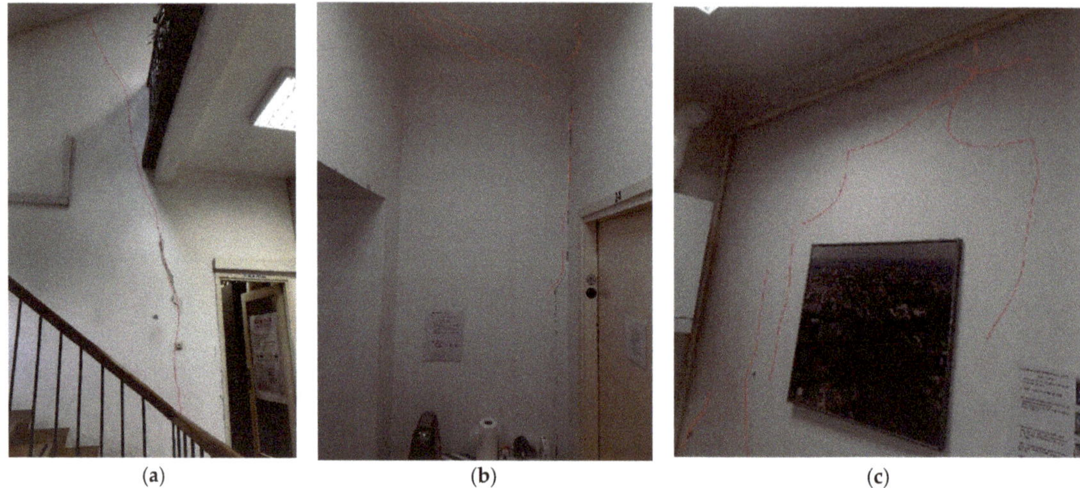

(a)   (b)   (c)

Figure 8. (**a**) Damage to the stair system; (**b**) Damage at the connections of load-bearing and partition walls; (**c**) Slight damage to the load-bearing walls.

3. Condition Assessment and Moderately Destructive Testing

The flat-jack method determined the vertical stress, the modulus of elasticity and the masonry's shear strength. The test was conducted on the ground and first floors on the same wall (Figure 9).

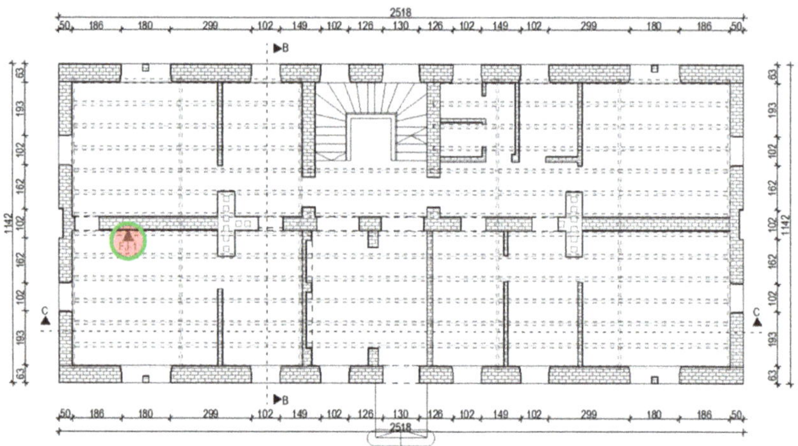

Figure 9. Ground floor plan with a marked test site.

The procedure for testing the vertical stress of the masonry was as follows:

- Removal of mortar from the horizontal joint of the masonry to partially release the masonry from compressive stress.
- Inserting a flat jack into the hole.
- Establishing the initial state of stress and strain by increasing the pressure in flat jacks.

It should be noted that the results obtained by this test are the average value of the masonry stress in the vicinity of the opening. Therefore, the obtained results can be assumed as representative stress for the whole tested wall when the wall is completely homogeneous, and the load is not eccentric.

The test procedure for masonry elasticity modulus was as follows:

- The test is performed in the same place as the vertical stress test.
- A second hole is made above the existing opening into which a flat jack is inserted.
- Both openings are horizontal, and they are vertically spaced by 5–7 rows of bricks.
- Inserted flat jacks are connected to one hydraulic pump.
- Displacement and relative deformation measuring devices are placed between flat jacks.
- Simultaneous application of vertical pressure to flat jacks and measurement of relative deformation using the device allows determining the modulus of elasticity.

The shear strength test procedure for masonry was as follows:

- The test is performed in the same place as the test of the modulus of elasticity of the masonry.
- One horizontal brick is removed to install the hydraulic press.
- Mortar was removed from the vertical joint of the horizontal test brick.
- A device for measuring displacements and relative deformations is installed over the test brick and the adjacent horizontal brick.
- Using a hydraulic press, horizontal pressure was applied to the test brick to move.
- Flat jacks enable the control of vertical stress to obtain the values of the coefficient of friction and the initial shear strength from the values of the ratio of shear strength and vertical stress.

The vertical stress of the masonry is determined by the following expression (ASTM C1196-14a):

$$\sigma_0 = K_m \cdot K_a \cdot p \tag{1}$$

where K_m is a dimensionless coefficient depending on the geometry and stiffness of the flat jack. The calibration of the flat jack determines it. K_a is a dimensionless coefficient determined from the ratio of the area of the flat jack and the area of the opening, and p is the pressure in the flat jack required to return the wall to its initial state of stress and strain.

According to the tests (Figure 10), the following values were obtained: compressive stress state in masonry at test location $\sigma_0 = 0.46$ N/mm^2 (used for model calibration regarding weight distribution), modulus of elasticity $E = 1469.5$ N/mm^2 (used for wall stiffness definition), initial shear strength $f_{v0} = 0.323$ N/mm^2 (used for wall shear resistance definition) and coefficient of friction $\mu = 0.447$.

(a)   (b)   (c)

Figure 10. (**a**) Connected hydraulic system and flat jack; (**b**) Determination of stress and deformation dependence of masonry; (**c**) Testing of shear strength of masonry with control of vertical stress by flat jacks. (photo credit: Luka Lulić).

Additionally, the so-called Masonry Quality Index (MQI) [31] was calculated. The MQI method is a simple and systematic qualitative approach appropriate for numerical estimation of the mechanical parameters of masonry. This method can be useful when in situ tests are not viable or for results validation when in situ tests are performed. More details on the mentioned method can be found in [31]. Table 1 shows the mechanical properties of masonry according to the method of MQI.

Table 1. Mechanical properties of masonry according to the Masonry Quality Index method (values in N/mm^2).

E_{min}	E_{max}	$f_{m,min}$	$f_{m,max}$	$\tau_{0,min}$	$\tau_{0,max}$	G_{min}	G_{max}	$f_{v0,min}$	$f_{v0,max}$
1786	2520	4.07	6.44	0.06	0.10	440	648	0.14	0.27

For all timber elements, the class of C22 is assumed, where the "C" letter implies softwood, e.g., spruce or pine, and the number "22" represents the major axis bending strength of timber. From the archives, it was concluded that softwood was used. The building was regularly maintained, and a value lower than the assumed (C24 or C27) was taken to be conservative. Seismic load analysis was performed according to EN 1998-1 [32] and the national annex [33]. The soil class is C, according to the latest geological research of the city of Zagreb.

4. Numerical Modeling of the Case Study Building

The building's choice of modeling and design method affects the accuracy and reliability of the results themselves. For example, simpler calculation methods give conservative results that can deviate greatly from the actual damage. On the other hand, more complex calculation methods give more accurate and reliable results even though they require more time. For this paper, for comparison, a seismic calculation was performed using two methods: the equivalent static load method and the pushover method. The modeling was performed using the 3Muri software.

Modeling the building in the 3Muri software is performed by inserting walls, columns and beams, which are then discretized into macroelements. There are two types of macroelements. These are the piers and spandrels in which all the damage is concentrated. Parts of the wall that are often undamaged are defined as rigid nodes, and they connect the former two [34]. The mathematical concept behind the use of macroelements makes it possible to find the mechanism of collapse, i.e., the mechanism of damage. Damage can be due to shear in the central part of the macroelements or due to combined compression and bending at the peripheral parts of the macroelements [34,35].

Horizontal diaphragms are modeled using floor elements connected by three-dimensional nodes. The loads on the horizontal diaphragms (used only for mass calculation and distribution) are perpendicular to the floor level, and the seismic action is in the direction of the floor level. For this reason, the horizontal diaphragms can be modeled as axially rigid or flexible but without bending stiffness. Such shaping of horizontal diaphragms is allowed because their main task is the acceptance of horizontal action due to seismic action and their further distribution to vertical load-bearing elements. 3Muri assumes good wall-to-wall and wall-to-floor connections, i.e., box behavior that is desirable but often unrealistic in the existing structures. Hence, during the modeling itself, it is assumed that the damaged masonry was restored to its original undamaged state by methods such as grouting and that the necessary measures were taken to ensure the box behavior of the observed structure. Good connection of walls and floors can be achieved by adding ties and anchors, as well as stiffening the floor structure. Additionally, 3Muri allows out-of-plane failure analysis of local mechanisms in a separate module. This is extremely useful since box behavior can accommodate only in-plane failure of the masonry. More on the analysis of local mechanisms in 3Muri can be found in [2].

Figure 11 shows the ground floor plan, and Figure 12 shows a 3D model of the building. Again, the floor plans of the floors roughly coincide, except that sometimes the layout of the partition walls is different.

Table 2 shows the legend of the material used to model the building.

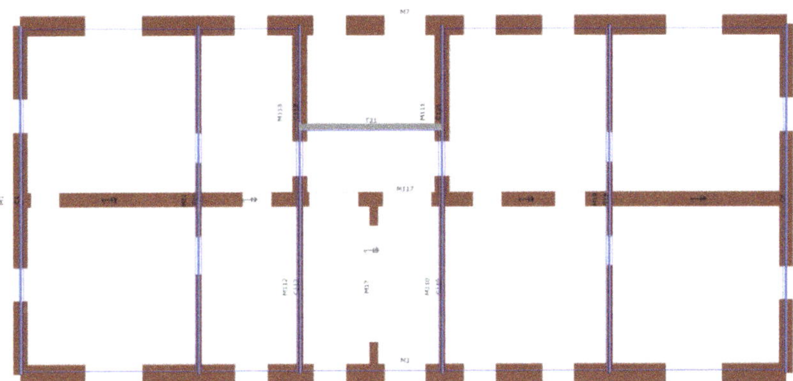

Figure 11. Ground floor plan.

Figure 12. Three-dimensional (3D) model—view of the eastern façade.

Table 2. Materials used in 3Muri model.

Material	Color	Norm
Masonry		According to the experimental results
Reinforced concrete		EN 1992-1-1:2005
Structural steel		EN1993-1-1:2005
Timber		EN 338:2002

Seismic action is determined by the equivalent static load method. To be able to apply the method of equivalent static load, the basic period of the first mode shape must be less than or equal to $4 \cdot T_C$ ($T_C = 0.6$ s for soil type C) and 2 s so that it satisfies the criterion of regularity in the vertical section. Therefore, the basic period of the first mode shape is 0.29, and the building satisfies the regularity criterion in the vertical section. The first mode shape was calculated by the following expression:

$$T_1 = C_t \cdot H^{\frac{3}{4}} \tag{2}$$

where C_t is a coefficient dependent on the structural system, and H is the building's height.

For the building in question, the importance class of II has been determined according to EN 1998. Importance class II corresponds to regular buildings.

Figure 13 shows the values of peak ground acceleration for the location of the building.

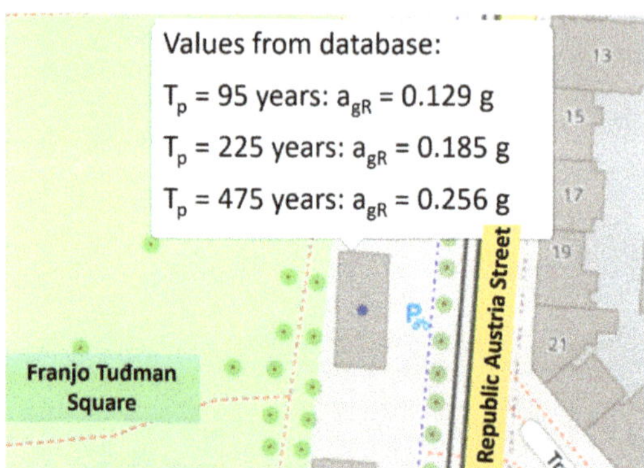

Figure 13. Peak ground acceleration values for the location of the building (from [36]—Croatian National Annex of EN1998).

For old unconfined masonry, the value of the behavior factor is set as $q = 1.00$. Seismic base shear force for each horizontal direction can be determined by the following expression:

$$F_b = S_d(T_1) \cdot m \cdot \lambda \tag{3}$$

where $S_d(T_1)$ is spectral acceleration for the first period of the building for the observed direction, m is the building's mass and λ is a correction factor dependent on the building's height. The values of the design spectrum and base shear force are given in Table 3.

Table 3. Design spectrum and base shear forces for different return periods.

Return Period [Years]	$S_e(T)$	F_b [kN]
95	0.373	3012
225	0.518	4183
475	0.748	6040

According to the EN 1998-1, depending on the local seismic hazard and the number of stories of the observed building, the minimum percentages of the cross-sectional area of the load-bearing walls in relation to the total floor area are given for the x- and y-directions (3% in our case). This check is the first step to establish the state of the existing building in terms of meeting the basic requirement used in the new building design process and to see if new walls should be added. Therefore, load-bearing walls in both directions meet the requirement of a minimum total area of load bearing walls for simple masonry buildings ($x = 7.79\%$, $y = 4.20\%$).

For the design purposes, the following mechanical characteristics and coefficients are taken based on in situ tests, code recommendations and literature review:

- Partial safety factor, $\gamma_M = 1.50$.
- Modulus of elasticity $E = 1470 \text{ N/mm}^2$
- Initial shear strength of masonry obtained from in situ testing, $f_{v0} = 0.323 \text{ N/mm}^2$.
- Confidence factor value, $FP = 1.20$ according to knowledge level 2.
- Diagonal tensile strength of masonry, $f_t = 0.114 \text{ N/mm}^2$.
- Local coefficient of friction of the joint, $\mu_j = 0.60$.
- Clamping coefficient, $\phi = 1.00$.
- Mean compressive strength of the units, $f_b = 12.00 \text{ N/mm}^2$.
- Value for clay unit from group 1 and general-purpose mortar, $K = 0.55$.
- Mortar compressive strength, $f_{mortar} = 1.50 \text{ N/mm}^2$.

The confidence factor is used to determine the seismic design method and depends on the level of knowledge. To determine the level of knowledge, it is necessary to know the geometric relationships of the structural and nonstructural elements, details (masonry, the connection of floor structure and masonry, etc.) and mechanical properties of the material from which the structure is built.

Suppose the level of knowledge is determined to be 1. In that case, the typical values of the mechanical characteristics of the material are assumed following the construction time of the building, and the structural tests are limited. If the level of knowledge is 2, then the values of mechanical characteristics of the material are assumed according to the original design specification or according to the values obtained from extensive research. The level of knowledge 2 was taken for the case study. Calculated base shear force is distributed on each floor, increasing linearly, along the height of the building. Next, floor forces are further distributed to walls according to their stiffnesses. To derive capacity utilization, i.e., the ratio of capacity and demand of individual walls, the resistance of the walls is compared with distributed wall forces. For global verification, the sum of the resistances of all the ground floor walls in the same direction was compared with total base shear force. According to the manual calculation (lateral force method), the capacity/demand ratio in the x-direction was 0.92, and in the y-direction 0.44. Masonry can fail in several different modes. Hence, the resistance to bending, shear sliding and diagonal tension failure (straight and stepped) are calculated. Expressions for resistance calculation can be found in [7]. According to the calculation of the masonry resistance, the load-bearing capacity of the walls in the x- and y-directions is not sufficient to absorb the earthquake force of the return period of 95 years. Therefore, adding walls (or an equivalent system for absorbing horizontal forces) in both directions is necessary.

The 3Muri software has a module for conducting modal analysis. This module offers the calculation of all possible mode shapes. In this paper, 10 mode shapes are considered, and 3 are shown in Figure 14 and in Table 4. The first and third modes are predominantly translational, while in the second mode, slight torsion occurs.

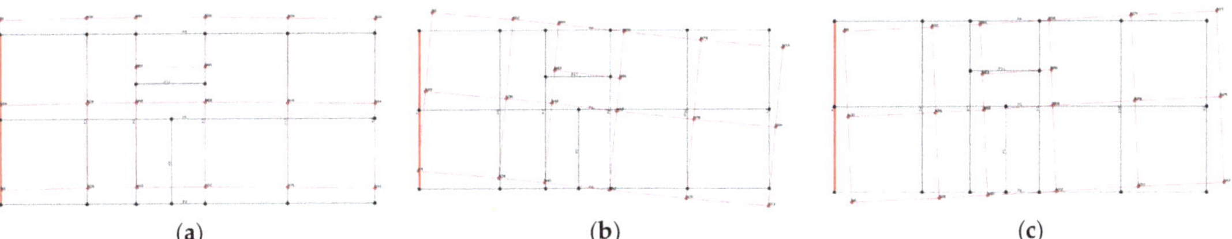

(a)   (b)   (c)

Figure 14. (**a**) First mode; (**b**) second mode; (**c**) third mode.

The pushover method in the 3Muri software is carried out depending on the distribution of lateral load on the structure. The lateral load distribution can be linearly increasing or uniform along the height of the building or in the form of the translational mode shape. Figure 15 shows a 3D model of the equivalent frames of the building developed in the 3Muri software. Four analyses were performed for the modal lateral load distribution, two in the x-direction (+X and −X) and two in the y-direction (+Y and −Y). Figures 16 and 17 show the capacity curves in the x- and y-directions (black) and their bilinear idealization (orange). Value "d_m" on the graph represents the near-collapse limit state. It is reached when the maximum value of the shear force drops by 20%. Figure 18 shows a 3D model with damage to the near-collapse limit state for critical analysis in the x- and y-directions.

Table 4. Modal analysis details.

Mode	T (s)	mx (kg)	Mx (%)	my (kg)	My (%)	mz (kg)	Mz (%)
1	0.2934	85	0.01	973,618	81.78	13	0.00
2	0.2293	209,897	17.63	150	0.01	0	0.00
3	0.2214	835,847	70.21	334	0.03	34	0.00

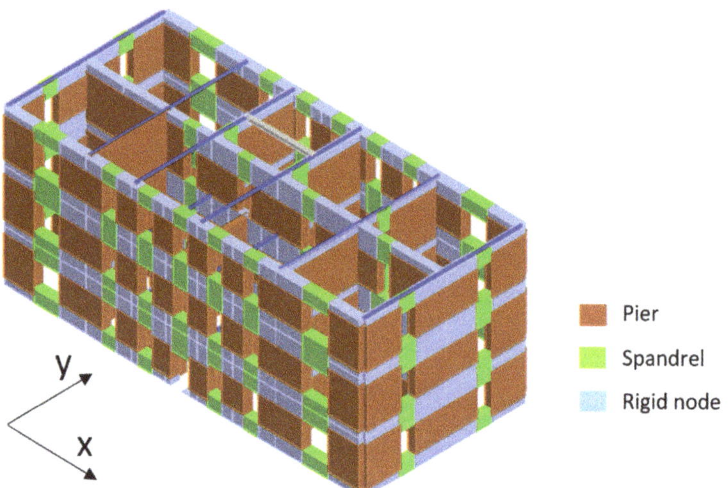

Figure 15. Presentation of 3D model of equivalent frames (macroelements) of the case study in the 3Muri software.

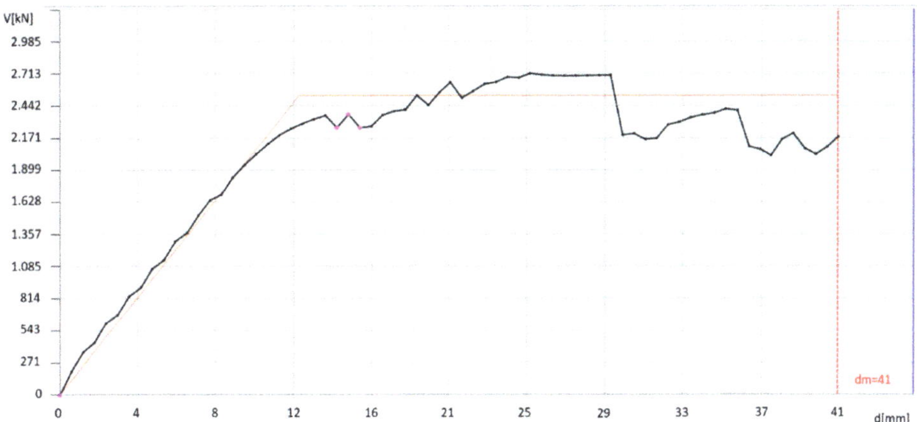

Figure 16. Capacity curve (black) and its bilinear idealization (orange) for *x*-direction.

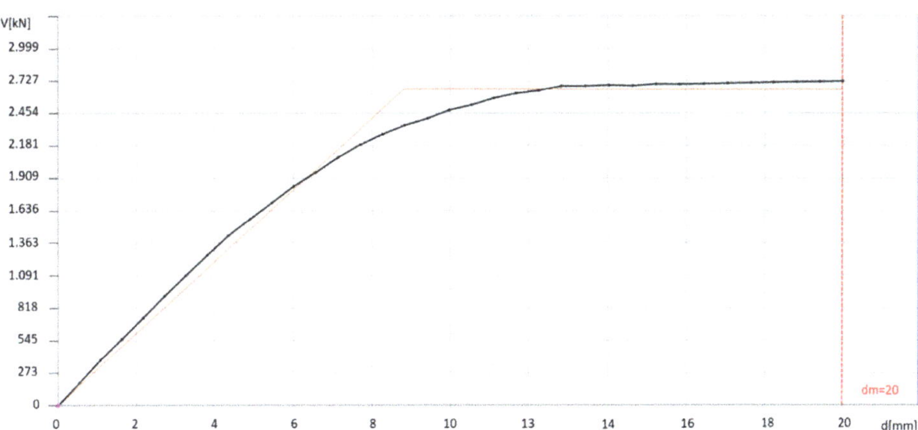

Figure 17. Capacity curve (black) and its bilinear idealization (orange) for *y*-direction.

The pushover analysis was performed for all three distributions of lateral load without random eccentricity. Finally, the capacity/demand ratio obtained according to the simplified calculations and the values of the safety indices for critical analyses obtained using the 3Muri software were compared, as shown in Table 5. According to the regulation, the safety index is the ratio of peak ground acceleration for which the structure reaches a certain limit state, i.e., capacity and peak ground acceleration (PGA), i.e., demand. PGA for a return period of 95 years was used, and it has a value of 0.13 g. The limit state of significant damage (SD) was observed.

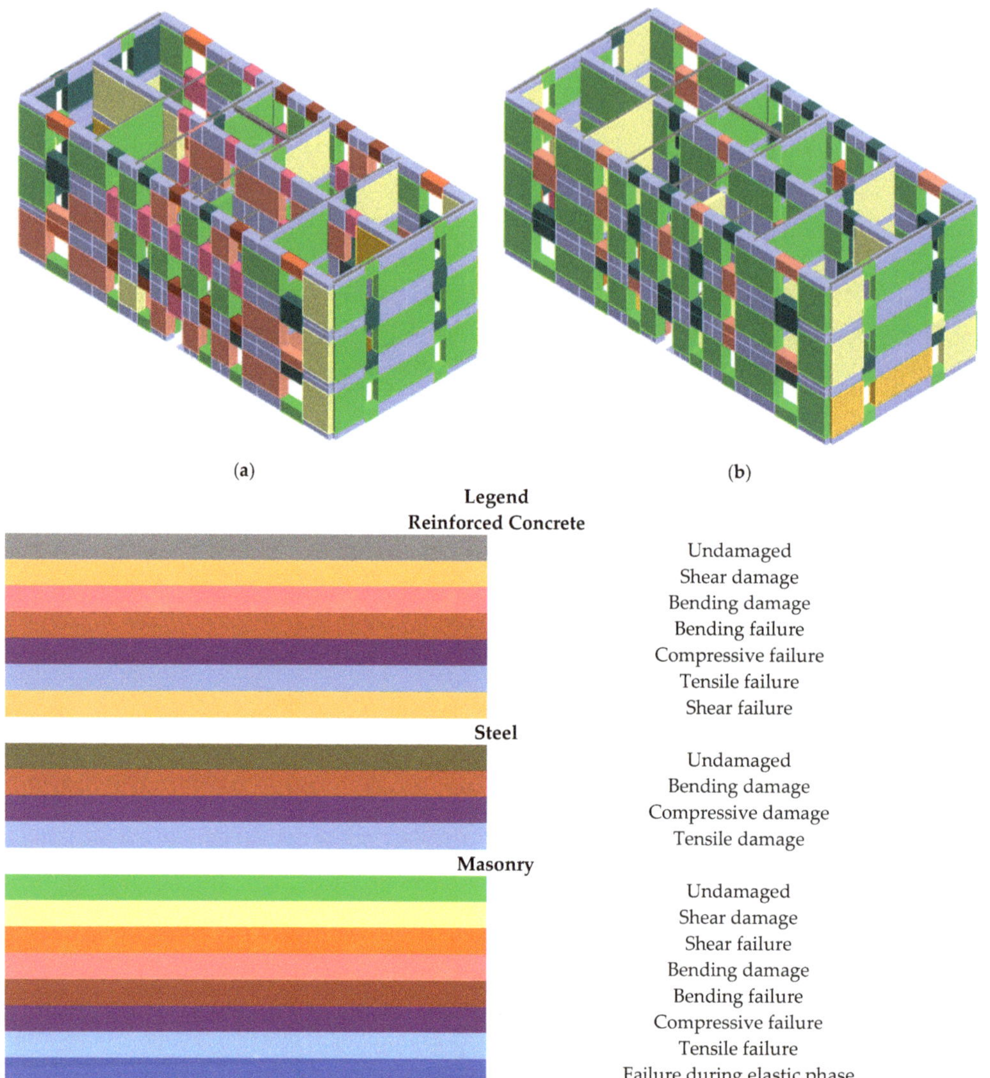

Figure 18. Three-dimensional (3D) model with damage for the near-collapse limit state: (**a**) x-direction; (**b**) y-direction.

Table 5. Capacity/demand ratios obtained by different methods of analysis.

Type of Analyses	x-Direction	y-Direction
Simplified hand calculation	92%	44%
Seismic load distribution according to equivalent static forces method	66%	71%
Modal distribution of seismic load	69%	78%
Uniform distribution of seismic load	81%	83%

According to the above, the simplified calculation visibly deviates from the calculation using the 3Muri software for calculating the capacity/demand ratio in the y-direction. The visible deviation occurred because the simplified calculation method has more geometric limitations due to the choice of the walls, such as the minimum wall thickness and the minimum wall length to height ratio. Thus, not all walls were considered when choosing load-bearing walls. Additionally, in the 3Muri software, in addition to partition walls, concrete and steel beams are modeled, contributing to the rigidity of the entire building, but mostly in the y-direction as can be seen in Figure 15 (steel beams in blue). Therefore, it can be concluded that the simplified design is more conservative for the y-direction compared to the design in the 3Muri software, which was expected. In the other direction, results are more similar, but the more conservative design is now reversed in favor of the 3Muri software.

To compare the actual damage and the damage obtained using the 3Muri software, a uniform lateral load distribution was selected. It is assumed that the peak ground acceleration of the earthquake in Zagreb in 2020 was about 0.18 g. The results of damage are shown in Figure 19, with locations of real damage shown in Figure 20a,b.

Figure 20a shows the actual damage to the right part of the load-bearing wall on the ground floor corresponding to the shear damage. Figure 20b shows the actual damage to the middle part of the load-bearing wall on the first floor, which corresponds to the damage due to bending. Similar damage is detected in height in the building itself, and damage corresponds to the model.

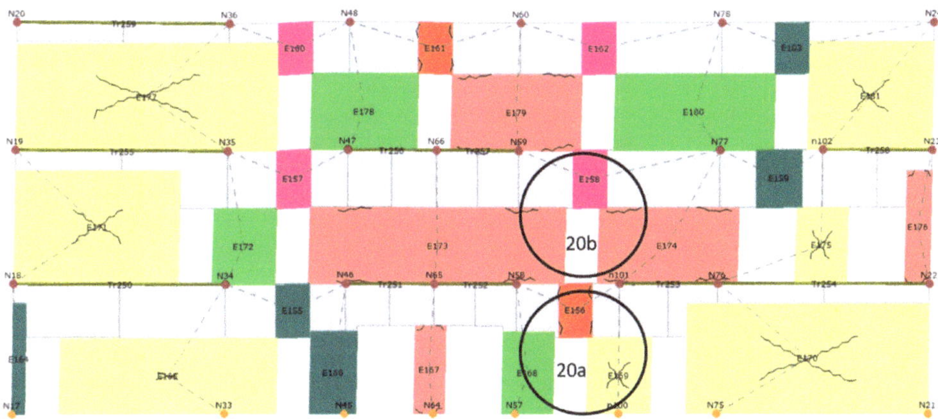

Figure 19. Damage to the central wall in the x-direction (3Muri).

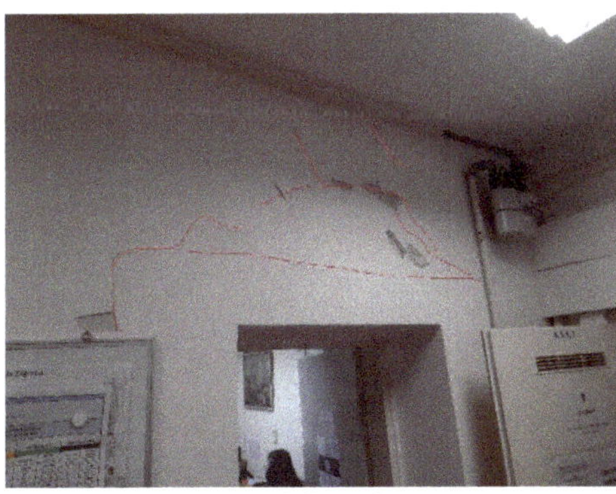

(**a**) Ground floor damage

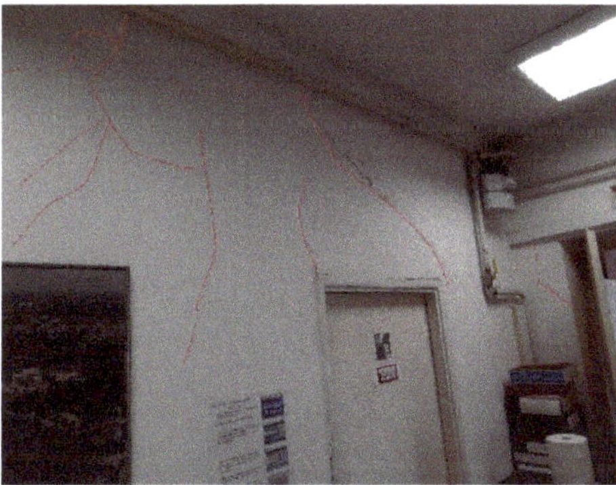

(**b**) First floor damage

Figure 20. Damage to the central wall in the x-direction.

5. Renovation Measures for Existing Masonry Buildings after the Earthquake(s) in Croatia

For the successful renovation of buildings damaged in the earthquake, it is necessary to apply appropriate measures for repair and strengthening of the building without compromising the mechanical characteristics of the material and the properties of the structure that contribute to the durability of the building [37].

After the earthquakes in Croatia, to create a legal framework for the faster, economical and easier reconstruction of earthquake-damaged areas, the Law on Reconstruction of Earthquake-Damaged Buildings in the city of Zagreb, Krapina-Zagorje County and Zagreb County [38] was passed. The Law defines the methods of reconstruction that depend on the degree of damage and purpose of the building. Additionally, an addendum to the technical regulations was issued [39], which defines the levels of renovation, which are:

- Level 1: repair of nonstructural elements.
- Level 2: structural repair to the return period of 95 years.
- Level 3: strengthening to the return period of 225 years.
- Level 4: Complete retrofitting to the return period of 475 years.

The Technical Regulation [39] defines the requirements, documentation, interventions and works, and the category of buildings that the renovated structure must meet for each level above. A proposal of measures for repair and reinforcement of buildings is given following the obtained results. Measures should follow the seismic design and be in line with the conservation and restoration rules. The minimum restoration level is level 2 for all structures with greater damage. However, in addition to the proposed minimum level of renovation, the building owner may request renovation to a higher level than the prescribed level of renovation at his own expense. For the building in question, the proposed level of renovation is level 2, but at the request of the owner of the building, renovation level 3 is selected.

As a measure of repair and reinforcement of the walls of the building, it is recommended to reinforce load-bearing walls by, e.g., FRCM system or concrete jacketing. Figure 21 shows a proposal for reinforcing load-bearing walls. To obtain good resistance in the transverse direction (y-direction), it is proposed to add new load-bearing walls with a minimum thickness of 38 cm. In addition, it is proposed to remove the brick partition walls and replace them with a drywall system. Figures 22 and 23 show a proposal for the position of the new load-bearing walls and a proposal for the removal and replacement of partition walls.

In addition to mentioned methods, it is necessary to strengthen the ceiling structure. Therefore, as a measure of repair and reinforcement of the wooden ceiling structure, a thin reinforced concrete compression slab is proposed to increase the load-bearing capacity and stiffen the structure (rigid diaphragm). All arched elements, vaults in the basement are planned to be kept in the original design with the possibility of strengthening with carbon fibers, maintaining the original proportion of the vaults in order to preserve the original construction and design characteristics of the building.

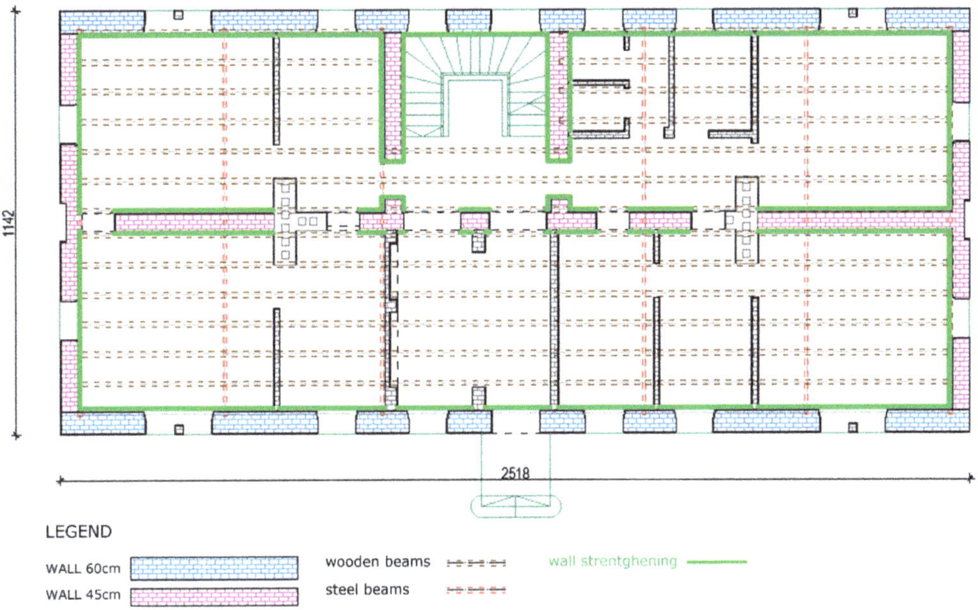

Figure 21. Proposal for reinforcement of load-bearing walls.

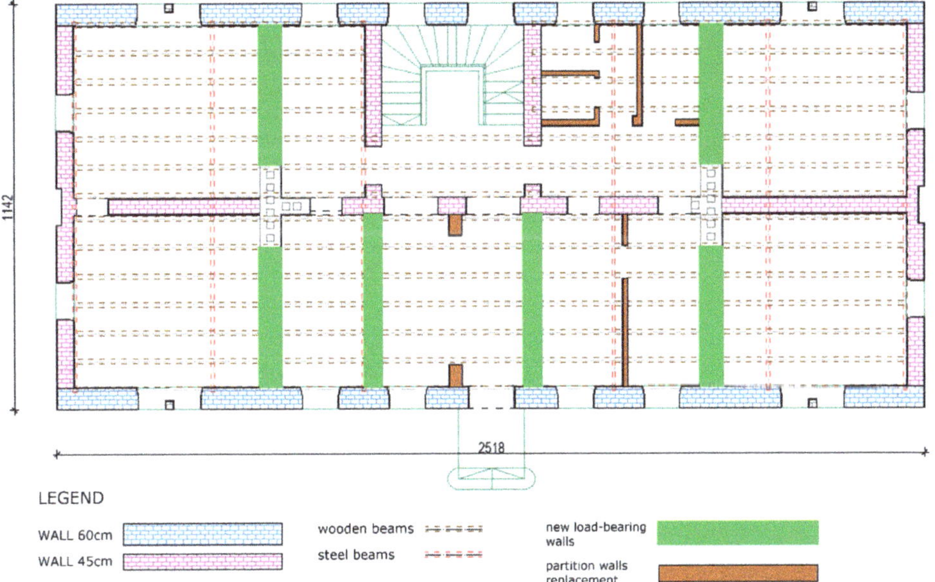

Figure 22. Proposal for the position of the new load-bearing walls and proposal for the removal and replacement of partition walls.

Figure 23. Proposal for the position of the new load-bearing walls and proposal for the removal and replacement of partition walls (cross-section).

Additionally, this paper analyzed the possibility to perform an equivalent system for an alternative retrofitting strategy to take over horizontal forces. An example of an idea for an equivalent system for taking over horizontal forces can be found in Figure 24. The idea of an equivalent system for taking over horizontal forces consists of using existing steel beams (crossbeams) in the floor structure. New steel beams would be added to the existing steel beams, which would end outside the structure itself. The new steel beams would be externally connected to the steel rope, as shown in Figure 24. Bracing elements could be placed to ensure the common behavior of the whole system and the building and ensure sufficient transverse stiffness.

This approach allows a clear differentiation of the old structure and the new-seismic one. The old structure becomes easier to read and more visible, due to the fact that the new seismic elements are mostly connected to the existing one and thus in some way additionally mark it. They also damage it less since no drastic interventions are needed for their installation or execution.

The proposed solution gives freedom in a case where interventions cannot be obtained from the interior. The existing horizontal frames steel beams can efficiently be connected to the exterior bracing system and throughout those beams transfer horizontal forces. In that way, the interior design can be saved, and the layout can be pretty much intact. The exterior vertical bracing system with a tension diagonal in general has a good dynamic response with the unreinforced masonry building, as those lateral systems are not too stiff as for example shear walls from omitted masonry or RC. This example can be also used for educational purposes to provide different solutions and different aspects in seismic retrofitting.

Another possible renovation method, i.e., seismic isolation, is used for the rehabilitation of buildings of special cultural and historical importance. The building should be separated from the ground, thus constructing new foundations on which insulating units are placed, and on them a new construction that will transfer loads from the existing building to the insulators. In addition, the biggest advantage of seismic insulation as a remedial measure is that the building does not require additional interventions and elements that could damage the façade or interior. This method is, on the other hand, considerably more expensive than others.

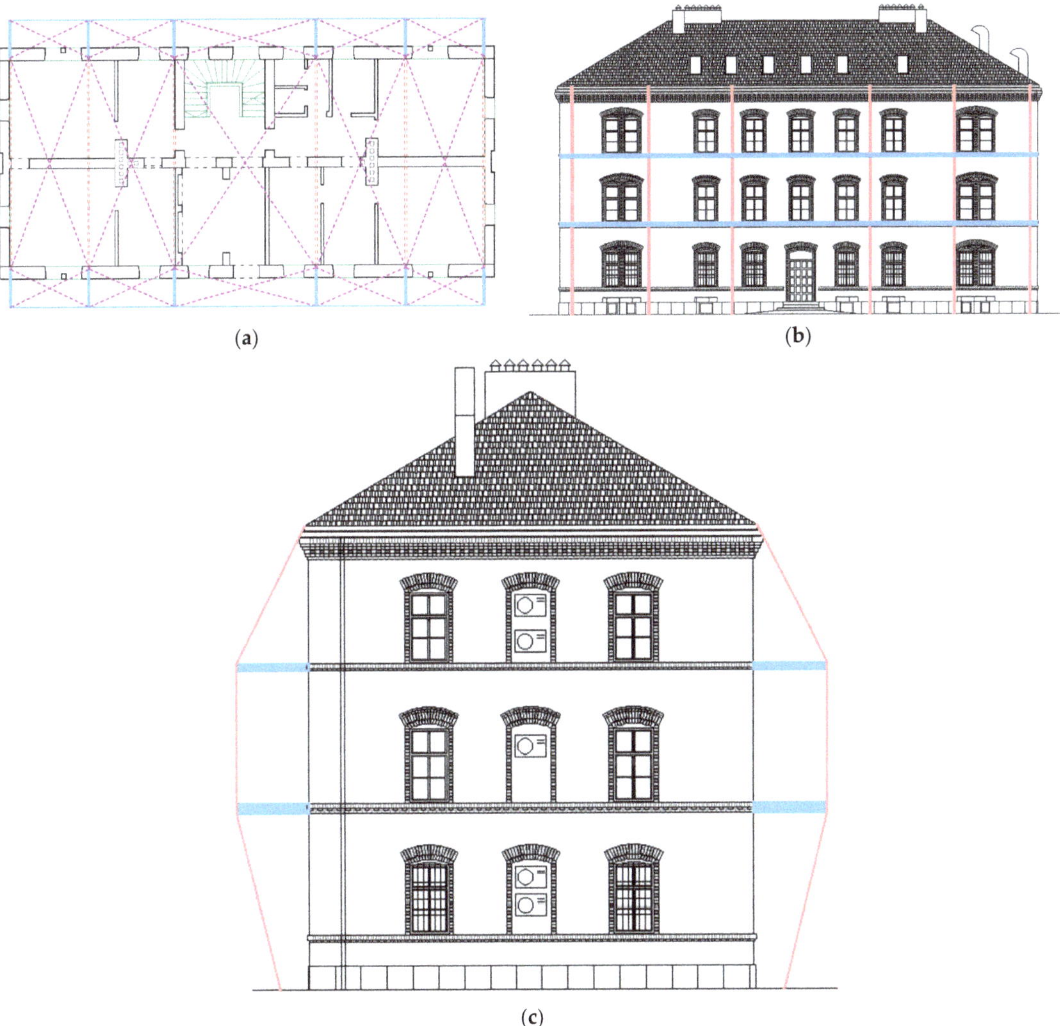

(a)

(b)

(c)

Figure 24. Alternative solution for the retrofitting: (**a**) ground floor plan; (**b**) longitudinal section; (**c**) transversal section.

6. Conclusions

The Republic of Croatia is one of the most seismically endangered countries in Europe, especially the Mediterranean area and northwestern Croatia. However, the city of Zagreb is a seismically active area due to the Žumberak-Medvednica fault and the Zagreb fault, which consists of a series of smaller faults. After the two earthquakes in Croatia, about 70,000 buildings were damaged, 25,000 in the Zagreb earthquake and about 45,000 in an earthquake whose epicenter was 70 km from Zagreb.

Most of the damaged buildings are of an older date of construction, mostly built in the period before the existence of the first earthquake regulations, and are built of brick with wooden floor structures. Such structures are characterized by uneven stiffness distribution, inappropriate or nonexistent connections between the walls and poor connection to the roof and floor structure. Many of these buildings are under cultural heritage protection, and such an example is the building presented in this paper.

The building was fully inspected after the earthquake by engineers in accordance with a pre-established methodology (EMS-98). Subsequently, for further potential restoration work, a digital "twin" of the building has been created with the Leica BLK360 device. Laser scanning resulted in a point cloud with a precision of 3 mm, which was processed in the Cyclone Register 360 software. This way, the original façade with external geometric contours and details was preserved for the future. Additionally, on-site investigative and moderately destructive tests were carried out using the flat-jack system. The tests provided an important insight into the material characteristics such as modulus of elasticity, compressive stress state, coefficient of friction and initial shear strength without the contribution of vertical stress, which are required for modeling. Since the standards recommend nonlinear methods of analysis for existing masonry structures, the pushover method integrated into the 3Muri program was used. Several different vertical distributions of seismic loads were considered. In addition, simplified manual calculations were performed. Finally, all methods were compared, where a significant deviation of the results of the manual method was observed.

The results obtained using the 3Muri software and the simplified method show that the case study building does not meet the states of limited damage, significant damage and near collapse, with return periods of 95 years, 225 years and 475 years. Therefore, in addition to the condition assessment and seismic design of the structure, a proposal of measures for repairs and strengthening of the structure was given in accordance with applicable laws and new regulations.

When designing a technical solution for the renovation and reinforcement of seismic resistance of the protected heritage building, it is necessary to envisage strengthening methods that are minimally invasive for historic structures and space utilization, using appropriate materials and methods, to enable preservation and presentation of original exterior and interior building characteristics.

In the process of strengthening, it is necessary to integrate and enhance the energy efficiency of the structure, as well as to preserve the architectural and historical values of the protected heritage while ensuring the safe and functional use of the building. Aseismic measures, elements whether exposed, visible or not, should respect the character and integrity of the cultural heritage and be visually in harmony with it. The seismic system should be reversible as much as possible so that it can be replaced by more advanced seismic measures in the future.

Author Contributions: Conceptualization, M.S.; methodology, M.M. and M.S.; software, M.M., B.Č.Š. and L.L.; validation, M.S., I.M. and L.L.; formal analysis, M.M., N.I., B.Č.Š. and L.L.; investigation, M.M. and N.I.; resources, M.S., I.M. and N.I.; data curation, M.M., N.I., B.Č.Š., L.L. and M.S.; writing—original draft preparation, M.S. and M.M.; writing—review and editing, M.S., Y.E., N.I., B.Č.Š. and I.M.; visualization, M.M., N.I. and M.S.; supervision, M.S., I.M. and Y.E.; project administration, M.S. and N.I.; funding acquisition, M.S. All authors have read and agreed to the published version of the manuscript.

Funding: This research was funded by the Croatian Science Foundation, grant number UIP-2019-04-3749 (ARES project—Assessment and rehabilitation of existing structures—development of contemporary methods for masonry and timber structures), project leader: Mislav Stepinac.

Data Availability Statement: All of the data shown in the paper, from laser scanning, visual inspection, non-destructive testing to structural modeling and strengthening proposals, was done by the authors. The data presented in this study are available on request from the corresponding author. The data are not publicly available due to privacy reasons.

Conflicts of Interest: The authors declare no conflict of interest.

References

1. *World Bank Report: Croatia Earthquake–Rapid Damage And Needs Assessment, 2020, June 2020*; Government of Croatia: Zagreb, Croatia, 2020.
2. Lulić, L.; Ožić, K.; Kišiček, T.; Hafner, I.; Stepinac, M. Post-Earthquake Damage Assessment-Case Study of the Educational Building after the Zagreb Earthquake. *Sustainability* **2021**, *13*, 6353. [CrossRef]
3. Saloustros, S.; Pelà, L.; Contrafatto, F.R.; Roca, P.; Petromichelakis, I. Analytical Derivation of Seismic Fragility Curves for Historical Masonry Structures Based on Stochastic Analysis of Uncertain Material Parameters. *Int. J. Archit. Herit.* **2019**, *13*, 1142–1164. [CrossRef]
4. Stepinac, M.; Kisicek, T.; Renić, T.; Hafner, I.; Bedon, C. Methods for the Assessment of Critical Properties in Existing Masonry Structures under Seismic Loads-the ARES Project. *Appl. Sci.* **2020**, *10*, 1576. [CrossRef]
5. Stepinac, M.; Lourenço, P.B.; Atalić, J.; Kišiček, T.; Uroš, M.; Baniček, M.; Šavor Novak, M. Damage Classification of Residential Buildings in Historical Downtown after the ML5.5 Earthquake in Zagreb, Croatia in 2020. *Int. J. Disaster Risk Reduct.* **2021**, *56*, 102140. [CrossRef]
6. Novak, M.Š.; Uroš, M.; Atalić, J.; Herak, M.; Demšić, M.; Baniček, M.; Lazarević, D.; Bijelić, N.; Crnogorac, M.; Todorić, M. Zagreb Earthquake of 22 March 2020–Preliminary Report on Seismologic Aspects and Damage to Buildings. *Gradjevinar* **2020**, *72*, 843–867. [CrossRef]
7. Kišiček, T.; Stepinac, M.; Renić, T.; Hafner, I.; Lulić, L. Strengthening of Masonry Walls with FRP or TRM. *Gradjevinar* **2020**, *72*, 937–953. [CrossRef]
8. *The Database of Usability Classification, Croatian Centre of Earthquake Engineering (HCPI-Hrvatski Centar Za Potresno Inženjerstvo)*; Faculty of Civil Engineering, University of Zagreb and The City of Zagreb: Zagreb, Croatia, 2020.
9. Forsyth, M. *Structures and Construction in Historic Building Conservation*; Blackwell Publishing Ltd.: Hoboken, NJ, USA, 2008. [CrossRef]
10. Vlachakis, G.; Vlachaki, E.; Lourenço, P.B. Learning from Failure: Damage and Failure of Masonry Structures, after the 2017 Lesvos Earthquake (Greece). *Eng. Fail. Anal.* **2020**, *117*, 104803. [CrossRef]
11. Cruz, H.; Yeomans, D.; Tsakanika, E.; Macchioni, N.; Jorissen, A.; Touza, M.; Mannucci, M.; Lourenço, P.B. Guidelines for On-Site Assessment of Historic Timber Structures. *Int. J. Archit. Herit.* **2015**, *9*, 277–289. [CrossRef]
12. Sánchez Rodríguez, A.; Riveiro Rodríguez, B.; Soilán Rodríguez, M.; Arias Sánchez, P. Laser Scanning and Its Applications to Damage Detection and Monitoring in Masonry Structures. In *Long-Term Performance and Durability of Masonry Structures. Degradation Mechanisms. Health Monitoring and Service Life Design*; Ghiassi, B., Lourenço, P.B., Eds.; Woodhead Publishing: Cambridge, UK, 2019; pp. 265–285. [CrossRef]
13. Funari, M.F.; Mehrotra, A.; Lourenço, P.B. A Tool for the Rapid Seismic Assessment of Historic Masonry Structures Based on Limit Analysis Optimisation and Rocking Dynamics. *Appl. Sci.* **2021**, *11*, 942. [CrossRef]
14. Endo, Y.; Pelà, L.; Roca, P. Review of Different Pushover Analysis Methods Applied to Masonry Buildings and Comparison with Nonlinear Dynamic Analysis. *J. Earthq. Eng.* **2017**, *21*, 1234–1255. [CrossRef]
15. Ortega, J.; Vasconcelos, G.; Rodrigues, H.; Correia, M. Assessment of the Influence of Horizontal Diaphragms on the Seismic Performance of Vernacular Buildings. *Bull. Earthq. Eng.* **2018**, *16*, 3871–3904. [CrossRef]
16. Stepinac, M.; Rajčić, V.; Barbalić, J. Inspection and Condition Assessment of Existing Timber Structures. *Gradjevinar* **2017**, *69*, 861–873. [CrossRef]
17. Fortunato, G.; Funari, M.F.; Lonetti, P. Survey and Seismic Vulnerability Assessment of the Baptistery of San Giovanni in Tumba (Italy). *J. Cult. Herit.* **2017**, *26*, 64–78. [CrossRef]
18. Malcata, M.; Ponte, M.; Tiberti, S.; Bento, R.; Milani, G. Failure Analysis of a Portuguese Cultural Heritage Masterpiece: Bonet Building in Sintra. *Eng. Fail. Anal.* **2020**, *115*, 104636. [CrossRef]
19. CIB 335. *Guide for the Structural Rehabilitation of Heritage Buildings*; CIB Commission W023: Ottawa, ON, Canada, 2010.
20. Lourenço, P.B. Conservation of Cultural Heritage Buildings: Methodology and Application to Case Studies. *Rev. ALCONPAT* **2013**, *3*, 98–110. [CrossRef]
21. Betti, M.; Bonora, V.; Galano, L.; Pellis, E.; Tucci, G.; Vignoli, A. An Integrated Geometric and Material Survey for the Conservation of Heritage Masonry Structures. *Heritage* **2021**, *4*, 585–611. [CrossRef]
22. Stepinac, M.; Gašparović, M. A Review of Emerging Technologies for an Assessment of Safety and Seismic Vulnerability and Damage Detection of Existing Masonry Structures. *Appl. Sci.* **2020**, *10*, 5060. [CrossRef]
23. Ortega, J.; Vasconcelos, G.; Rodrigues, H.; Correia, M. Seismic Vulnerability and Loss Assessment of Vila Real de Santo António, Portugal: Application of a Novel Method. *Int. J. Archit. Herit.* **2020**, *15*, 1585–1607. [CrossRef]
24. Saloustros, S.; Pelà, L.; Roca, P.; Portal, J. Numerical Analysis of Structural Damage in the Church of the Poblet Monastery. *Eng. Fail. Anal.* **2015**, *48*, 41–61. [CrossRef]
25. Giordano, E.; Ferrante, A.; Clementi, F.; Milani, G.; Formisano, A. Cultural Heritage and Earthquake: The Case Study of San Francesco's Church in Amandola (Central Italy). In Proceedings of the 2019 IMEKO TC4 International Conference on Metrology for Archaeology and Cultural Heritage, MetroArchaeo 2019, Florence, Italy, 4–6 December 2019.

26. Formisano, A.; Vaiano, G.; Fabbrocino, F. A Seismic-Energetic-Economic Combined Procedure for Retrofitting Residential Buildings: A Case Study in the Province of Avellino (Italy). In *AIP Conference Proceedings*; AIP Publishing LLC: Melville, NY, USA, 2019. [CrossRef]
27. Capanna, I.; Aloisio, A.; Di Fabio, F.; Fragiacomo, M. Sensitivity Assessment of the Seismic Response of a Masonry Palace via Non-Linear Static Analysis: A Case Study in l'aquila (Italy). *Infrastructures* **2021**, *6*, 8. [CrossRef]
28. Kneževicć, S. *Zagrebaćke Planirane Vojarne Iz Doba Habsburške Monarhije (Zagreb Planned Barracks From the Age of the Habsburg Monarchy)*; Institut Za Povijest Umjetnosti: Zagreb, Croatia, 2007.
29. Knežević, S. Povijest Područja Bivše Rudolfove Vojarne i Trga Francuske Republike u Zagrebu (History of the Area of the Former Rudolf Barracks and the Square of the French Republic in Zagreb). *Godišnjak Zaštite Spomenika Kult. Hrvat.* **1996–1997**, *22–23*, 57–72.
30. Digital Collections of the National and University Library in Zagreb [Internet]. Available online: Https://Digitalna.Nsk.Hr/Pb/?Object=list&mr%5B553362%5D=a (accessed on 28 June 2021).
31. Borri, A.; Corradi, M.; De Maria, A.; Sisti, R. Calibration of a Visual Method for the Analysis of the Mechanical Properties of Historic Masonry. *Procedia Struct. Integr.* **2018**, *11*, 418–427. [CrossRef]
32. EN 1998-1. Eurocode 8: Design of Structures for Earthquake Resistance—Part 1: General Rules, Seismic Actions and Rules for Buildings. *Eur. Comm. Norm. Brussels* **2004**. Available online: https://www.phd.eng.br/wp-content/uploads/2015/02/en.1998.1.2004.pdf (accessed on 17 September 2021).
33. HRN EN 1998-3:2011 Eurocode 8: Design of Structures for Earthquake Resistance—Part 3: Assessment and Retrofitting of Buildings (EN 1998-3:2005+AC:2010). Available online: https://www.phd.eng.br/wp-content/uploads/2014/07/en.1998.3.2005.pdf (accessed on 17 September 2021).
34. Lagomarsino, S.; Penna, A.; Galasco, A.; Cattari, S. TREMURI Program: An Equivalent Frame Model for the Nonlinear Seismic Analysis of Masonry Buildings. *Eng. Struct.* **2013**, *56*, 1787–1799. [CrossRef]
35. S.T.A. DATA, 3Muri Program 12.5.0.2. Available online: http://www.stadata.com/ (accessed on 14 September 2021).
36. Herak, M.; Allegretti, I.; Herak, D.; Kuk, V.; Marić, K.; Markušić, S.; Sović, I. Maps of Seismic Areas of the Republic of Croatia. Available online: http://seizkarta.gfz.hr/hazmap/ (accessed on 21 September 2021).
37. Uroš, M.; Todorić, M.; Crnogorac, M.; Atalić, J.; Šavor Novak, M.; Lakušić, S. (Eds.) *Potresno Inženjerstvo-Obnova Zidanih Zgrada*; Građevinski Fakultet, Sveučilišta u Zagrebu: Zagreb, Croatia, 2021.
38. Law on the Reconstruction of Earthquake-Damaged Buildings in the City of Zagreb, Krapina-Zagorje County and Zagreb County (NN 102/2020). Available online: https://www.zakon.hr/z/2656/Zakon-o-obnovi-zgrada-o%C5%A1te%C4%87enih-potresom-na-podru%C4%8Dju-Grada-Zagreba,-Krapinsko-zagorske-%C5%BEupanije,-Zagreba%C4%8Dke-%C5%BEupanije,-Sisa%C4%8Dko-moslava%C4%8Dke-%C5%BEupanije-i-Karlova%C4%8Dke-%C5%BEupanije (accessed on 10 September 2021).
39. Tehnički Propis o Izmjeni i Dopunama Tehničkog Propisa Za Građevinske Konstrukcije (Technical Regulation on Amendments to the Technical Regulation for Building Structures) [Internet]. Available online: Https://Narodne-Novine.Nn.Hr/Clanci/Sluzbeni/2020_07_75_1448.Html (accessed on 29 June 2021).

MDPI
St. Alban-Anlage 66
4052 Basel
Switzerland
www.mdpi.com

Buildings Editorial Office
E-mail: buildings@mdpi.com
www.mdpi.com/journal/buildings